S0-BVO-929

Basic Acoustics

Basic Acoustics

Donald E. Hall
California State University

HARPER & ROW, PUBLISHERS, New York
Cambridge, Philadelphia, San Francisco, Washington
London, Mexico City, São Paulo, Singapore, Sydney

Sponsoring Editor: *Lisa Berger*
Project Editor: *Ellen MacElree*
Cover Design: *Wanda Lubelska Design*
Text Art: *Fineline Illustrations, Inc.*
Production: *Debra Forrest*
Compositor: *Tapsco, Inc.*
Printer and Binder: *R. R. Donnelley & Sons Company*

Basic Acoustics

Library of Congress Cataloging-in-Publication Data

Hall, Donald E.
Basic acoustics.
Bibliography: p.
1. Sound. 2. Acoustical engineering. I. Title.
QC225.15.H35 1987 620.2 86-11971
ISBN 0-06-042611-X

86 87 88 89 9 8 7 6 5 4 3 2 1

Contents

Metric Units and Prefixes

A	ampere	= C/s	current
C	coulomb		charge
c	centi	$= 10^{-2}$	
dB	decibel		factor $\sqrt[10]{10} = 1.26$
F	farad	= C/V	capacitance
g	gram		mass
H	henry	= V-s/A	inductance
Hz	hertz	= cyc/s	frequency
h	hour		time
J	joule	= N-m	energy
K	kelvin		temperature
k	kilo	$= 10^3$	
kg	kilogram		mass
M	mega	$= 10^6$	
m	milli	$= 10^{-3}$	
m	meter		length
min	minute		time
μ	micro	$= 10^{-6}$	
N	newton	$= \text{kg-m/s}^2$	force
Pa	pascal	$= \text{N/m}^2$	pressure
p	pico	$= 10^{-12}$	
rad	radian	= 57.3°	angle
s	second		time
T	tesla	= N/A-m	magnetic field
V	volt		electric potential
W	watt	= J/s	power
Ω	ohm	= V/A	electrical ohm

List of Symbols

Numbers in right column denote chapters where used or introduced.

Symbol	Meaning	Units	Chapters
A, B, C, D	real wave amplitudes		
A, B, C, D	frequency weightings for SPL		2, 13
A	area	m^2	
A_e	effective absorption area	m^2	12, 13
a	acceleration	m/s^2	
a	complex wave amplitude		3
a, b	radius of disk, pipe, or sphere	m	
B	susceptance = Im{admittance}		6
B	bulk modulus	Pa	9
B	amplitude reflection coefficient		14
B	baffle radius	m	16
b	bar thickness	m	8
C	heat capacity per unit mass	J/kg-K	9
C_a	acoustic compliance	m^4s^2/kg	14, 18
C_e	capacitance	F	
CF	crest factor		2
CNEL	community noise equivalent level	dB	13
c	speed of sound	m/s	1, 9
c	speed of transverse waves	m/s	7, 8
D	dipole moment	m^4/s	11
D, d	distance or diameter	m	
Df	average mode spacing	Hz	12
E	total energy	J	
E_k, E_p	kinetic and potential energy	J	
e	2.72, base of natural logarithms		
e	energy density	J/m^3	
e_k	kinetic energy density	J/m^3	
e_p	potential or internal energy density	J/m^3	
$\mathcal{E}$	emf = electromotive force	V	
F	force	N	
FFT	fast Fourier transform		3
f	frequency	Hz	
f, g, h	arbitrary function or trial solution		
G	conductance = Re{admittance}		6, 14
g	acceleration of gravity	m/s^2	1
H	barrier height	m	4
H	angular amplitude factor		11, 17
I	intensity	W/m^2	1
IL	insertion loss	dB	4

i	electric current	A	
J_m	Bessel function of order m		8, 11, C
j	$\sqrt{-1}$		
j_{mn}	nth root of order-m Bessel function		8, C
K	combination of bar parameters	m^2/s	8
K	thermal conductivity	W/m-K	9
K	a real constant of proportionality		
k	wave number	rad/m	1
L	length or distance	m	
L	electrical inductance	H	
L	loudness	sone	2
l, m, n, p	integer indices		
l, l_e	pipe length and effective length	m	14
l	length of wire in a coil	m	15, 16
M	molecular weight		9
M	acoustic inertance	kg/m^4	14, 16
M	microphone sensitivity	V/Pa	2, 17, 18
ML	microphone sensitivity level	dB	17
m	mass	kg	
N	total number of objects, cycles, or modes		
N_F	Fresnel number		4
NR	noise reduction	dB	13
n	any integer or harmonic number		
P	power	W	
PBL	pressure band level	dB	3, 13
PSL	pressure spectrum level		3
p	pressure, especially wave amplitude	Pa	
p_0	equilibrium or atmospheric pressure	Pa	
Q	quality factor of resonant system		5
Q	quantity of heat	J	9
Q	flow source strength (see U)	m^3/s	11
q	charge on a capacitor	C	6, 14
R	ideal gas constant	J/kg-K	9
R	early/late sound ratio		12
R	power-reflection coefficient		14
R_a	acoustic resistance	kg/m^4s	14
R_e	electrical resistance	ohm	6
R_m	mechanical resistance	kg/s	5, 6
R_{rad}	radiation resistance (R_a or R_m)		11, 14, 15
R_1	dimensionless piston resistance		11
r	radius in spherical or cylindrical coordinates		
r_a	specific acoustic resistance	kg/m^2s	10
rms	root mean square average		2
S	speaker output ratio	Pa/V	2, 15, 18
S	acoustic stiffness $= 1/C_a$	kg/m^4s^2	14

S	standing-wave ratio		14
$S_x(f)$	spectral density of quantity x	x^2/Hz	3, 5, 13
SIL	sound intensity level	dB	1
SPL	sound pressure level	dB	1
s	mechanical stiffness	N/m	5
s	radial coordinate	m	11
s	capacitor plate separation	m	17
T	period of vibration	s	1
T	averaging time	s	2
T	string tension	N	7
T	temperature	K	9
T	power-transmission coefficient		14
T_r	reverberation time	s	12
TL	transmission loss	dB	13
t	time	s	
$\mathcal{T}$	membrane tension	N/m	8, 17
U	internal energy	J	9
U	volume flow rate	m^3/s	11
u	internal energy density	J/m^3	9
$u(x)$	initial string position	m	7
V	voltage = potential difference	V	
V	volume	m^3	
v	velocity	m/s	
$v(x)$	initial string velocity	m/s	7
W	work	J	
W	width	m	
w	$ct \pm x$		7, 8
w	complex number $u + jv$		B
X, Y, Z	room dimensions	m	12
X_a	acoustic reactance	kg/m^4s	14
X_e	electrical reactance	ohm	6
X_m	mechanical reactance	kg/s	6
X_{rad}	radiation reactance (X_a or X_m)		11, 14, 15
X_1	dimensionless piston reactance		11
x, y, z	position, cartesian coordinates	m	
x_a	specific acoustic reactance	kg/m^2s	10
$Y_{a,e,m}$	complex admittance = $1/Z$		6, 14
Y	Young's modulus for solid stiffness	Pa	8
Z	complex impedance $Z = R + jX$		
Z_a	complex acoustic impedance	kg/m^4s	11, 14
Z_e	complex electrical impedance	ohm	6
Z_m	complex mechanical impedance	kg/s	6
Z_{mot}	motional electrical impedance	ohm	15
Z_{rad}	radiation impedance (Z_a or Z_m)		11, 14, 15
Z_1	dimensionless piston impedance		11
z	complex number $x + jy$		B
z_a	specific acoustic impedance	kg/m^2s	10
z_c	ρc = characteristic z_a of fluid	kg/m^2s	10

α, β, γ	angles in complex plane	rad	B
α	attenuation per unit distance	dB/m	1
$\alpha, \bar{\alpha}$	material absorptivity, and average		12
α	constant multiplying factor		8, 9, 14
β	duty factor = fraction of period		2, 3
β	fraction of string length		7
β, γ	exponential damping rates	/s	5, 12
β	nonlinearity coefficient		17
γ	radian frequency like ω	rad/s	3
γ	ratio of specific heats		9
γ	plane polar coordinate angle	rad	11
δ	string or membrane slope angle	rad	7, 8
δ	phase of Z	rad	10, 18
Δ	finite difference		
Δf	bandwidth	Hz	
ϵ	membrane slope angle	rad	8
ϵ	electric permittivity	F/m	17
η	exponent of loudness law		2
η	0.16 s/m		12
η	viscosity	kg/m-s	14
θ	angle between wave direction and surface normal	rad	9, 12
θ	plane polar angle	rad	8
θ	spherical polar angle	rad	
λ	wavelength	m	
μ	mass per unit length	kg/m	7
ξ	displacement from equilibrium	m	7
π	3.14159		
ρ	mass per unit volume	kg/m^3	
σ	mass per unit area	kg/m^2	8, 13
τ	exponential-average time scale	s	2
τ	pulse width	s	3
τ	exponential damping time scale	s	5
τ	early/late arrival time boundary	s	12
ϕ	plane or spherical azimuthal angle	rad	
ϕ (ϕ_x)	phase angle (of quantity x) relative to $t = 0$		
ϕ_{xy}	phase lag of x with respect to y	rad	5, 14
Ω	radian frequency of natural mode	rad/s	5
Ω_d	damped natural frequency	rad/s	5
$d\Omega$	differential solid angle	rad^2	12
ω	frequency of any signal or motion	rad/s	
ω	frequency of a driving force	rad/s	5
ω_1	fundamental frequency of harmonic series		
$\Delta\omega$	half-power bandwidth	rad/s	5
∂	partial derivative		
∇	gradient operator	/m	

Preface for the Student

Sometimes it seems that neither physicists nor engineers are quite sure who should claim the subject of acoustics as their own. On the one hand, sound remains one of the most fundamental physical phenomena for which physics is committed to supplying clear and detailed explanations. On the other hand, the basic laws governing sound have been well established for a long enough time that most physicists move on to other things, assuming that acoustics can be turned over to the engineers for practical application. The positive aspect of all this is that the study of basic acoustics can help you gain familiarity with concepts and techniques that have much wider use in other areas of science, while at the same time having practical relevance in our daily lives.

This book will not make you an expert in acoustics; one should simply not expect to get to that level in a single step. So I have not felt obligated to touch every topic that might eventually be important to you. For some readers this will be a prelude and preparation for more specialized studies. For others it will be their only study of acoustics, and I hope it not only will serve as an enjoyable elective but will also make a specific contribution to their professional preparation. Even though they may never have primary responsibility for solving acoustical problems themselves, it is certainly an asset for any physicist, engineer, or architect to be able to talk intelligently with those of their colleagues who specialize in sound.

In the first few chapters we will try to learn a little about sound and its measurement from a relatively practical point of view. This should enable you to proceed with many kinds of measurements and laboratory experiments. Afterward, we will develop the basic theory of vibrations and waves that underlies all sound phenomena and instrumentation, and use that theory to show the basis for practical work with room acoustics, environmental noise, and sound-reproduction systems.

Do not forget that acoustics means real physical phenomena, not just elegant theories. Use every opportunity to build your practical experience with sound generation and measurement. This includes not only getting acquainted with the latest in electronic aids but also educating your own ears through thoughtful listening. Remember that you have mounted right in your own head two of the most sensitive and versatile sound detectors ever built, complete with software for extremely subtle and sophisticated signal analysis.

Your study of this book should put you in a position to appreciate many others that are concerned with individual subfields of acoustics. Here are a few of the possibilities:

Noise Control for Engineers, H. W. Lord, W. S. Gatley, and H. A. Evenson (McGraw-Hill, 1980)

Handbook of Noise Control, C. M. Harris, editor, 2d ed. (McGraw-Hill, 1979)

Principles of Underwater Sound, R. J. Urick, 3d ed. (McGraw-Hill, 1983)

Room Acoustics, H. Kuttruff, 2d ed. (Applied Science Publishers, 1979)

High-Performance Loudspeakers, M. Colloms, 2d ed. (Pentech Press, London, 1980)

An Introduction to Hearing, D. M. Green (Halsted, 1976)

Musical Acoustics: An Introduction, D. E. Hall (Wadsworth, 1980)

For more detailed exposition of all the fundamental material of acoustics, any aspiring acoustician should eventually delve into one or more of these excellent graduate-level texts:

Elements of Acoustics, S. Temkin (McGraw-Hill, 1981)

Acoustics, A. D. Pierce (McGraw-Hill, 1981)

Theoretical Acoustics, P. M. Morse and K. U. Ingard (McGraw-Hill, 1968)

Ongoing research is reported in several journals, of which the most important for beginning your acquaintance with the field are *Journal of the Acoustical Society of America* (*JASA*), *Acustica, Journal of Sound and Vibration,* and *Journal of the Audio Engineering Society* (*JAES*). You may also find it interesting to look at practical trade journals such as *Sound and Vibration* or *Sound and Video Contractor,* which are supported by advertising and sent free to people working in the field.

Donald E. Hall

Preface for the Teacher

The number of textbooks in acoustics has been much smaller in recent decades than in some other areas of physics. The best of them have tended to be at the level of beginning graduate study. I have felt the need of a book aimed more directly at the junior/senior level, one truly accessible to the student whose sophistication does not yet go much beyond a standard introductory course in university physics. My own classes include a mix of physics and engineering majors. I hope you will find that this book fits the stereotype of neither physics texts (in being too predominantly theoretical to hold the interests of engineers) nor engineering texts (in being too practical to spend much time trying to develop real understanding of the basic phenomena). Some readers may feel that parts of this book (for instance, some of the material on instrumentation and signal averaging in Chap. 2) are not true to the title; but even if they are not strictly acoustics, they are there because in my own experience I found that these were things my students and I wanted and needed to know in order to do acoustics.

I cover most of this material in a one-semester course that includes laboratory periods. I feel that lab experience is a very important aspect of the course and would urge that in any case where it is impossible to have the lab component there should at least be extensive demonstrations in the lectures. The suggestions for lab exercises at the ends of chapters are necessarily quite vague because of the differences in equipment that will be available in different schools. All experienced teachers will recognize that they need to give much more specific and extensive instructions to their students. A number of interesting things can be done in the lab with very simple equipment that is common even in small schools—a few cheap microphones and speakers, function generators, and small sound-level meters. A great deal more can be done with the addition of a digital spectrum analyzer, and fortunately these can now be obtained for a mere few thousand dollars. If your department does not yet have one, I would suggest that it will be one of the best educational investments you can make.

I believe there is some flexibility in the way this book may be used. To begin with, how much is done with Chap. 3 must depend a great deal on the role your own course has relative to others in the curriculum. One could in principle begin the fundamental theory of Chaps. 5 to 11 before completing coverage of the four introductory chapters. The applications in Chaps. 12 to 17 do not necessarily have to be taken in that order. Chapter 13, for instance, could be at the very end since later chapters do not depend on it; on the other hand, it could follow Chap. 3 if an earlier emphasis on practical problems is desired. Sections in several chapters, as well as all of Chap. 18, have been marked with an asterisk as optional, and in my view these are the prime candidates for omission where time is short.

Donald E. Hall

Acknowledgments

I am grateful to the students who tolerated this book in manuscript form for two years. Their positive attitude was a great help in seeing that the problems were workable, in correcting errors, and most of all in making me aware of changes in style of presentation that would make the book more useful to them. The manuscript was also reviewed by J. Gerard Anderson, Stanley Christensen, Murray Korman, William Savage, William J. Strong, and Arnold Tubis, whom I thank for their many helpful critical comments. Since they occasionally disagreed with each other on how some topics should be presented, I naturally bear full responsibility for the choices that appear in the final product.

Wadsworth Publishing Company has graciously extended permission to reproduce Figs. 1.1, 2.2, 8.4, 9.3, 9.4, 12.6, 12.8, 17.6, 17.7, and 17.10, which originally were drawn for my textbook on musical acoustics. I am grateful to the many people at Harper & Row who worked on this project. Heidi Udell was physics editor during most of the time I worked on this manuscript, and she provided much valuable assistance. I also owe words of appreciation to my department chair, Edward Gibson, for his support and encouragement, and to my wife, Carol Maxwell, who remains my best friend in spite of all my hours with the word processor.

Chapter 1

Acoustics: A Preliminary Acquaintance

What is acoustics? Historically, this meant simply the scientific study of sound. But the word *sound* itself is used both for the physical waves that travel through the air and for the psychological sensations they cause inside your head. This book is concerned primarily with the physical aspect of sound, but we will also briefly mention, where appropriate, a few fundamental facts about human perception of sound.

We will develop in several stages the tools needed for a detailed and quantitative description of sound waves and how they behave. This includes definitions of fundamental wave properties, as well as arguments justifying the basic equations of motion and showing techniques for finding their many solutions. Part of our understanding of sound clearly must come from studying the vibrating objects which create that sound. We all have an intuitive awareness that this connection is important—think about the motion of a guitar string or a drumhead, or simply press your fingers lightly against the front of your neck to feel the vibrations in your voice box while you talk or sing.

We must also become familiar with the devices and methods that are used to measure sounds. This means both awareness of available instrumentation and its practical operation as well as some theoretical understanding of what is being measured and how. Finally, several chapters aim to give you some feeling for widespread applications of acoustics that affect our experiences with sound in everyday life. Although examples most often involve sound in air, keep in mind that the theory to be formulated should be applicable to similar waves traveling in other substances as well.

This chapter is devoted mainly to reminding you of the terminology with which simple waves are described and to making some preliminary practical definitions of sound strength. But first let us consider briefly the variety of activities to which the study of sound may lead.

1.1 THE SCOPE OF ACOUSTICS

During the twentieth century, our view of acoustics has gradually broadened to encompass a number of areas more or less related to audible sound. Try to remember as you proceed that a wide variety of practical reasons may motivate our study of these things. At a typical meeting of the Acoustical Society of America, you could hear research papers presented in all of the following areas:

- Vibration of mechanical structures.
- Acoustical properties of materials.
- Sound measurement and instrumentation.
- Noise effects and noise control.
- Infrasonics and ultrasonics (sound inaudible to human ears because the vibration frequencies are too low or too high).
- Shock waves and sonic booms.
- Underwater sound.
- Signal-processing methods and theory.
- Architectural acoustics.
- Physiological acoustics and hearing disorders.
- Psychological acoustics: perception, cognition, and judgment of sound.
- Speech communication, speech disorders, and artificial speech generation.
- Music and musical instruments.

In this book we make no attempt to cover all these areas. We concentrate on audible sound in air, on how vibrating bodies radiate such sound, and on how these sounds can be measured. This will provide a general foundation that will put you in a position to profit from studying other books devoted to each specific area, such as those listed in the Preface.

Acoustics as a profession is enriched by the variety of backgrounds with which people enter it. Most have been officially educated in electrical or mechanical engineering, physics, experimental psychology, or biomedical sciences and have chosen relatively late in their training to specialize in acoustics. The research activities that continue to generate new fundamental knowledge are naturally centered often in universities. But manufacturing and sales of consumer audio equipment, as well as scientific laboratory apparatus, also support a large amount of engineering, design, and development. These contribute to our knowledge of new materials, structures, and devices and their integration into useful sound systems. A significant number of acousticians are active in consulting work, where they may assist in the design of new auditoriums and theaters as well as solve problems with environmental noise and vibration in offices, factories, and residential areas.

1.2 THE NATURE OF SOUND

Let us leave many of the details until later and mention here just a bare minimum about sound needed to proceed with practical questions. First, *sound is a wave phenomenon;* that is, each little parcel of air (or water, or concrete, or whatever else carries the wave) vibrates in some fashion and passes on the disturbance to its neighbors. But while this disturbance carries both information and energy to distance places, each bit of air remains always in the vicinity of its original position. In a gas or liquid the local vibration is always parallel to the direction of wave travel; so sound waves are classified as **longitudinal waves** (Fig. 1.1*a*). This is unlike radio waves or guitar-string vibrations, which are **transverse waves** (Fig. 1.1*b*), or water surface waves, which fit neither category (Fig. 1.1*c*). A solid material can transmit shearing or bending stresses, so can support transverse as well as longitudinal vibrations; but generally we limit the word *sound* to refer only to the longitudinal waves.

The speed of sound in air at room temperature is roughly 340 meters per second; we will be able to show in Chap. 9 why it has this value. The speed of sound in several other substances is listed for easy reference inside the front cover of the book. You must understand clearly the difference between **speed** (how *fast* a signal goes from one place to another) and **frequency** (how *often* the oscillating motion repeats at a single place). While speed is measured in meters per second, frequency requires cycles per second, which are also designated as

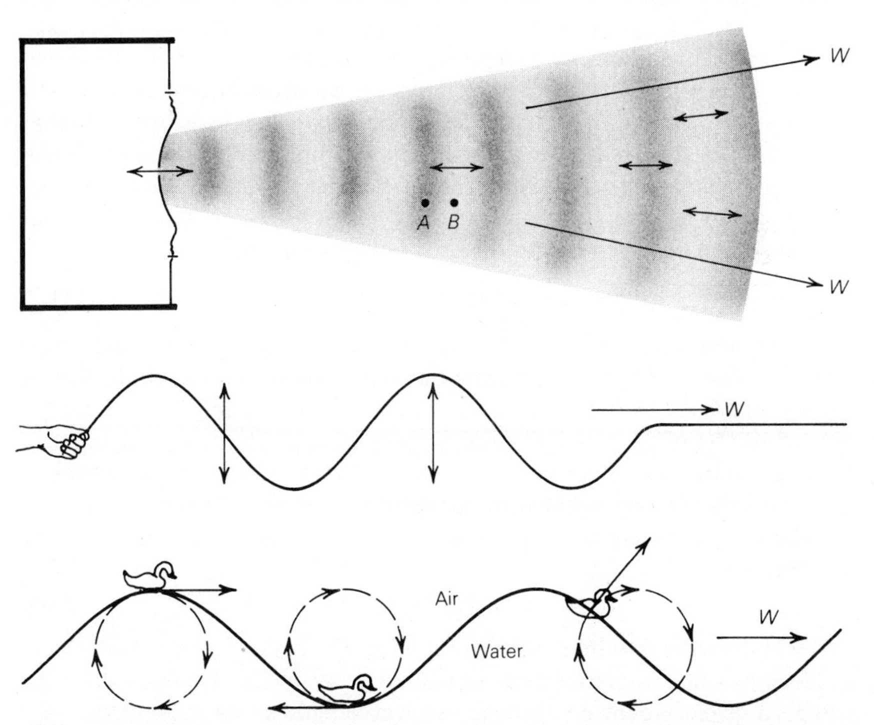

Figure 1.1 Different wave types. (*a*) Sound from a loudspeaker, with compression at *A* and rarefaction at *B*, and double-ended arrows showing longitudinal air motion parallel to the direction of wave travel *W*. (*b*) A rope shaken at one end, with each piece moving transversely to the direction of wave travel. (*c*) A water surface wave carrying some sitting ducks in circular paths with both longitudinal and transverse components. (From D. E. Hall, *Musical Acoustics: An Introduction,* Wadsworth, 1980.)

hertz. It is sounds whose frequency of vibration lies between about 20 and 20,000 cycles per second (20 Hz to 20 kHz) that are audible to human ears. You should be accustomed to describing frequencies with either cycles or radians. We will always use f for cycles/second and ω for radians/second, so that

$$\boxed{\omega = 2\pi f = 2\pi/T.} \tag{1.1}$$

Here $T = 1/f$ is the **period,** or length of time for one cycle of vibration.

Another important quantity is the **wavelength** λ, or crest-to-crest distance in space along the direction of wave travel. (Think of a snapshot, with the motion frozen.) By regarding the units of λ as meters per cycle, you may see that frequency f, wavelength λ, and speed c are related by

$$\boxed{c = f\lambda.} \tag{1.2}$$

This relation holds true for wave phenomena of *any* kind and is not unique to sound. Using $c = 340$ m/s, we find that audible sounds in air have wavelengths $\lambda = c/f$ between roughly 17 mm (at 20 kHz) and 17 m (for 20 Hz). Thus audible sound waves are comparable in size with people and doorways and loudspeakers, and that is why their interaction with these objects is sometimes complicated.

The simplest kind of sound waves involve sinusoidal motion, as illustrated in Fig. 1.2. (Keep in mind that such graphs are an abstract representation of information about the motion, not a literal picture of the disturbance itself as they would be for transverse waves.) Although we think only of these sinusoidal waves for now, there are certainly other kinds as well; in Chaps. 2 and 3 we will see how to deal with them.

You should be familiar with the use of expressions such as

$$y - y_0 = A \cos \omega t \quad \text{or} \quad A \sin \omega t \tag{1.3}$$

to represent sinusoidal variations of any physical quantity y above and below its average value y_0. Here A represents the **amplitude** of the vibration, that is, the maximum value ever attained by $y - y_0$. For sound, y might represent the density of one bit of air, or its pressure, temperature, displacement, or velocity. Notice that an increase of 2π in ωt, that is, $\Delta t = 2\pi/\omega$, always brings y back to its previous value in agreement with our definition of period above.

When air in different locations is at different stages in its vibration, we need terms like

$$y - y_0 = A \cos (\omega t - kx) \tag{1.4}$$

to represent the wave motion everywhere at once. The cosine function still returns to its previous value whenever its argument increases by 2π. That is, $k\,\Delta x = 2\pi$ represents a displacement Δx through one wavelength; so we must have

$$\boxed{k = 2\pi/\lambda.} \tag{1.5}$$

The quantity k is called the **wave number** (or spatial frequency); it has units of radian/meter and is closely analogous to ω, as you can see by comparing (1.1)

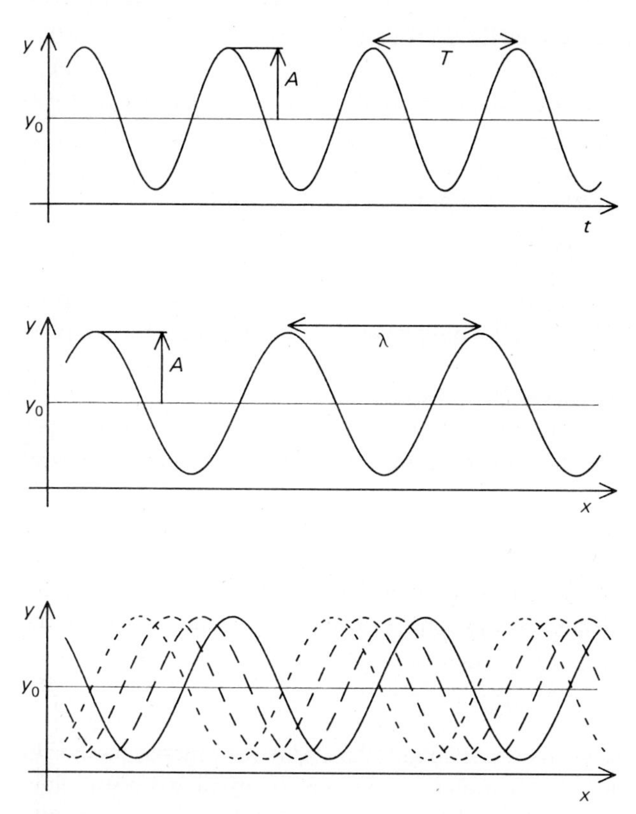

Figure 1.2 Representations of sinusoidal motion with period T, wavelength λ, and amplitude A. (a) Time history of disturbance at one point in space for some quantity y with equilibrium value y_0. (b) Variation of y in space at a single moment in time. (c) Multiple exposure indicating progress of a traveling wave through space, described by Eq. 1.4.

and (1.5). As an example, the low bass sounds with $f = 20$ Hz and $\lambda = 17$ m mentioned above have $\omega = (6.28)(20) = 126$ rad/s and $k = (6.28)/(17) = 0.38$ rad/m.

Many physical phenomena have a property called **linearity.** This means that the forces acting on the medium produce motions in exact proportion; doubling the cause will double the effect. Although nonlinearity (and resulting distortion of wave shapes) does occur for sound waves of very large amplitude, the approximation that sound waves behave linearly is generally an excellent one for any sound you would willingly allow to enter your ears. That is, sound in everyday life represents an exceedingly small disturbance of the average properties of the air. Whenever linearity holds, we are entitled to use the extremely powerful concept of **superposition.** This means that when many waves pass through the same region, their effects are simply additive; each continues to behave just as if the others were not present. This justifies our devoting most of our study to simple sine waves, for whatever we learn can still be applied when many such waves are put together as "components" of a more complicated disturbance.

1.3 DESCRIPTIONS OF SOUND STRENGTH

We most often describe the strength of a sound wave in terms of its **pressure amplitude,** meaning the size of the pressure fluctuations produced by the wave. Air, like any fluid, already has a steady equilibrium pressure p_0. The sound wave is represented by the small disturbance p_1 in the total pressure

$$p(x, t) = p_0 + p_1(x, t). \tag{1.6}$$

Wherever p_1 is positive or negative, we may say there is a **condensation** or **rarefaction,** respectively. It is p_1 that would have the form (1.4) for the simplest sound waves. Most of the time we drop the subscript and use the symbol p to stand only for the acoustic part of the pressure.

Since pressure means force per unit area, it is measured in newtons per square meter. It has become increasingly common to use the designation Pascal for this combination of units:

$$1 \text{ Pa} = 1 \text{ N/m}^2 = 10 \text{ dynes/cm}^2. \tag{1.7}$$

Leaving aside day-to-day variations because of weather conditions (on the order of 1 to 2 percent), the steady pressure at sea level due to the weight of the atmosphere above us is

$$p_0 = 1 \text{ atm} = 1.013 \times 10^5 \text{ Pa} = 14.7 \text{ lb/in}^2. \tag{1.8}$$

For moderate altitudes, p_0 decreases at about 12.5 Pa/m with increasing height above sea level. The sounds we encounter in everyday life involve pressure fluctuations on the order of 0.001 to 10 Pa, and thus fractional changes of only about 10^{-8} to 10^{-4} atm.

Another way to characterize the strength of a sound wave is by the energy it carries. But we do not ordinarily measure the total energy E emitted by a source over all time; instead, we are usually concerned with the *rate* of emission, or **power** P:

$$\boxed{P = dE/dt.} \tag{1.9}$$

Furthermore, although the total power P is a reasonable way to describe a sound source, we do not ordinarily collect this total power by completely surrounding the source with detectors. The strength of each portion of a sound wave ought to be measured in terms of the power density arriving at a detector in one particular region of space. So the local strength of a sound wave is given by its **intensity** I:

$$\boxed{I = dP/dA.} \tag{1.10}$$

Here dA represents the area of a surface oriented perpendicular to the direction in which the wave travels, so that it intercepts that portion dP of the power. For sufficiently small dA, doubling dA would also double dP, so that I defined in this way is a property of the wave and not of the detector. The units of intensity are watts per square meter, and sounds we ordinarily hear have I in the range from about 10^{-8} to 10^{-1} W/m^2.

If you could collect all the sound energy leaving a football stadium, how much would you have? For a rough estimate, an imaginary dome over the stadium would have a diameter of 100 m or so and an area on the order of 10^4 m^2. If the average intensity of sound passing outward through this surface is something like 10^{-3} W/m^2, the total power is about

$$P = IA = (10^{-3}\ \mathrm{W/m^2})(10^4\ \mathrm{m^2}) = 10\ \mathrm{W}.$$

Even if it continues for 3 hours, the accumulated total energy is still only about

$$E = Pt = (10\ \mathrm{W})(10^4\ \mathrm{s}) = 10^5\ \mathrm{J},$$

or enough to run a TV set for a few minutes. Your ears are extremely sensitive detectors that need only tiny amounts of energy to do their job.

It is true in many situations that energy depends on the square of displacement, velocity, or some similar amplitude. You may recall such examples as $mv^2/2$ for kinetic energy or $kx^2/2$ for potential energy in a spring, as well as $CV^2/2$ for electrical energy stored in a capacitor. So we also have

$$I = bp^2 \tag{1.11}$$

for the relation between intensity I and pressure amplitude p of a simple sound wave. In Chap. 9 we will give a proof of this equation, which will also show that the constant b should be $1/\rho c$, where ρ is the mass density of the fluid.

1.4 SOUND LEVELS AND DECIBEL SCALES

Since we encounter a vast range of sound-wave amplitudes, it is very common to represent their strength on a logarithmic scale called **sound level.** (Logarithmic scales are also used by astronomers in describing starlight and by geologists in their Richter scale of earthquake magnitudes.) This scale is defined so that every additional power of 10 in the amount of energy carried by the wave is represented as one **Bel,** named after Alexander Graham Bell. But the Bel is an inconveniently large step; so we introduce the **decibel,** which is one-tenth of a Bel. Thus the **sound intensity level** is given by

$$\mathrm{SIL(dB)} = 10 \log_{10} (I/I_{\mathrm{ref}}), \tag{1.12}$$

where the factor 10 converts Bels to decibels. Notice that dB, like radians, are dimensionless—they are not "units" in the usual sense, since they do not involve mass, length, or time. By mutual agreement, the reference intensity for sound in air is

$$I_{\mathrm{ref}} = 10^{-12}\ \mathrm{W/m^2}. \tag{1.13}$$

The intensity $I = 10^{-3}$ W/m^2 in the example above, then, has SIL = 10 $\times \log(10^9) = 90$ dB. Very roughly speaking, human ears may serve to detect a range of sound levels from a minimum of 0 dB (barely audible to a few people with the most sensitive ears) to a maximum of 130 dB (intolerably loud). Inexpensive sound-level meters may typically handle a range such as 40 to 140 dB. Table 1.1 will give you a rough idea of what intensity and what sound level correspond to various sounds in everyday life.

Another version of this logarithmic scale is called **sound pressure level:**

$$\boxed{\text{SPL(dB)} = 20 \log_{10} (p/p_{\text{ref}}),} \tag{1.14}$$

where p represents the amplitude of the wave and p_{ref} is a reference standard. The proportionality between I and p^2 requires that we use the factor 20 instead of 10 in order to make the two sound-level scales similar. For sound in air it is customary to use a reference level of 20 micropascals:

$$p_{\text{ref}}(\text{air}) = 2 \times 10^{-5} \text{ Pa}. \tag{1.15}$$

For example, amplitude $p = 0.4$ Pa corresponds to a sound level

$$\text{SPL} = 20 \log (2 \times 10^4) = 86 \text{ dB}.$$

Going the other way, SPL = 58 dB means

$$p = (2 \times 10^{-5})10^{2.9} = 0.016 \text{ Pa}.$$

For sound in water, 1 micropascal is now the preferred reference pressure. But you may also encounter older sources that use 20 μPa or 0.1 Pa; the latter is about 10^{-6} atm, so is often called 1 microbar. Therefore, it is always important in discussing underwater sound to specify clearly which reference level is used.

TABLE 1.1 SOME TYPICAL SOUND LEVELS*

Sound source	SIL or SPL (dB)	I(W/m^2)	p(Pa)
Jet engine at 30 m	140	10^2	200
Car stereo contest winner	130		
SST takeoff at 500 m	120	1	20
Amplified rock concert	110		
Heavy machine shop	100	10^{-2}	2
Subway train	90		
Factory	80	10^{-4}	0.2
City traffic	70		
Subdued conversation	60	10^{-6}	2×10^{-2}
Quiet auto interior	50		
Library	40	10^{-8}	2×10^{-3}
Empty auditorium	30		
Whisper at 1 m	20	10^{-10}	2×10^{-4}
Falling pin	10		
	0	10^{-12}	2×10^{-5}

* While these are typical of levels that might be encountered in a variety of situations, individual examples could easily be 10 dB higher or lower.

As an illustration, consider a sound in water with effective pressure amplitude $p = 0.04$ Pa. Its sound level is

$$\text{SPL} = 20 \log (4 \times 10^4) = 92 \text{ dB } re \ 1 \ \mu\text{Pa},$$

but it is also

$$\begin{aligned} \text{SPL} &= 20 \log (2 \times 10^3) = 66 \text{ dB } re \ 20 \ \mu\text{Pa} \\ &= 20 \log (0.4) = -8 \text{ dB } re \ 1 \ \mu\text{bar}. \end{aligned}$$

Although SIL is at first sight easier to understand intuitively, SPL is almost universally used for two reasons. First, nearly all the instruments with which we measure sound strength respond directly to pressure rather than energy; so it is technically more correct to describe experimental results with SPL. Second, when many waves are traveling in all directions past the detector, it is much more difficult to make a sensible definition of intensity or SIL, while pressure and SPL remain well defined. Therefore, we will make little further mention of SIL.

Often our concern may be to compare two sounds with one another, in which case the reference pressure is not really relevant. We may show this explicitly by using (1.14) to write

$$\text{SPL}_1 - \text{SPL}_2 = 20 \log_{10} (p_1/p_2). \tag{1.16}$$

For example, an SPL difference of 46 dB would always imply a pressure ratio $p_1/p_2 = 200$ and intensity ratio $I_1/I_2 = 4 \times 10^4$, regardless of how far either sound exceeds the reference level.

1.5 WAVE STRENGTH AND GEOMETRY

There are some simple conditions in which it is very easy to predict how the strength of a sound wave will vary with position. First is the trivial case of **plane waves,** such as those radiated by a very large flat plate oscillating in a direction perpendicular to its own plane (Fig. 1.3*a*). Here the waves are not spreading; so if we believe that they neither gain nor lose energy as they go, the wave intensity and amplitude must simply be the same at all distances from the plate. Since in

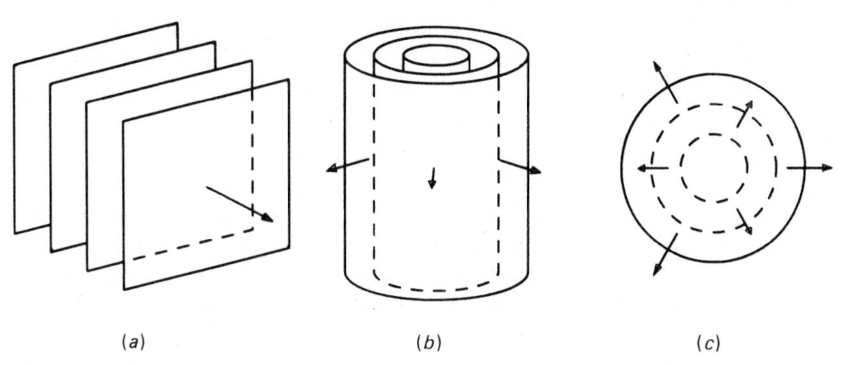

Figure 1.3 Successive wave fronts or crests for idealized plane, cylindrical, and spherical waves.

real life we have no truly infinite plates, we would have to limit such statements to distances from the plate not larger than its own dimensions.

If sound waves are created uniformly along a straight line rather than a flat surface, they will spread out in the form of concentric cylinders (Fig. 1.3*b*). Now if we consider cylinders with length L and radii r, a constant amount of energy is being spread over a larger and larger area $A(r) = 2\pi rL$ as r increases. Thus the intensity and amplitude of these **cylindrical waves** must drop off with distance according to

$$I(r) \propto 1/r, \qquad p(r) \propto 1/\sqrt{r}, \tag{1.17}$$

so that

$$\text{SPL}(r_1) - \text{SPL}(r_2) = 10 \log (r_2/r_1). \tag{1.18}$$

If we make a similar argument for spheres of radius r around a point source, the **spherical waves** (Fig. 1.3*c*) are spreading to cover area $A(r) = 4\pi r^2$, so that

$$I(r) \propto 1/r^2, \qquad p(r) \propto 1/r, \tag{1.19}$$

and

$$\text{SPL}(r_1) - \text{SPL}(r_2) = 20 \log (r_2/r_1). \tag{1.20}$$

We will find later that these arguments should be applied only when the distance r is much greater than the wavelength λ.

But what about those situations where a significant amount of sound energy is literally lost through **absorption** processes which convert it irreversibly into heat? This can be caused by both the viscosity and thermal conductivity of the fluid, as well as by resonant interactions of individual molecules at particular frequencies. We do not present the theory of any of those processes in this book, but you may find it in advanced texts. It suffices here to know that all these processes produce exponential decay of the sound energy, that is, energy loss per unit distance of travel which is always the same fraction of the energy present.

Thus the strength of a plane wave will in fact fall off with distance x as

$$I(x) = I(0)\ 10^{-\alpha x/10}, \qquad p(x) = p(0)\ 10^{-\alpha x/20}, \tag{1.21}$$

or equivalently

$$\text{SPL}(x_1) - \text{SPL}(x_2) = \alpha(x_2 - x_1). \tag{1.22}$$

The only reason for writing this form instead of $I(0) \exp(-\alpha x)$ is so that α can be expressed in decibels per meter. In cylindrical or spherical geometry, the correct result will be obtained by simply multiplying (1.17) or (1.19) by the same factor that appears in (1.21). Equivalently, the decrease in level given by (1.22) can be added on to that in (1.18) or (1.20).

For instance, in glycerin at 1 MHz (with $\alpha \simeq 10$ dB/m), sound from a line source traveling from $r_1 = 20$ cm to $r_2 = 60$ cm would lose $10 \log (r_2/r_1) = 10 \log (3) = 5$ dB to spreading and another $\alpha(r_2 - r_1) = 10 \times 0.4 = 4$ dB to absorption, for a total 9 dB reduction in signal strength.

Figure 1.4 gives values of α for air, based on both theory and measurement. You will see that absorption rapidly becomes more important as frequency in-

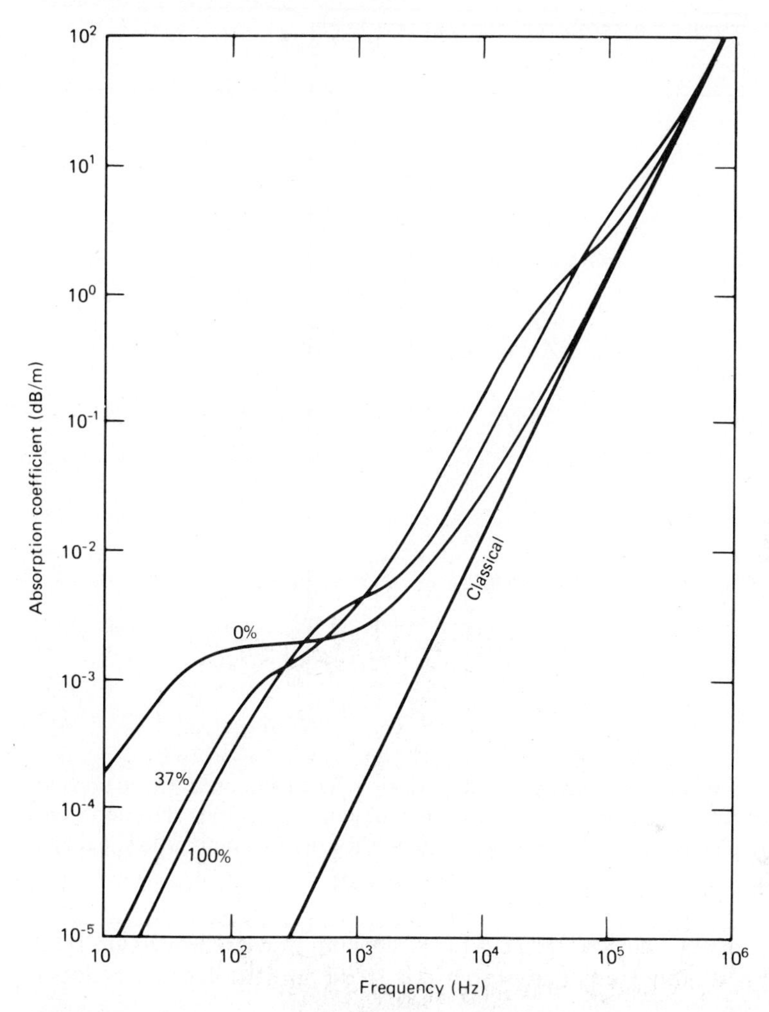

Figure 1.4 Absorption coefficient α (dB/m) for sound in air of temperature 20°C as a function of frequency. Irregular curves represent data for different relative humidities. Straight line represents "classical" absorption, that portion due to viscosity and thermal conduction alone, which is proportional to f^2. (Reproduced by permission from Kinsler et al., *Fundamentals of Acoustics,* John Wiley & Sons, New York, 3d ed., 1982, p. 154. Based on results from Bass et al., *JASA,* **52,** 821, 1972.)

creases. For ultrasonic signals at 100 kHz, for instance, several decibels of signal level are lost for every meter traveled. In the audible range the attenuation is small for most purposes but depends in quite a complicated way on the relative humidity of the air. Figure 1.5 shows some sample values of α for water. The dissolved salts cause seawater to absorb sound much faster than fresh water. We must expect that differences in salinity in different parts of the ocean, as well as schools of fish or rafts of seaweed, may cause further differences in α. Thus practical situations often require information about absorption to be sought in empirical data rather than in theory.

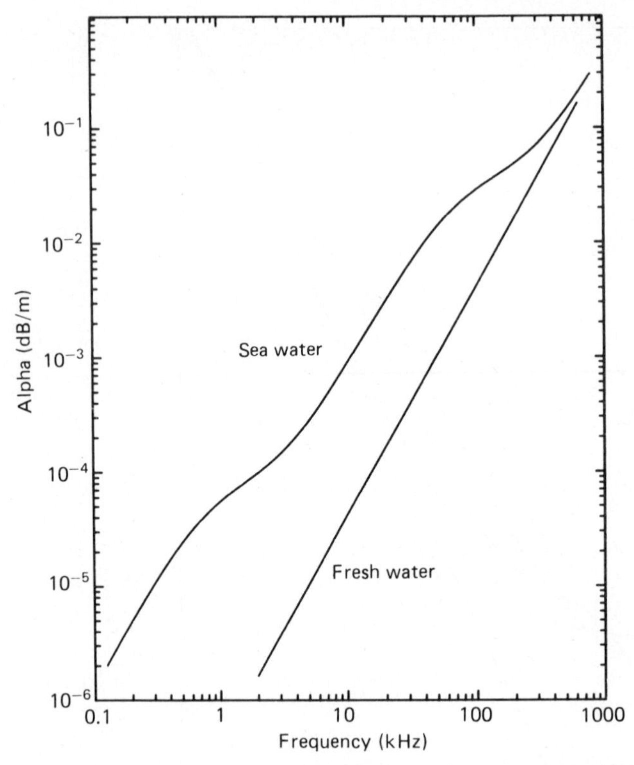

Figure 1.5 Absorption coefficient for sound in water with temperature 5°C and pressure 1 atm. Seawater salinity (in this case 35 parts per thousand) and resulting absorption vary considerably in different parts of the ocean. (After Fisher and Simmons, *JASA*, **62,** 558, 1977.)

For a final example, consider a 20-kHz signal in water-saturated air. If the SPL at $r_1 = 30$ m from a point source is 70 dB, at what distance r_2 does it drop below 40 dB? This requires

$$\mathrm{SPL}(r_1) - \mathrm{SPL}(r_2) = 30 = 20 \log (r_2/r_1) + \alpha(r_2 - r_1).$$

According to Fig. 1.4, $\alpha = 0.2$ dB/m, so that $\alpha r_1 = 6$ dB. It clarifies the mathematical problem to think of the ratio

$$z = r_2/r_1$$

as being the unknown rather than r_2 alone. Then the equation to be solved becomes

$$30 = 20 \log z + 6(z - 1)$$

or, after transposing the last term and dividing by 20,

$$1.8 - 0.3z = \log z.$$

You should sketch a rough graph of each side of this transcendental equation, whose solution is represented by the intersection of the two curves.

The procedures discussed in Appendix A lead to the value $z = 4.0$. Thus the answer is

$$r_2 = 4 \times 30 = 120 \text{ m}.$$

Spherical spreading provided $20 \log(4) = 12$ dB reduction in signal strength, and absorption provided the other $0.2 \times (120 - 30) = 18$ dB.

PROBLEMS

1.1. **(a)** The speed of light in vacuum is 3×10^8 m/s. What fraction of this is the speed of sound?
(b) Visible light has wavelengths between about 0.4 and 0.7 μm. Compare these with audible sound, by finding ratios of sound wavelength to light wavelength.
(c) What then are the frequencies for visible light? Compare with those for sound.

1.2. 500 Hz is, in hi-fi terminology, a midrange frequency. What period and what wavelength does such a sound have? What are ω and k?

1.3. Show that $c = f\lambda$ is equivalent to $\omega = ck$. (The latter form is often very useful in theoretical derivations.)

1.4. If a sound wave traveling through water in the x direction has wavelength 20 cm, what is its period? If the greatest speed produced by the wave motion is 2 m/s, write an expression of the form (1.4) to describe this wave. If you use $A = 2$ m/s, what physical interpretation must y have? If you wish y to represent displacement, what must you use instead for A? How is the direction of the oscillating velocity of each bit of water related to the x direction?

1.5. Atmospheric pressure represents simply the weight of all the air above you, where weight is mass times acceleration of gravity.
(a) If $g = 9.8$ m/s^2, what mass of air lies in the vertical column standing above each square meter of the Earth's surface?
(b) If the density of air remained constant at $\rho = 1.21$ kg/m^3 all the way to the top of the atmosphere and then abruptly dropped to zero, what would be the altitude of that upper boundary?
(c) If, instead, the density diminished gradually with altitude y according to $\rho(y) = 1.21 \exp(-y/H)$ (as it would if the temperature were the same at all altitudes), what value would the *scale height* H have? What would the density be at $y = 10\,H$?

1.6. We mentioned 0 dB and 130 dB as extreme examples of soft and loud sounds in air. In each of these two cases, by what fraction is normal atmospheric pressure changed by the presence of such a wave? If you had a wave whose pressure fluctuations were comparable with 1 atmosphere, where would it rate on the decibel scale?

1.7. For a sound in air with SPL = 84 dB, what is the pressure amplitude? For $p = 4$ Pa, what is the sound level?

1.8. **(a)** Show that a SPL *re* 1 μbar can always be converted to SPL *re* 1 μPa by simply adding 100 dB.
(b) State a similar rule for changing from 20 μPa reference to 1 μPa.
(c) For a sound in water with amplitude $p = 500$ Pa, what is the SPL with respect to each of the three reference levels?

1.9. If sound A has twice the intensity of sound B, how much will their SPLs differ? Is prolonged exposure to 115 dB at a rock concert "only a little worse" than 110? In what sense is that difference vastly greater than the difference between 55 and 50 dB?

1.10. **(a)** Starting at a distance of 20 m from an air-raid siren, how far away would you have to go to cut the sound intensity in half? How far to cut the pressure amplitude in half?

(b) Answer the same questions for freeway noise, considering the source to be spread uniformly along the road. Assume that absorption losses are negligible.

1.11. Suppose you compare SPL readings at distances 100 m and 200 m from a point source. In each of the following cases, how many decibels decrease do you expect because of geometrical spreading and how many decibels because of absorption?

(a) Air, 37 percent humidity, 1 kHz.
(b) Air, 37 percent humidity, 10 kHz.
(c) Seawater, 10 kHz.
(d) Seawater, 100 kHz.

1.12. If you begin at a distance 2 km from a point source of sound in seawater, to what distance can you go before the SPL drops 20 dB if the frequency is **(a)** 1 kHz, **(b)** 10 kHz, or **(c)** 100 kHz? (*Hint:* This problem involves a transcendental equation and gives you opportunity to apply the technique discussed in Appendix A.)

Chapter 2

Detecting and Measuring Sound

When we encounter an interesting sound, we would like to be able to make measurements that will precisely describe its properties. But the full power of modern instrumentation is most readily available when analyzing signals in electrical form. So our usual strategy calls for an assortment of devices that will detect sound waves and, in so doing, convert them into electric-circuit vibrations. We describe in the first section several ways of doing this.

In the case of a simple sine wave, a complete description requires exactly three pieces of information: the amplitude, the frequency, and the phase. The last of these has very little physical importance, since it only relates the wave to a time $t = 0$, which is arbitrarily chosen anyhow. In the second section below we introduce the simple instruments with which amplitude and frequency can easily be measured. In the third section, we comment briefly on the human perception of sound frequency and strength.

Most sound waves, of course, are not sinusoidal. Some, like a steady note played on a flute or a sung vowel, are **periodic** signals: an oscilloscope display will show the same unit being repeated over and over (Fig. 2.1*a*). After a specific length of time T has passed, the same disturbance occurs again; so period and frequency are still well defined for such a sound. But we must tell (in Secs. 2.2 and 2.5) how the nonsinusoidal shape of the repeating waveform will affect our attempts to measure frequency and strength. Indeed, we will have to more carefully define what we mean by "amplitude" for such a sound.

Many sounds do not remain steady for a long time. So we must also comment (in Sec. 2.4) on how well our measuring instruments can respond to rapid changes in sound strength. Finally, we encounter **broadband noise,** which is steady in the sense that its statistical properties remain constant, even though the waveform never repeats itself (Fig. 2.1*b*). Such a sound has no single well-defined period or frequency, but should be thought of as a mixture of many

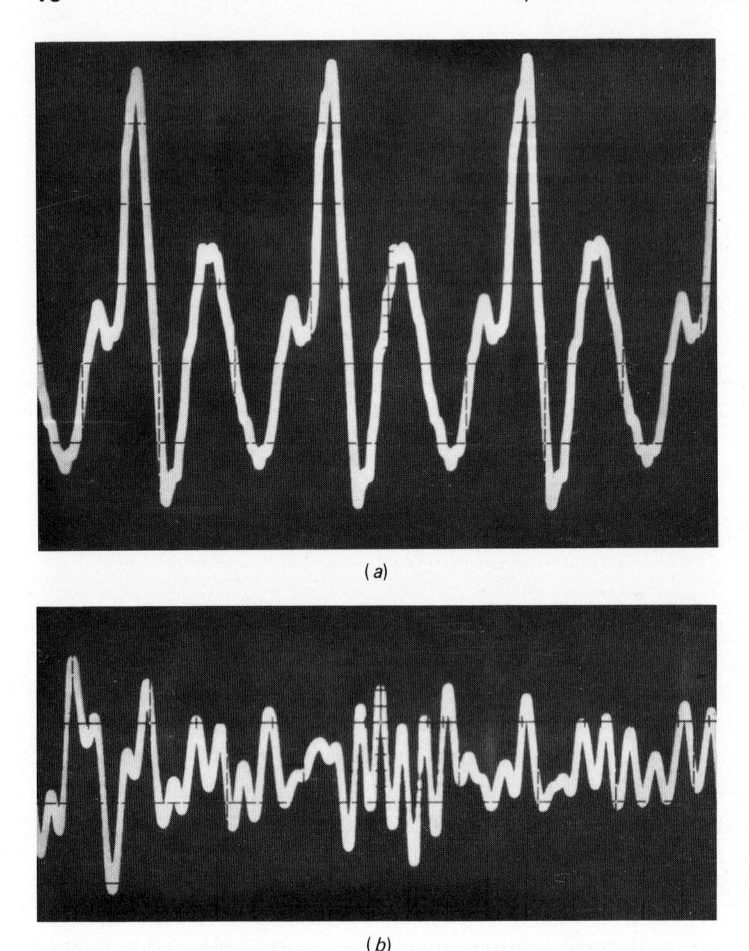

(*a*)

(*b*)

Figure 2.1 Two oscilloscope traces produced by the author's voice. (*a*) Period signal with frequency 200 Hz, singing a steady vowel *ah*. (*b*) Nonperiodic signal, a hissing sound.

sounds covering a whole range of frequencies. Section 2.4 describes how our simplest measuring devices deal with that problem. But we wait until Chap. 3 to develop the idea of frequency analysis and to describe instruments for measuring the detailed makeup of all sounds.

2.1 SOUND TRANSDUCERS

Generation, manipulation, and measurement of information-carrying signals is generally easiest when these signals take the form of vibrations in electric circuits. So one of our prime interests is in how to convert these electrical signals into sound, and vice versa. A device that converts one kind of vibration energy into another is called a **transducer.** Loudspeakers accept electrical input and produce acoustic output, and microphones turn sound input into electrical output. Both, then, may be called electroacoustic transducers, since they translate signals between electric and acoustic forms. Similarly, a phonograph cartridge can be used to detect the mechanical vibrations of any solid object and may be called an electromechanical transducer.

The appropriate measure of strength for an electrical oscillation is usually taken to be its voltage, or how great a push acts on the electrons. This is to be distinguished from the associated current, or how much flow of electrons results from this push. Similarly, the best measure of sound-wave strength is generally the acoustic pressure, which is associated with, but not identical to, the amount of air motion taking place. We can wait until later (Chap. 9) to study the precise connection between the sound pressure and the air displacement in the sound wave.

A typical microphone detecting an ordinary sound is a rather small device that can extract only tiny amounts of energy from the passing wave. Its output signal may be only a few millivolts and unable to drive more than microamperes of current. So the signal is invariably strengthened by an electronic amplifier before being put to any other use. A loudspeaker, on the other hand, must move enough air to fill an entire room with sound. So it is physically larger and is supplied (by a power amplifier) with an electrical signal measured in volts rather than millivolts, generally delivering several watts of power.

The sensitivity of a microphone is best characterized by the amount of voltage output per unit pressure acting on it:

$$M = V_{\text{out}}/p(0). \tag{2.1}$$

The effectiveness of a speaker can be described analogously in terms of the acoustic pressure created at a specified distance d away, such as 3 m (but *not* right at the speaker), per unit driving signal:

$$S = p(d)/V_{\text{in}}. \tag{2.2}$$

In later chapters we also consider some other measures of speaker output.

A particularly important property for both loudspeakers and microphones is their **frequency response.** This means the variation of M or S as a function of frequency, measured ideally by slowly sweeping the frequency of a sine-wave input signal while holding its amplitude constant. The simplest kind (and usually the most desirable as well) is **flat response,** for which the output-to-input ratio does not depend on frequency. You must, however, be careful never to assume that any microphone you use has flat response unless that is justified in detailed manufacturer's specifications or by your own calibration measurement. Even more so, you must expect as a matter of course that no real loudspeaker will achieve flat response over any very wide range of frequencies.

Most common electroacoustic transducers are reversible. You can easily take any speaker and connect it to an oscilloscope input and verify that it is simply a giant microphone. Similarly, many microphones are simply miniature loudspeakers operating backward. So parts of the theory we develop later will be applicable equally to both. For now, let us give a qualitative description of the principal types of transducers, using the simplified sketches of Fig. 2.2 to emphasize the basic physical processes involved. Suppose for the moment that these are all designed to interact with sound in air.

The **electrostatic** transducer is simply a capacitor (Fig. 2.2*a*), so has often been known as the condenser microphone. One plate of the capacitor is a very thin membrane exposed to the air, whose pressure fluctuations then cause variations in the capacitor plate spacing. The resulting capacitance changes cause corresponding voltage variations across the large series resistor, and these serve as input to a signal amplifier. Electrostatic microphones are often superior to others in the flatness of their frequency response, so are commonly chosen for laboratory measurement work; some details of their theory will be presented in Chap. 17. For operation as a loudspeaker, the input voltage drives fluctuating amounts of charge onto the capacitor, thus varying the amount of electrical force attracting the two plates together and causing the flexible membrane to move against the adjacent air.

The **piezoelectric** (crystal) or **electrostrictive** (ceramic) transducers rely on the fact that deformation of many solid materials causes them to become electrically polarized. Conversely, application of an alternating polarizing voltage will cause mechanical vibration. Here a light but stiff diaphragm is needed to interact with the air and translate its pressure into stress on the solid material; Fig. 2.2*b* suggests a version in which a thin sheet is rigidly mounted on two sides and bent by pushing on the free corner. Single-crystal types may use substances such as quartz, Rochelle salt (which cannot be exposed to high humidity or temperatures above 46°C), or ADP (ammonium dihydrogen phosphate). Ferroelectric ceramic (polycrystalline) materials such as barium titanate or lead zirconate will also serve if they have been given a permanent polarization by cooling from above their Curie temperature while in a strong electric field. Ceramics have the advantage that they can be molded into any desired shapes. Crystal transducers have relatively large values of M and have received wide use as microphones in public-address systems and hearing aids, and as speakers in headphones.

Dynamic transducers take advantage of electromagnetic effects on a coil of wire placed in the radial magnetic field in the narrow gap between poles of a permanent magnet, as shown in Fig. 2.2*c*. Air-pressure fluctuations on the attached diaphragm move the coil through the magnetic field to create an induced emf for microphone operation. It is actually immaterial whether the magnet is stationary and the coil moves or vice versa. (Indeed, dynamic phonograph cartridges have been made both ways by using very small magnets.) This kind of microphone is widely used in recording and in live performance of popular music. It is relatively rugged in careless hands but also heavy because of the magnet; it has relatively small sensitivity M. When the microphone is operated in reverse, an electric current sent through the coil experiences a magnetic force and so moves the diaphragm. Dynamic loudspeakers are by far the most common type in home hi-fi systems, and we consider them at length in Chap. 15. The ribbon microphone of Fig. 2.2*d* also belongs to the dynamic type, with the single strip of current-carrying foil taking the place of the coil.

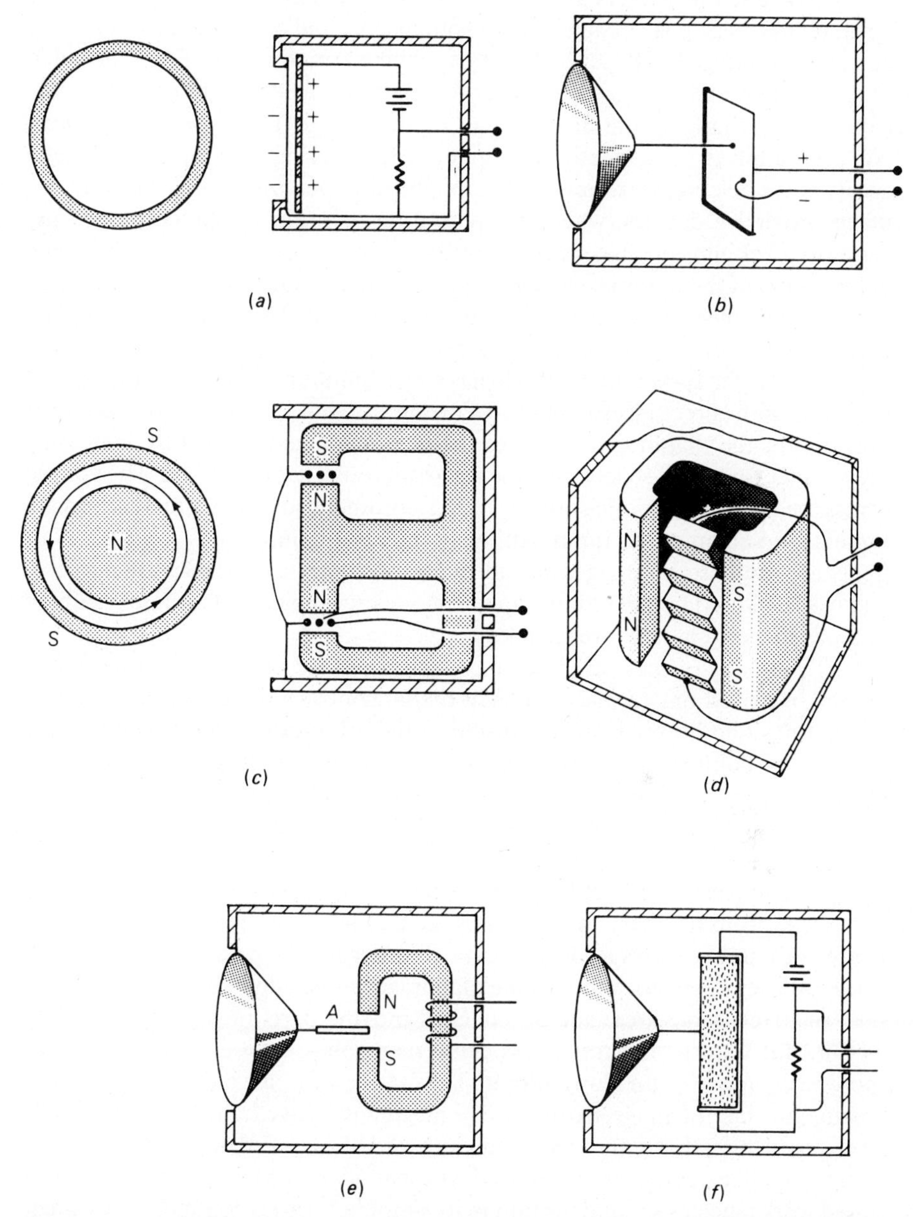

Figure 2.2 Idealized versions of common transducers. (*a*) Electrostatic, with axis of rotational symmetry from left to right. (*b*) Piezoelectric, with electrical leads to opposite faces of a thin crystal. (*c*) Dynamic, again with an axis of rotational symmetry from right to left; the coil of wire passing through the plane of the paper shows as a row of dots. (*d*) Ribbon dynamic. (*e*) Magnetic, with armature *A* moving in gap between magnet poles. (*f*) Carbon button. (From D. E. Hall, *Musical Acoustics: An Introduction,* Wadsworth, 1980.)

Magnetic transducers are closely related to dynamic, in that they involve coils and magnets. It is possible to keep both magnet and coil stationary, however, and still get induced voltages by having a moving armature *A* in Fig. 2.2*e* which makes it alternately harder and easier for magnetic field flux to cross the pole gap, thereby changing the flux through the fixed coil. This type of transducer has been applied in magnetic microphones for hearing aids and in magnetic speakers for telephone receivers. We might also put under this heading the **magnetostrictive** devices, where magnetization of some substances (especially certain nickel alloys) causes significant expansion or contraction. Conversely, deformation of the material can cause a change in its magnetization, which then is detected through induced emf in a surrounding coil. These have been applied mainly in dealing with underwater sound.

Finally, the **carbon** microphone has small granules of carbon tightly packed in a box, with direct electric current flowing through it (Fig. 2.2*f*). Excess air pressure on the diaphragm squeezes the granules together, slightly increasing their contact area and so lowering the electrical resistance of the box. The voltage across the unchanging series resistor rises accordingly, and these fluctuations are presented to an amplifier input. Although the carbon button has relatively poor frequency response, it is rugged and adequate for speech communication purposes and has been used by the millions in telephone mouthpieces. This is an example of a transducer that is *not* reversible.

To use any of these as electromechanical instead of electroacoustic transducers, one need only replace the light diaphragm by a direct connection with a vibrating solid body. Thus, for instance, the information above can provide the starting point for discussion of phonograph needles (crystal, dynamic, or magnetic) or electric-guitar pickups (crystal mounted in the bridge with the string pressing directly on it, or magnetic mounted farther out and detecting nearby motion of the steel string without touching it).

Transducers intended for underwater use are called **hydrophones.** Their differences from ordinary microphones are due largely to the fact that water is nearly a thousand times as dense as air. In order for the pressure of thin air to cause sufficient detector motion, the diaphragm must allow that pressure to act on a relatively large area, and all attached moving parts must be very light in weight. But water is not much less dense than solid materials; so it can much more easily move them along with it. Thus (aside from a waterproof protective coating) the face of an electrostrictive or magnetostrictive transducer can simply be exposed directly to the water pressure with little need for a separate diaphragm. Unlike audio work in air for the benefit of human ears, some underwater sound work (such as sonar detection in its simplest form) is conducted at a single fixed frequency. In that case flat frequency response is unimportant, and resonant transducers designed for strong response near that one frequency may be used.

2.2 FREQUENCY AND AMPLITUDE MEASUREMENT

When a microphone is connected to an oscilloscope, all the details of the sound it received can be seen. But we usually want to use only a small portion of that information; much of the detail is insignificant and confusing. The period and amplitude of the wave can be read directly from the scope screen, but other instruments designed for those specific purposes can provide the information much more conveniently.

It is quite easy to measure the frequency of simple waveforms such as sine or square waves with a **frequency counter.** Every time the input voltage from the microphone passes a set threshold, a short pulse is generated to advance a digital counter. Commercial frequency counters contain their own internal clocks, based for instance on 10-MHz oscillations of a quartz crystal. Thus they can be set to accept sound signals only during a 1-second window in time and then display the total number of cycles counted. Since there is no attempt to estimate fractions of a cycle, the actual frequency is effectively rounded to an adjacent integer—but not necessarily the nearest one. For instance, if the actual frequency of a signal is 931.4 Hz, the counter will sometimes display 931 and other times 932. In measuring radio frequencies in the megahertz range, rounding to the nearest hertz would be more than adequate. But audio frequencies may be less than 100 Hz, in which case the counter may be inherently unable to attain 1 percent accuracy when allowed to count for only 1 second. Most models offer you the option of counting for 10 seconds, so that the result will be correct within 0.1 Hz.

There is an alternative way of achieving high precision on low-frequency signals without requiring a long time to complete the measurement. Let the roles be reversed: The incoming audio signal opens and closes the gate to begin and end the count, while it is the internal clock signal that is counted. If, for instance, during one audio cycle 32,763 pulses were counted from a 10-MHz clock, it would mean that the period $T = 3.2763$ ms for that sound. Thus its frequency, $f = 1/T = 305.22$ Hz, is accurately determined within a small fraction of a second.

If a waveform is complex enough to have more than one peak within the repeating unit, a frequency counter may well display a multiple of the actual frequency. In fact, as you vary its trigger threshold setting, you may obtain several different readings depending on how many of the peaks intercept the trigger level. To avoid this, such signals should generally be sent through a sharp low-pass filter before their frequencies are measured. Remember this further limitation: for many sounds, a unique frequency or period is not even defined. So for broadband noise you cannot expect a frequency counter to produce any consistent or meaningful reading. Even for an orchestra, each of whose instruments is playing a note of definite frequency, the combined sound is nonperiodic; so it is inappropriate to attempt a frequency measurement.

The usual tool for quick measurement of sound strength is the hand-held, battery-powered **sound-level meter** (Fig. 2.3). It contains a microphone, circuits to process and measure its output-voltage signal, range and function switches, and either a digital or a moving-needle display for the results. In order that the reading may represent sound level, rather than pressure amplitude, a logarithmic conversion is included either in the electronics preceding a digital output or in the way the scale is marked behind a moving needle. Since sounds occur with such vastly differing strengths, a range switch is often used so that only a 10- or

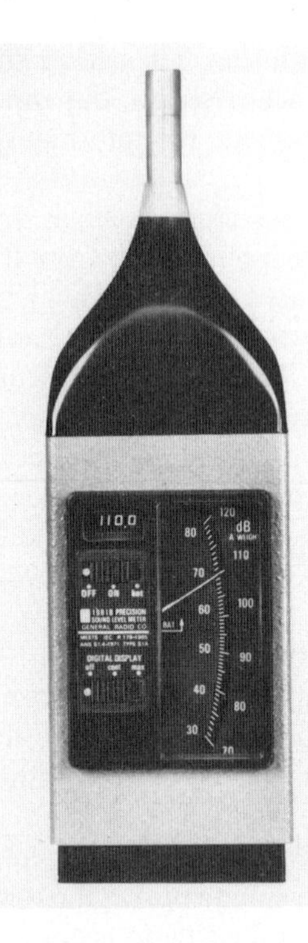

Figure 2.3 A precision sound-level meter, with both analog and digital displays. (Courtesy of GenRad, Inc.)

20-dB range need be covered for each switch position. Whenever important measurements are being made, the calibration of the meter should be checked with a sound source of known strength; such calibrators come in the form of small battery-powered devices that fit snugly over the microphone.

2.3 PSYCHOACOUSTICS

Whenever we are concerned with human responses to sound, it is important to distinguish those responses from the physical stimuli that caused them. Let us mention just enough about sound perception so that we can be careful to use the correct terminology.

First, **pitch** designates the sensation of low or high in the sense of bass and treble. That sensation is produced primarily by the frequency of the stimulus sound, but the terms pitch and frequency should not be used interchangeably. Pitch is conveniently measured in a unit borrowed from music, the **octave.** One octave is the amount of pitch change that produces the unique perceptual response where the second tone sounds so closely related to the first that we give it the same name, as with middle C and high C. This turns out always to correspond closely to a doubling of frequency. A piano keyboard, then, represents a logarithmic frequency scale: 100, 200, 400, 800, 1600 Hz are all equally spaced, at one-octave intervals. Human singing is done mostly in the three octaves between 100 and 800 Hz.

Second, let us reserve the word **loudness** for perceptual judgments of sound strength. If we need a physical measure of sound strength, we may use either wave amplitude or intensity (SPL being merely a coded form of that information). The distinction must be maintained because our ears do not respond on a linear scale: for source A to "sound twice as loud" as source B, it must provide significantly more than twice as much physical energy.

In fact, for sounds of midrange frequency and moderate strength, perceived loudness L is well approximated by a power law:

$$L/L_1 = (I/I_1)^\eta = (p/p_1)^{2\eta}. \tag{2.3}$$

The most commonly used value of η is 0.3, though some people believe the data are better represented by values as large as 0.5; but then η changes somewhat with frequency anyhow. If the loudness of a 1-kHz tone with $\text{SPL}_1 = 40$ dB is defined to be $L_1 = 1$ **sone,** then using 40 dB as a reference level (i.e., $I_1 = 10^{-8}$ W/m^2 or $p_1 = 0.002$ Pa) in Eq. 2.3 will give L in sones for other tones with different intensities. As you may show (see Prob. 2.4), the common belief that the decibel scale represents the way our ears judge sound is not correct.

Our sensitivity to sound decreases toward both ends of the audible range of frequencies; so Eq. 2.3 will overestimate the loudness of signals much below 500 Hz or above 5 kHz. At 200 Hz, for instance, it takes a SPL of about 52 dB to produce one sone (i.e., to sound equally loud with a 40-dB, 1-kHz tone), and the exponent η is about 50 percent higher than at 1 kHz. Even at midrange frequencies, (2.3) somewhat overestimates L for SPL below 30 or above 100 dB.

Besides pitch and loudness, we also perceive the **timbre** or tone color of sounds. This property is what makes it possible still to tell some difference between a violin and a clarinet when both play the same note on a musical scale equally loudly. The sensation of timbre depends on the physical waveform of steady periodic sounds, and we will see in the following chapter that this is intimately related to their "harmonic content." Timbre is also influenced by the transients with which musical notes begin and end, and these attacks and decays are important cues in our ability to recognize different instruments.

2.4. TIME CONSTANTS AND WEIGHTING NETWORKS

In measuring sound signals, we must deal with several important questions involving the careful definition of what is being measured and how. These questions arise for ac signals of any kind and are usually discussed in connection with electric circuits and electronic measurements. Let us ignore any corrections for transducer behavior and suppose we are concerned with measuring an electrical signal that is a perfect replica of the sound pressure waveform.

One important issue arises when we measure signals that do not remain steady for a long time. Any real meter must have a finite response time and will not register the full strength of short transient sounds. In effect, the meter can only indicate some running average of the actual signal $X(t)$. You may naturally think first of a simple **linear average** extending over the time interval τ just past:

$$X_{av}(t) = (1/\tau) \int_{t-\tau}^{t} X(y)\, dy. \tag{2.4}$$

Here X is intended to represent the amplitude envelope of a sound disturbance (changing perhaps on a time scale of tenths of a second), not every individual vibration.

But in practice a more important type of smoothing is the **exponential average,**

$$X_{av}(t) = (1/\tau) \int_{-\infty}^{t} X(y) \exp[-(t-y)/\tau]\, dy. \tag{2.5}$$

Notice how X_{av} at any time t depends on X at all earlier times $y < t$, with the most recent values having the greatest weight. The parameter τ characterizes how long each input continues to strongly influence the average.

Take, for example, the signal shown in Fig. 2.4 with amplitude

$$X(t) = V(1 - e^{-t/T}).$$

Eq. 2.5 gives

$$\begin{aligned} X_{av}(t) &= (V/\tau)e^{-t/\tau} \int_{0}^{t} (1 - e^{-y/T})e^{y/\tau}\, dy \\ &= V\{1 - [(\tau/T)e^{-t/\tau} - e^{-t/T}]/(\tau/T - 1)\}. \end{aligned}$$

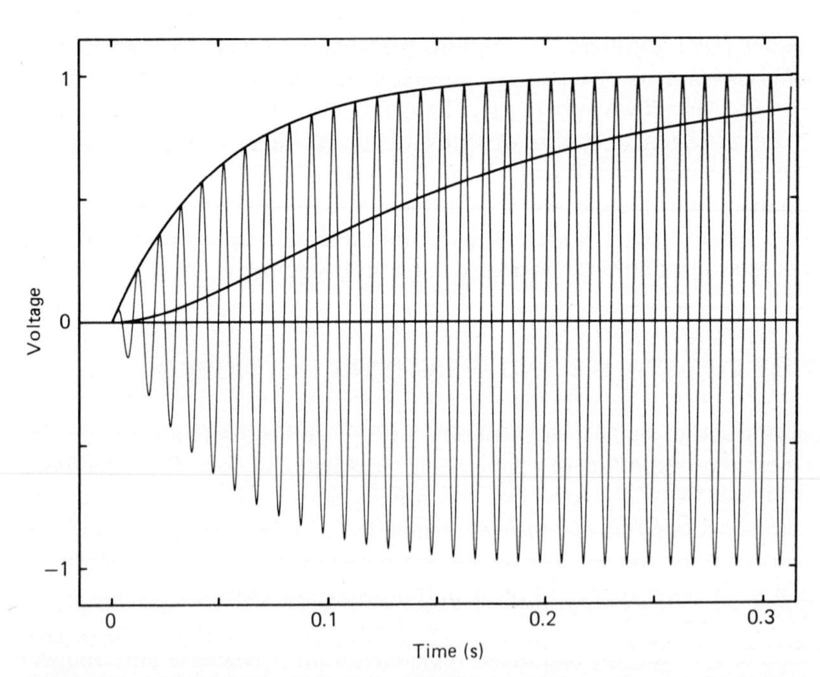

Figure 2.4 A signal of frequency $f = 100$ Hz turned on at time $t = 0$ whose amplitude envelope has an exponential rise time $T = 0.05$ s, and its exponential average with response time $\tau = 0.125$ s.

Figure 2.4 shows how this response lags behind the rise in strength of the signal.

A good sound-level meter will offer you a choice of two or three agreed standard response characteristics. Though the technical definitions are a bit more complicated, **Slow** response on the meter in my lab means roughly an exponential time constant τ of 1.0 s, and **Fast** averages over 0.125 s. The fast response most nearly approximates human perceptions, since our aural processing includes an effective integration time of about 0.1 to 0.2 s. **Impulse** response uses a time constant of 0.035 s. In the last case your eye could hardly follow such rapid swings of the needle or digital display, so the meter shows only rapid rises of signal strength; its reading will take several seconds to decay even if the signal is suddenly cut off. An even more extreme version of this is available on some meters as **Maximum Hold:** the reading for a single isolated impulse, or the peak value occurring during any sequence of sounds, will remain on the meter for many minutes. In this case, a Reset button is needed to prepare the meter for another reading.

When making laboratory measurements for purely physical purposes, we usually prefer that a sound-level meter be equally sensitive to sounds of all frequencies—even those beyond the range audible to humans. The meter control labeled Flat or Linear will carry out this choice. Any real meter, of course, still has some inherent circuit limitations on its ability to do this—its amplifier might operate linearly from 2 Hz to 70 kHz, for instance. Merely selecting this control would not ensure response up to 70 kHz, though, unless the particular microphone involved would respond at such high frequencies.

But many sound measurements are motivated by the effects these sounds will have on people—highway or aircraft noise, for instance. In these cases we would prefer that the meter response mimic the human ear's unequal sensitivity to signals of different frequencies. Several standard filter characteristics have been defined for this purpose, and they are shown in Fig. 2.5. The A, B, and C networks approximate human responses for soft, medium, and loud sounds (SPL near 40, 70, and 100 dB, respectively). The D filter is used only for specialized measurements of aircraft noise, where it is intended to represent perceived "noisiness."

Whenever these filters are used to measure a mixture of sounds of widely differing frequencies, the resulting number is a weighted average of the individual sound levels. We must always indicate clearly when this is done, by statements such as "the C-weighted sound level was found to be 57 dB" or SPL(C) = 57 dB, or SPL = 57 dBC. Notwithstanding the historical rationale for each weighting network, community noise criteria are nearly always stated in terms of A-weighted sound levels. This is because the A weighting has turned out to correlate well with the effectiveness of environmental sounds in causing annoyance. Thus, unless you have specific reasons for doing otherwise, you should generally use the A scale for such measurements and report the results as dBA.

If we want to understand more detail about the origins and effects of complex sounds, it is reasonable that we ask for separate information about how

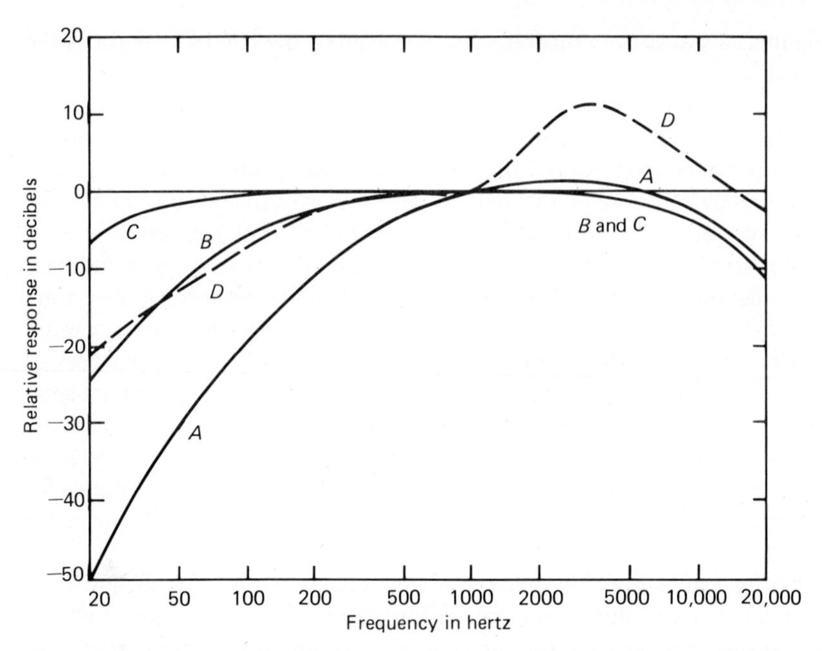

Figure 2.5 The A, B, C, and D weighting curves. (Courtesy of Bruel and Kjaer.)

much sound energy is present at various frequencies instead of having only a single weighted average. We describe several ways to do that in Chap. 3.

2.5 SIGNAL PROCESSING AND AVERAGING

We encounter another difficulty when we try to measure the strength of various waveforms, even when their amplitudes remain steady. If we dealt exclusively with square waves, it would be quite clear what we mean by the pressure (or voltage) associated with the presence of the wave; aside from sign, it always has the same value. But a sinusoidal wave involves continuous changes; so we must specify whether we wish to describe it by its peak value or by some average value of the changing disturbance. Please note that we are now concerned with averaging of each individual oscillation and thus with a much shorter time scale than in the preceding section.

The most useful kind of average is generally the one most closely related to the amount of energy the wave carries. This is the **root-mean-square** average, or rms amplitude, and is defined by

$$V_{\text{rms}} = \sqrt{(V^2)_{\text{av}}} = \left[\int_0^T V^2(t)\, dt/T\right]^{1/2}. \tag{2.6}$$

Here T may be either a very long time or a single period of a repeating waveform. For a sine wave, $V(t) = V_0 \sin \omega t$, you may easily show that $V_{\text{rms}} = 0.707 V_0$. It is most important to realize that this relation does not hold for other waveforms;

for instance, a square wave has $V_{rms} = V_0$, a triangular wave has $V_{rms} = 0.577V_0$, etc. (see Prob. 2.8).

The simplest way to build an ac voltmeter is to connect four diodes with a dc meter (as in Fig. 2.6) so that positive and negative voltages will both drive the meter in the same direction. If a sinusoidal signal is sent to such an instrument, the time average of the rectified waveform that actually drives the meter is $2V_0/\pi = 0.636V_0$ (see Prob. 2.9). The instrument is customarily calibrated to include a factor $0.707/0.636 = 1.11$ so that the output reading for a sine wave is V_{rms}. If such an instrument is used with a nonsinusoidal waveform, however, its reading has no such simple interpretation. For a square wave, for instance, the reading is $1.11V_{rms}$, while for a triangular wave it is $0.96V_{rms}$, etc.

An inexpensive volt-ohm meter is likely to operate that way, but a good sound-level meter will use different circuits to achieve **true-rms** operation, that is, to register V_{rms} for a wide range of waveforms. Without concerning ourselves

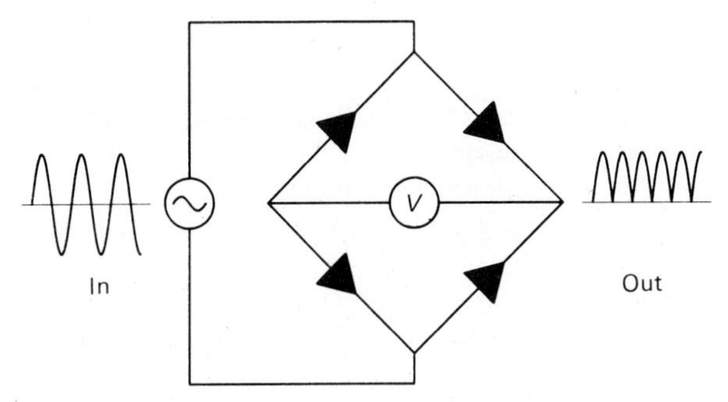

Figure 2.6 A simple full-wave rectifier circuit. Each diode allows current to flow through it only in the direction in which its symbol points.

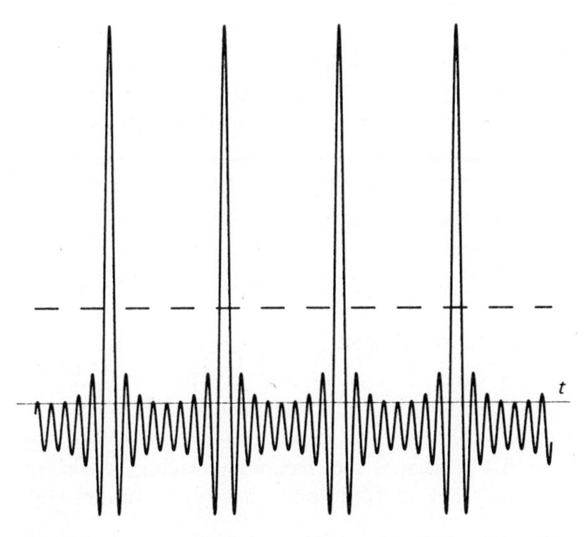

Figure 2.7 A periodic waveform with CF $= V_{max}/V_{rms} = 4$. Dashed line shows rms average.

with exactly how that is done, we may comment that there is still a limitation if you must measure extremely jagged waveforms. A sound-level meter can be trusted to achieve rms operation only if the **crest factor** of the waveform does not exceed some maximum value (for instance, the manufacturer may specify true-rms operation for CF < 5). Figure 2.7 illustrates the definition of crest factor as the ratio

$$CF = V_{max}/V_{rms}. \tag{2.7}$$

All of our succeeding discussion will be based on the assumption that every sound measurement is made with a true-rms meter and that all waveforms measured are sufficiently smooth that they do not exceed the crest-factor capabilities of the apparatus. In particular, it is important from here on that the reference pressure $p_{ref} = 2 \times 10^{-5}$ Pa used in defining the decibel scale is to be understood as an rms average.

PROBLEMS

2.1. **(a)** If a microphone whose sensitivity is $M = 0.3$ mV/Pa gives an output of 6 mV, what is the SPL of the sound it is receiving?
(b) If it receives sound with SPL = 70 dB, what is its output?

2.2. **(a)** If a speaker with $S(d = 3\text{ m}) = 0.06$ Pa/V is driven by a 0.4-V signal, what SPL is created at the distance d?
(b) What input voltage would be needed to make SPL = 100 dB?
(c, d) If the speaker has 8 Ω of resistance, what input power $P = V^2/R$ is delivered in both cases?

2.3. **(a)** How many cycles must pass for a frequency counter to give a reading guaranteed to be accurate within 0.01 percent?
(b) How long will that take if f is around 100 Hz?
(c) How long for 10 kHz?
(d) If this counter has a 4-MHz internal clock, how long must it count to determine a time interval within 0.01 percent?
(e) Is one cycle of the input signal enough for accurate period determination for the 100-Hz signal?
(f) How about 10 kHz?

2.4. Use Eqs. 1.16 and 2.3 to express loudness L in terms of SPL and thus to prove that they are not linearly proportional.

2.5. For 1-kHz tones with SPL = **(a)** 70, **(b)** 80, and **(c)** 85 dB, use Eq. 2.3 to estimate their loudness in sones.
(d) What SPL would it take to produce 0.5 sone?

2.6. Suppose the actual history of a signal is given by $X(t) = 0$ for $t < 0$, $X = A$ (a constant) for $0 < t < t_1$, and $X = 0$ again for $t > t_1$. **(a)** Use (2.5) to calculate $X_{av}(t)$ for both $t < t_1$ and $t > t_1$, and show that its maximum value is at $t = t_1$. If the signal duration is $t_1 = 0.1$ s, what is the maximum reading you expect for an averaged signal on **(b)** Slow, **(c)** Fast, **(d)** Impulse, and **(e)** Maximum-Hold settings?

2.7. If you have a sound-level meter offering only A, B, and C weighting, which of these would you choose if you wanted to approximate flat response?

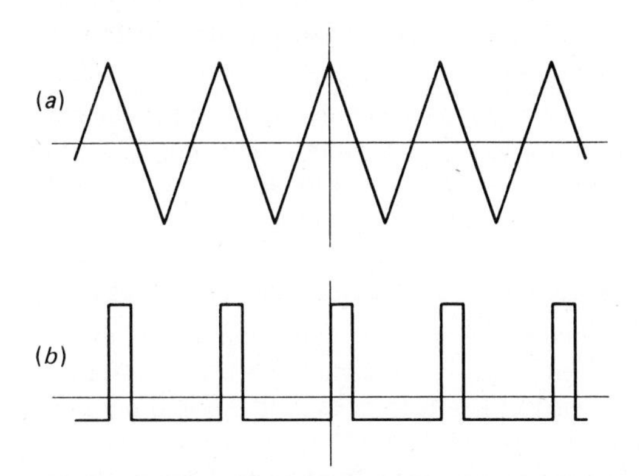

Figure 2.8 Definitions of triangular and pulse waves.

2.8. **(a)** Show that the rms average for a triangular wave (Fig. 2.8) is 0.577 times its peak value.

(b) Let us define a pulse wave of duty factor β to mean a wave that has constant positive voltage for a fraction β of its period and constant negative voltage for the remaining time: $V(t) = +V_0$ for $0 < t < \beta T$ and $V(t) = -\beta V_0/(1-\beta)$ for $\beta T < t < T$. Show that V_{rms}/V_0 is $\sqrt{\beta/(1-\beta)}$.

2.9. **(a)** Compute the average of the absolute value of sin ωt, to show the origin of the factor 0.636 used in Sec. 2.5.

(b) Prove that the reading on a simple rectifying meter will be 1.11 times the rms average for a square wave but 0.96 times rms for a triangular wave.

(c) Using the definition and result from part *b* of the preceding problem, show that the corresponding factor for a pulse wave of duty factor β is $2.22\sqrt{\beta(1-\beta)}$.

2.10. **(a)** What is the crest factor of a sine wave?

(b) A triangular wave?

(c) A pulse wave with duty factor $\beta = 10$ percent?

(d) With $\beta = 1$ percent? (See Prob. 2.8 for definitions.)

LABORATORY EXERCISES

All of the following exercises are suggested with the idea of being primarily to gain experience connecting and turning the knobs of the test equipment available in your laboratory. With this preparation you can more efficiently undertake more careful and detailed versions of these experiments as the motivating theory develops in Chaps. 12 to 17.

2.A. Familiarize yourself with the operation of whatever electronic signal generators, frequency counters, and oscilloscopes are available to you. These are tools with which you should become so familiar that you can use them effortlessly in creating and diagnosing acoustical signals.

2.B. Check out a sound-level meter and familiarize yourself with its operating instructions. Carry it around for a day or two, and measure as wide a variety of sounds as you can discover or create. Note that reflections from your body can modify the sound field, so it is best to hold the meter at arm's length. When writing down each measurement, be sure to record how the meter was located relative to the source, since signal strength usually depends on position.

2.C. Place a microphone beside a sound-level meter so that both should receive the same acoustic pressure, and display the microphone output on an oscilloscope. Use this to determine M for this microphone. Determine S for a loudspeaker similarly by placing the sound-level meter at a stated distance away and measuring the driving voltage supplied by an electronic oscillator. Think about what measures you can take to eliminate the confusing effect of reflected sound from the walls of the room.

2.D. *Loudspeaker frequency response:* Obtain a microphone that is known to have flat frequency response over some specified range. (That might mean the manufacturer has provided a certified calibration curve.) Use it to measure the sound produced by a loudspeaker driven by a sine-wave generator. "Measure" might be taken to mean reading the amplitude of the microphone signal from an oscilloscope display for each of several different signal frequencies, or it might mean sending the microphone signal to a strip-chart recorder while the frequency scans continuously, if you have such equipment available. How will you meet the possible objection that what you have measured is at least as much due to room response as it is to loudspeaker output? If you do not have an anechoic chamber in which to do the measurement, think about how you might best approximate one.

2.E. *Microphone frequency response:* Use a microphone with known frequency response as a standard to determine the frequency response of a second microphone. Put them side by side so that both will receive the same signal from a speaker driven as in Lab Exercise 2.D above. Insofar as both microphone signals vary in unison from one frequency to another, you will blame that on the loudspeaker (or the room). But a change in strength of one signal relative to the other will tell you about a difference in frequency response of the two microphones.
Improved version: If you have such equipment, use the output of the standard microphone in a feedback loop to control the oscillator signal strength. That is, let the speaker be driven harder at those frequencies where it is less efficient, so that this automatic compensation will always ensure that the same actual sound strength arrives at the standard microphone. Then the output of the second microphone will give its frequency response directly.

Chapter 3

Spectral Analysis

Sound-level measurements, as we presented them in the preceding chapter, are aimed at obtaining a single number to represent sound strength. This number may be a weighted average, in which sounds of differing frequencies have unequal importance. But sometimes we want to know more detail about the origins and effects of complex sounds. It is reasonable to ask how we may obtain separate information about how much sound energy is present at various frequencies.

In this chapter we describe several methods or devices for making such measurements. Laboratory instruments for spectral analysis are extremely powerful and have application in all branches of physics and engineering. To give a deeper understanding of their meaning, we also present a brief summary of the underlying mathematical theory called Fourier analysis. Though that theory will not be explicitly required for most of the material in this book, it is offered here in hopes that it can give better insight into the meaning of spectrum measurements.

3.1 BAND-LEVEL MEASUREMENTS

The first logical step is to send the microphone signal through a filter circuit. A bandpass filter can effectively allow signals within a certain range of frequencies to pass through, while blocking all lower or higher frequencies. Though this is easy to say and to appreciate intuitively, it is still imprecise. Only after considering Fourier theory later in the chapter will we have a clear definition of the sense in which all signals may be decomposed into components of definite frequency. Nevertheless, let us proceed to consider an input signal that is a mixture of sine waves.

Suppose then we use a sound-level meter in conjunction with a whole set of filters, which offer the ability to select any one of a number of narrow frequency ranges for a passband. Some modern sound-level meters even have such a filter set completely built in. Then the set of measurements on the filtered signal will give more information about it than would a single overall sound level.

If we use passive filters, we may say we are simply measuring the sound level (i.e., the rms pressure) of that portion of the original signal that is allowed to pass through. We may call this a **pressure band level:**

$$\text{PBL (dB)} = 20 \log_{10} (p_{\text{band}}/p_{\text{ref}}). \tag{3.1}$$

Here p_{band} is the rms average of the filtered signal and p_{ref} is 20 μPa for sound in air, as before. If we use a filter set that spans all frequencies, but with no overlap, the total strength in all bands must be that of the original signal. But note that this does not mean simple addition either of p_{band} or of PBL. As shown in Prob. 3.12, it is p^2 that is additive for signals of different frequencies:

$$p_{\text{rms}}^2 = \sum p_{\text{band}}^2 \tag{3.2}$$

so that

$$\begin{aligned} \text{SPL} &= 10 \log_{10} (p_{\text{rms}}^2/p_{\text{ref}}^2) \\ &= 10 \log_{10} \sum (p_{\text{band}}/p_{\text{ref}})^2 \\ &= 10 \log_{10} (\textstyle\sum 10^{\text{PBL}/10}). \end{aligned} \tag{3.3}$$

For example, suppose measurements in three different frequency bands give PBLs of 46, 48, and 40 dB, for which the corresponding p_{band}'s are 4, 5, and 2 mPa. The combined signal in all three bands does *not* have a PBL of 46 + 48 + 40 = 134 dB, nor is its effective pressure 4 + 5 + 2 = 11 mPa. Rather, p_{rms}^2 is 16 + 25 + 4 = 45 mPa^2, so that p_{rms} is 6.7 mPa and the corresponding level is

$$20 \log (6.7 \times 10^{-3}/2 \times 10^{-5}) = 50.5 \text{ dB}.$$

This is exactly the same answer that would result from direct use of (3.3):

$$10 \log (10^{4.6} + 10^{4.8} + 10^{4.0}) = 50.5 \text{ dB}.$$

The most commonly used filters have a bandwidth of one octave. This means the highest frequency allowed to pass must be exactly twice the lowest. An octave-band filter is labeled with the center frequency of its passband, but that is defined as the geometric, not the arithmetic, mean of the upper and lower band limits:

$$f_c = \sqrt{f_u f_l}, \tag{3.4}$$

not $(f_u + f_l)/2$. Thus the 1-kHz octave-band filter, in order to have $10^6 = f_u f_l$ and $f_u/f_l = 2$, must have its cutoffs at $f_l = 707$ and $f_u = 1414$ Hz. Any real filter will not have an infinitely steep cutoff, of course; but for most purposes that can be well approximated, as illustrated in Fig. 3.1. It is now a widely accepted standard to choose the center frequencies for such a filter set so that all are

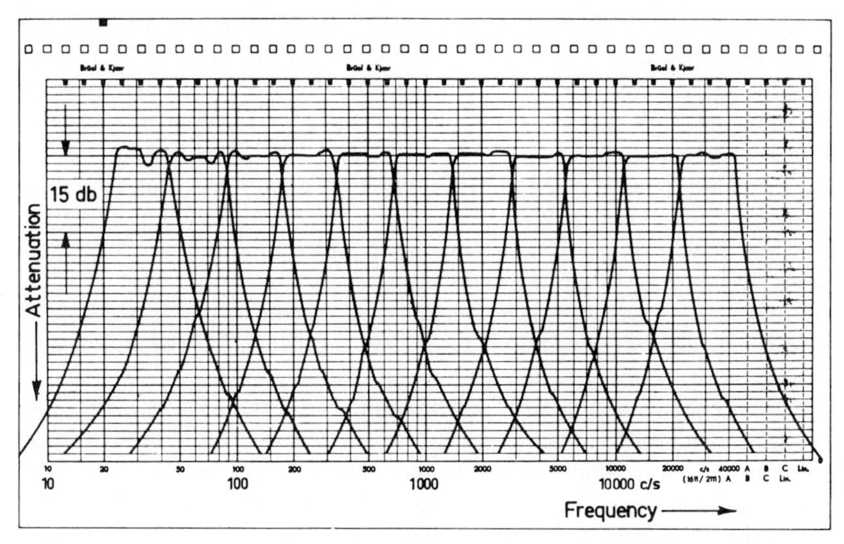

Figure 3.1 Transmission curves of a set of actual octave-band filters, showing slight overlap at boundaries. (Courtesy of Bruel and Kjaer Instruments, Inc., Marlborough, MA).

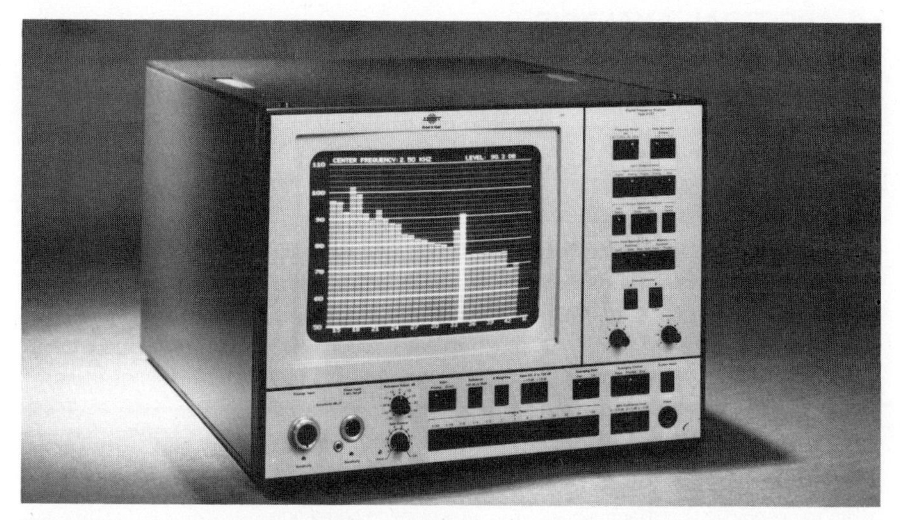

Figure 3.2 Graphical representation of band levels. Here a modern digital analyzer has simultaneously determined levels in many third-octave bands without requiring measurement through a physically separate filter for each one. (Courtesy of Bruel and Kjaer Instruments, Inc., Marlborough, MA.)

doublings or halvings of 1000 Hz: . . . 63, 125, 250, 500, 1K, 2K, 4K, 8K, ... Results of measurements with such a filter set would conveniently be represented by a bar graph, as in Fig. 3.2.

For still finer analysis it is common to use another set of filters, with third-octave bandwidth. Whatever the ratio f_u/f_l is for one-third octave, this ratio applied three times must raise the frequency by a whole octave, which is a factor of 2. Thus the separation of one center frequency from the next, and the width of each third-octave filter, are both given by a factor of $\sqrt[3]{2} = 1.26$ in frequency. For instance, the 1000-Hz octave band mentioned above is divided into three bands with center frequencies $f_c = 794$, 1000, and 1260 Hz (though you will probably find these labeled 800 and 1250). The 1000-Hz third-octave band must extend only from $f_l = 891$ to $f_u = 1122$ Hz, in order that these bounding frequencies may have a ratio 1.26 and a geometric mean 1000 Hz. You may even encounter equipment with tenth-octave filters.

If band-level measurements are available, it is possible to construct weighted averages where those were not already directly measured. To compute an A-weighted sound level, for instance, you would use the adjustment factors given in Table 3.1. (They actually repeat information shown graphically in Fig. 2.4.) After each individual PBL is adjusted, the weighted levels would then be combined in the same way as in Eq. 3.3. It may also be of interest to adjust band levels with A weighting simply to see which band is the largest contributor to the overall weighted SPL. Then noise-reduction measures can be concentrated on that most offensive band.

3.2 SPECTRAL ANALYSIS

The logical extension of the preceding discussion would be to make larger and larger numbers of measurements through filters of increasingly narrow bandwidth. In the limit where we attain essentially continuous measurements at all frequencies, with infinitesimal bandwidths, we would say we have determined the **spectrum** of the original signal.

As long as we had a finite bandwidth, it was still appropriate to characterize the finite signal strength for each filter by p_{rms} or by the corresponding PBL in dB. But as the bandwidth approaches zero, the signal strength passing through

TABLE 3.1 STANDARD OCTAVE-BAND FILTERS AND ADJUSTMENT FACTORS TO BE ADDED TO OCTAVE-BAND LEVELS TO OBTAIN A-WEIGHTED SOUND LEVEL

Center frequency (Hz)	Bandwidth (Hz)	A-weighting correction (dB)
31	22	−39.4
63	44	−26.2
125	88	−16.1
250	177	−8.6
500	354	−3.2
1000	707	0
2000	1414	+1.2
4000	2830	+1.0
8000	5660	−1.1

may also approach zero; it is then the **spectral density** of the signal strength that is of interest. By this we mean the signal strength per unit frequency range:

$$\boxed{S_p(f) = \Delta(p_{\text{rms}}^2 \text{ in } \Delta f)/\Delta f} \tag{3.5}$$

in the limit as Δf approaches zero. (Why Δp^2 rather than Δp? Because its definition as an rms average means p^2 is additive but p is not; see Eq. 3.2 and Prob. 3.12.) The units of this spectrum function $S_p(f)$ must be Pa^2/Hz. Like total sound levels, this is more commonly reported in coded form as a **pressure spectrum level** (in dB *re* S_{ref}):

$$\text{PSL}(f) = 10 \log [S_p(f)/S_{\text{ref}}]. \tag{3.6}$$

For sound in air, the reference level is generally chosen to be

$$S_{\text{ref}} = (20\ \mu\text{Pa})^2/(1\ \text{Hz}) = 4 \times 10^{-10}\ \text{Pa}^2/\text{Hz}, \tag{3.7}$$

that is, sound strong enough that a 1-Hz-wide filter would let through just enough to register a PBL of 0 dB.

As an example, a spectrum function

$$S(f) = (10^{-4}\ \text{Pa}^2/\text{Hz}) \exp(-f/1\ \text{kHz})$$

will have

$$\text{PSL}(f) = 54 - 4.34(f/1\ \text{kHz}),$$

where 54 comes from $10 \log_{10} [10^{-4}/(4 \times 10^{-10})]$ and 4.34 from $10 \log_{10} (e)$.

Note that overall SPL or band levels can be calculated from $S_p(f)$ by turning definition 3.5 around:

$$p_{\text{rms}}^2 = \int_0^\infty S_p(f)\, df \tag{3.8}$$

and

$$p_{\text{band}}^2 = \int_{f_l}^{f_u} S_p(f)\, df. \tag{3.9}$$

Over a band whose width $\Delta f = f_u - f_l$ is sufficiently narrow that $S_p(f)$ is nearly constant, (3.9) gives the approximations

$$p_{\text{band}}^2 \simeq S_p(f_c)\, \Delta f \tag{3.10}$$

and

$$\text{PBL} \simeq \text{PSL}(f_c) + 10 \log \Delta f. \tag{3.11}$$

For the example in the preceding paragraph

$$S_p(2\ \text{kHz}) = 10^{-4} e^{-2} = 1.35 \times 10^{-5}\ \text{Pa}^2/\text{Hz},$$

$$\text{PSL}(2\ \text{kHz}) = 54 - 4.34 \times 2 = 45.3\ \text{dB},$$

and the 2-kHz octave band has $\Delta f = 1414$ Hz. Thus we would estimate

$$p_{\text{band}} \simeq \sqrt{(1.35 \times 10^{-5})(1414)} = 0.138 \text{ Pa}$$

and
$$\text{PBL} \simeq 45.3 + 10 \log 1414 = 76.8 \text{ dB}.$$

For comparison, the exact result from (3.9) would be

$$p_{\text{band}}^2 = 0.1(e^{-1.414} - e^{-2.828}) = 0.0184,$$

$$p_{\text{band}} = 0.136 \text{ Pa} \quad \text{and} \quad \text{PBL} = 76.6 \text{ dB}.$$

3.3 AUTOMATIC SPECTRUM ANALYZERS

How is the measurement of a sound spectrum actually realized in the laboratory? Three approaches are available. First, we can have a set of many separate filters, each physically distinct (even if all enclosed in a single box), and make measurements through each one. If we are satisfied with only a rough approximation to the true spectrum, we may well do this with octave or third-octave filters, as described above. And that may be entirely adequate for some work with broadband noise sources. But when very fine discrimination in frequency is required, this multiple-filter approach is clearly not very practical.

Second, we might have a single **tunable filter** and make measurements continuously while sweeping the filter's center frequency. The most straightforward way to do this would be with a resonant "tank circuit" using a variable capacitor, just like a home radio tuning dial. If the frequency control is linked (by a belt, chain, or cable) with a strip-chart recorder, we have an electromechanical system that can with the single push of a button make automatic recordings of frequency spectra. The filtered microphone signal, probably in the form of PBL(dB), controls pen movement on the vertical axis while the chart paper moves in synchronism with the filter tuning to produce the horizontal axis.

There is another, purely electronic, way of simulating a tunable filter that is sometimes advantageous. This is the **heterodyne analyzer,** which is available in many laboratories as a plug-in unit for a storage oscilloscope. Here a sawtooth waveform from an internal oscillator controls the frequency of an internal comparison signal, while at the same time providing the horizontal drive for an oscilloscope display. Through heterodyne mixing, any component of the input signal whose frequency is close enough to that of the comparison signal will generate an output that can pass through a narrow-band filter and drive the vertical display. As the reference frequency sweeps, then, each component of the input in turn causes a peak to appear on the screen. Since there are no moving mechanical parts, these analyzers can in some cases perform spectrum measurement much faster than the electromechanical systems. Figure 3.3 shows a heterodyne analyzer spectrum and illustrates how one may control the resolving power by choosing wider or narrower bandwidths for the tunable filter. In simplest form this merely means substituting one resistor for another in the resonant

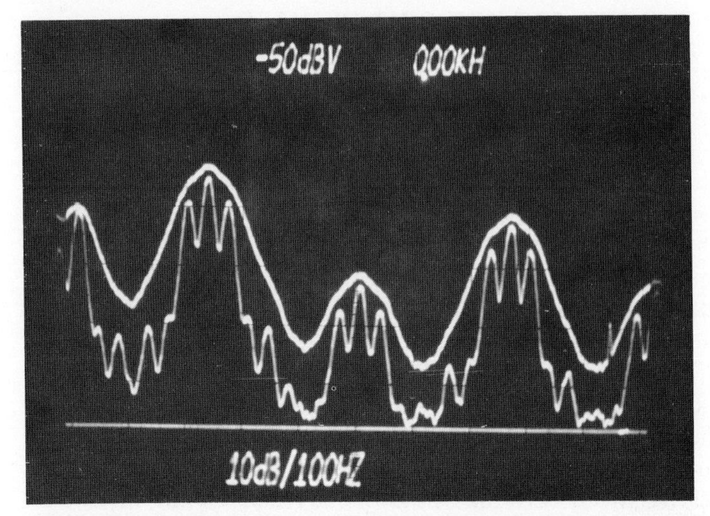

Figure 3.3 Oscilloscope traces showing heterodyne analyzer results for a 500-Hz rectangular wave whose amplitude was modulated at a rate of 100 Hz. Traces were taken with higher and lower resolution settings, and frequency range 0 to 2 kHz is displayed.

circuit. It is important to note that these filters generally maintain a constant bandwidth (expressed as a frequency difference in hertz) as they sweep, which is *not* the same as having always a constant fraction of an octave.

The third approach to spectrum measurement, which has become extremely powerful in the last decade, is **digital analysis.** Here the original waveform is stored in the memory of a small dedicated computer. Then this information can be operated on at leisure (which may mean only a fraction of a second) to extract information about its frequency content. There exist two distinct computing algorithms (or "digital filters") which are equivalent to the two types of bandwidth variation mentioned above. Recursive digital filters can analyze simultaneously in many octave bands (or third-octave bands), and that is particularly useful in measuring noise spectra. (It is this type that is illustrated in Fig. 3.2.) But FFT (fast Fourier transform) programs will perform constant-bandwidth narrow-band analysis; it is typical to resolve details whose separation is about 1 percent of the total frequency range being analyzed.

These digital algorithms may (or may not, depending on the parameters of operation) achieve **real-time analysis.** That would mean the spectrum of a changing source is continually updated during the change, with no incoming data ignored because of "busy signals" and no noticeable waiting time for the analysis to be performed. All the information in each waveform sample is used in determining the entire spectrum. In contrast, swept-filter analyzers derive successive parts of the spectrum from successively later samples of signal. So the ability of a digital analyzer to capture and store a waveform gives it a distinct advantage over both electromechanical and heterodyne analyzers if the signal to be studied is transient rather than steady.

*3.4 FOURIER THEORY FOR PERIODIC SIGNALS

Much of the time, it is enough to make empirical use of the electronic spectrum analyzers described above and simply accept their results. But it is also very useful to be aware of some extremely powerful mathematics that lies behind the concept of a spectrum. Without presenting proofs in every case, we summarize here a few of the most useful properties that stem from the discoveries of the young Frenchman Fourier in the early nineteenth century.

We may begin with some simple qualitative statements. **Fourier synthesis** points out that adding together many sinusoidal waves of different frequencies will produce a total disturbance with some more complicated shape. The principle of superposition (Sec. 1.2) assures us that the combined waveform is physically meaningful if we are dealing with a linear system. **Fourier analysis,** on the other hand, proposes the opposite: given any complex waveform, we may choose to view it as a mixture of simple sine waves. To be more specific, the mathematics of Fourier synthesis (relatively easy) takes as given the recipe for how many sine waves are to be added, and the frequency and strength of each, and considers the calculation of the resulting total waveform and description of its properties. Fourier analysis (much more sophisticated) takes any specified complex wave and shows explicitly how to calculate the strengths and frequencies of the components needed to construct it.

Fourier synthesis is represented symbolically by

$$g(t) = \sum_n C_n \cos(\omega_n t - \phi_n). \tag{3.12}$$

Here $g(t)$ is the total disturbance created by the individual components labeled with the index $n = 1, 2, 3, \ldots$; C_n is the amplitude of each, ω_n the frequency, and ϕ_n the phase lag. Here is an extremely important distinction that should be made: *If* all frequencies used are exact integral multiples of the first one ($\omega_n = n\omega_1$), then the result of synthesizing all these waves is guaranteed to be a periodic wave; that is, the signal will have the property $g(t + T) = g(t)$ for all values of t. Such a set of frequencies is called a **harmonic series** (borrowing musical terminology again), and the lowest frequency ω_1 is called the **fundamental** frequency. The period of repetition of the complex wave must be that of the fundamental component: $T = 2\pi/\omega_1$. The truth of this claim follows directly from observing that after time T (but no shorter time) every component has completed an integral number of cycles and is starting a repeat performance. *If,* on the other hand, synthesis is done with frequencies that do *not* form a harmonic series, then the total disturbance will *not* be periodic. In particular, if a continuous range of frequencies is all present, then the resulting mixture cannot be a repetitive waveform.

The same distinction works with equal importance in the other direction. *If* we perform Fourier analysis of a periodic signal that always repeats the same waveform after a time T (over the entire domain $-\infty < t < +\infty$), then the spectrum of this signal must consist of spikes at the discrete harmonic series of frequencies $\omega_n = 2\pi n/T$ and *nothing* in between. On the other hand, *if* we analyze a nonperiodic signal, then we *must* find that it has a nonharmonic spectrum. In particular, both finite-length transients and random-noise signals always have continuous spectra, meaning that the sum in (3.12) would become an integral with $C(\omega)$ existing over a continuous range of frequencies. We can say, then, that the first signal in Fig. 2.1 has a harmonic spectrum but the second has a continuous spectrum. Let us pursue the case of periodic signals and their discrete spectra here, and return to the problem of continuous spectra in the following section.

The basic task of Fourier-series analysis is to take a periodic function $g(t)$ as given and ask how to determine the amplitudes C_n and ϕ_n of its components. It becomes a little easier to see how to do this if we use the trigonometric identity

$$\cos(\omega t - \phi) = \cos \omega t \cos \phi + \sin \omega t \sin \phi$$

to write (3.12) as

$$g(t) = \sum_{n=1}^{\infty} (A_n \cos \omega_n t + B_n \sin \omega_n t). \tag{3.13}$$

Here we have defined

$$A_n = C_n \cos \phi_n, \qquad B_n = C_n \sin \phi_n, \tag{3.14}$$

and since for now $g(t)$ must be periodic with period T,

$$\omega_n = n2\pi/T. \tag{3.15}$$

The clever trick that accomplishes our purpose is to multiply both sides of (3.13) by $\cos \omega_m t$, where m is some fixed positive integer, and integrate over one full period of the wave. Integrating the right-hand side term by term, we find that only a single term survives, one for which n equals m:

$$\int_0^T g(t) \cos \omega_m t \, dt = (T/2)A_m. \tag{3.16}$$

But since we made no special assumptions about m, this must be true for every integer value of m. A similar calculation with sine instead of cosine gives

$$\int_0^T g(t) \sin \omega_m t \, dt = (T/2)B_m. \tag{3.17}$$

These two equations, read from right to left, assure us that we can always find all the A's and B's for any given $g(t)$. And, if we wish, we can go back to (3.12) instead of (3.13) by using

$$C_m = \sqrt{A_m^2 + B_m^2}, \qquad \phi_m = \arctan(B_m/A_m). \tag{3.18}$$

In any case where it might make the integration more convenient, the limits $-T/2$, $+T/2$ (or any others that cover one complete period) are just as good as 0, $+T$.

We have supposed that $g(t)$ is strictly an ac signal. It may be allowed to have a dc component as well if the summation in (3.12) or (3.13) is extended to include a term

$$A_0 = (1/T) \int_0^T g(t)\, dt. \tag{3.19}$$

A trivial first example, just to verify that the machinery works, could be $g(t) = \cos 2\pi t/T$. Equations 3.16 and 3.17 quickly produce $A_1 = 1$ and all other coefficients zero. A more interesting example is a square wave, represented by $g(t) = +V$ for $0 < t < T/2$ and $-V$ for $T/2 < t < T$, together with $g(t + T) = g(t)$ for all other t to describe its infinite repetition. Here we obtain all $A_m = 0$ from (3.16), and (3.17) gives $B_m = 4V/m\pi$ for all odd values of m, but $B_m = 0$ for even m. Figure 3.4 shows how the square wave shape is more and more nearly reproduced as larger numbers of terms are included in the series. Further examples are provided in the problems.

Consideration of symmetry can sometimes save grinding through these integrals. If a particular $g(t)$ happened to be an even function, that is, one for which $g(-t) = g(t)$, that would already guarantee that all $B_m = 0$ and only A_m need be considered. Similarly, an odd function $g(-t) = -g(t)$ will have all $A_m = 0$.

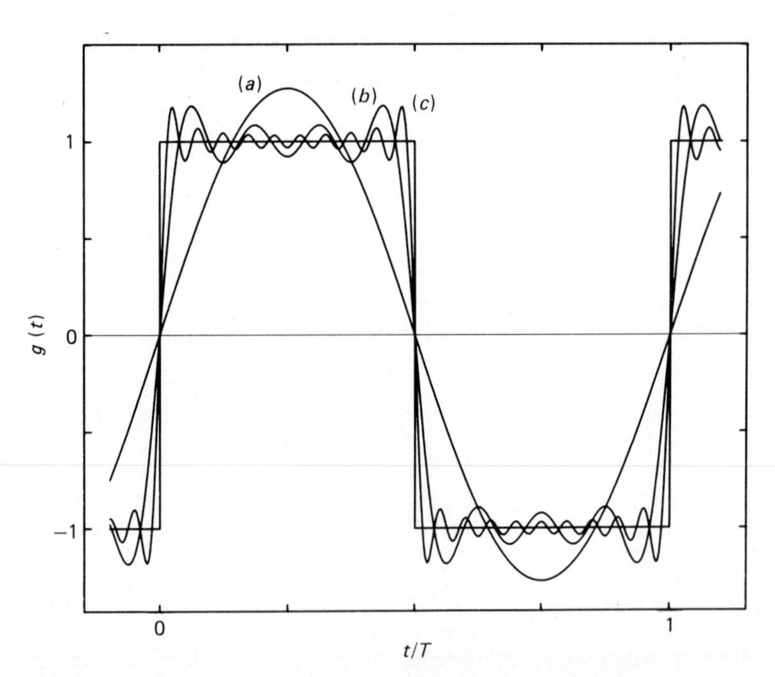

Figure 3.4 A square wave and its approximation by truncated Fourier series with (a) one term ($n = 1$), (b) four terms ($n = 1, 3, 5, 7$), and (c) ten terms ($n = 1$ through 19).

Take as an example the triangular wave defined by

$$g(t) = V(1 - 4|t|/T) \quad \text{for} \quad -T/2 < t < +T/2$$

and $g(t \pm T) = g(t)$ for all other values of t. [Please pause to substitute some values like $t = T/4$ and $T/2$ and be sure you are convinced what this formula for $g(t)$ says.] Since this is an even function, we can say immediately that all B_m are zero. In computing the A_m, it is easier to integrate from $-T/2$ to $+T/2$ than from 0 to T; in fact, since the integrand is even, it is easier still to integrate only from 0 to $T/2$ and double the result. Thus each coefficient is

$$A_m = (4V/T) \int_0^{T/2} (1 - 4t/T) \cos(m2\pi t/T)\, dt,$$

which is easily computed with the help of the substitution $y = m2\pi t/T$ to be

$$A_m = (2V/\pi m)[\sin y - (2/\pi m)(y \sin y + \cos y)]_0^{m\pi}.$$

This reduces to

$$A_m = 8V/\pi^2 m^2$$

for odd m and zero for even m.

It is intuitively appealing to suppose that a function that executes very rapid changes must require a recipe in which the high-frequency components have appreciable amplitudes. That is true and can be stated more precisely as follows. Any function that has discontinuous jumps will have A's and B's whose general trend is proportional to m^{-1}, as we saw for square waves above. Any function that is continuous but has jumps in its first derivative, such as a triangular wave, must have Fourier coefficients behaving asymptotically as m^{-2}. In general, if the pth derivative is the first one to be discontinuous, the component amplitudes will fall off for sufficiently large p in proportion to $m^{-(p+1)}$.

If we are willing to use complex-variable ("phasor") representations of oscillating quantities (see Appendix B), some of the statements of Fourier theory become even more simple and symmetrical, or easier to prove. Using the identity

$$\exp(jx) = \cos x + j \sin x,$$

where $j = \sqrt{-1}$, we may bring (3.12) or (3.13) into the form

$$\boxed{g(t) = \sum_{-\infty}^{+\infty} a_n\, e^{+jn\omega_1 t}.} \tag{3.20}$$

Now the sum includes negative integers n as well as positive and zero. The amplitude and phase of each component are both incorporated in the complex coefficient

$$\begin{aligned} a_n &= (A_n - jB_n)/2 \\ &= C_n(e^{-j\phi_n})/2 \qquad (n \neq 0). \end{aligned} \tag{3.21}$$

We must omit the factor 2, however, from $a_0 = A_0$.

Similarly, (3.16) and (3.17) are replaced by

$$\boxed{\int_0^T g(t)\, e^{-jm\omega_1 t}\, dt = Ta_m.} \tag{3.22}$$

The requirement that $g(t)$ is real imposes the restriction that a_{-m} must be the complex conjugate of a_m, so that a_{-m} does not actually represent any additional information. Problem 3.9 illustrates why the phase ϕ_m is often physically unimportant, so that we are concerned only with the magnitude of a_m.

An example where complex notation reduces computational labor is a half-rectified sine wave,

$$g(t) = V \sin 2\pi t/T \quad \text{for} \quad 0 < t < T/2$$

and 0 for $T/2 < t < T$. Equation 3.22 is easily integrated by using $(e^{jy} - e^{-jy})/2j$ in place of $\sin y$ and produces $a_1 = -jV/4$, $a_m = 0$ for all other odd m, and

$$a_m = -V/\pi(m^2 - 1)$$

for even m. These are equivalent to $A_0 = V/\pi$, $A_m = -2V/\pi(m^2 - 1)$ (nonzero even m only), and $B_1 = V/2$. Successive approximations are illustrated in Fig. 3.5.

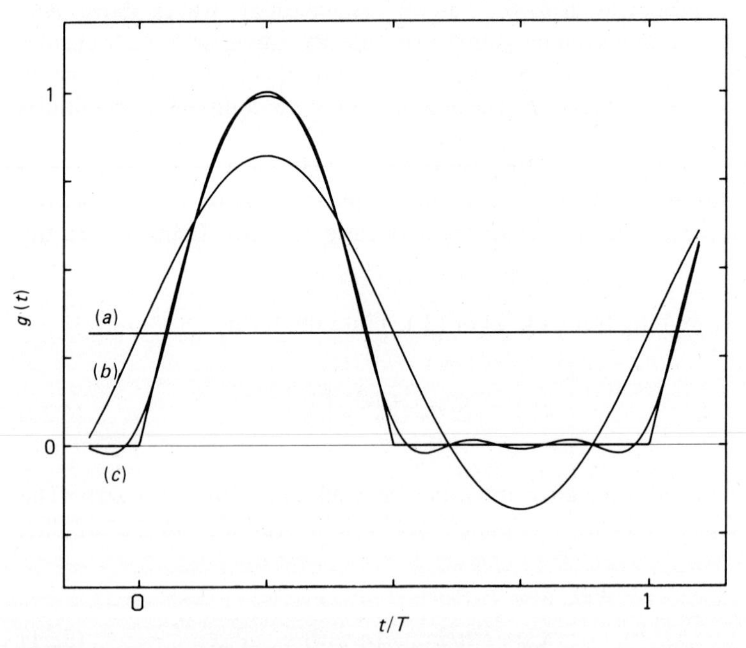

Figure 3.5 A half-rectified sine wave and its approximation by (*a*) one ($n = 0$), (*b*) two ($n = 0, 1$), and (*c*) four ($n = 0, 1, 2, 4$) Fourier terms.

*3.5 FOURIER INTEGRALS

The equations in the preceding section apply only in the case of a periodic signal. For a nonrepeating signal and its continuous spectrum let us replace the sum in (3.20) by an integral:

$$g(t) = \int_{-\infty}^{+\infty} a(\omega)\, e^{+j\omega t}\, d\omega. \tag{3.23}$$

Notice that the frequency ω here is a continuous variable, where in (3.20) above we had a harmonic series labeled by the discrete variable n. Instead of finite coefficients a_n we now have a continuous function $a(\omega)$ to represent the spectral amplitudes, and it is $a(\omega)\, d\omega$ rather than a_n that has the same physical units as $g(t)$. [It is a mathematical convenience to let the integration extend to $-\infty$, and you should not fret over giving a physical interpretation to negative frequencies; (3.25) will give an alternate form involving only positive frequencies.]

Again, the trick with which $a(\omega)$ can be determined from $g(t)$ is to multiply both sides by $\exp(-j\gamma t)$ and integrate over dt; but now the integration must extend over all time to get all the information that is in $g(t)$. Evaluating the right-hand side is a little tricky (see Prob. 3.13), but it results in a close analogy to Eq. 3.22:

$$\int_{-\infty}^{+\infty} g(t)\, e^{-j\omega t}\, dt = 2\pi\, a(\omega). \tag{3.24}$$

Again, if $g(t)$ is real, $a(-\omega)$ must be the complex conjugate of $a(+\omega)$.

As an example, consider a signal consisting of a single isolated pulse: $g(t) = V$ for $0 < t < \tau$ and zero otherwise. Equation 3.24 then gives

$$a(\omega) = (V/\pi\omega) \sin(\omega\tau/2) \exp(-j\omega\tau/2).$$

If we had taken $g(t)$ to be 1 between $-\tau/2$ and $+\tau/2$ instead, we would have obtained

$$a(\omega) = (V/\pi\omega) \sin(\omega\tau/2),$$

which has exactly the same magnitude for all ω as the previous expression, even though different phase.

The last example illustrates (as does Prob. 3.9) that the phases are of less physical interest than the magnitudes, because they depend on the arbitrary definition of time zero. In fact, by putting $t = t_1 + T$ in (3.24), it is easy to see in general that shifting the time origin by any amount T merely multiplies $a(\omega)$ by a factor $\exp(-j\omega T)$.

Those who prefer may express Fourier integral theory entirely in terms of real functions. As you may show (in Prob. 3.16), Eqs. 3.23 and 3.24 are then replaced by

$$g(t) = \int_0^{\infty} [A(\omega) \cos \omega t + B(\omega) \sin \omega t] \, d\omega \tag{3.25}$$

and

$$\int_{-\infty}^{+\infty} g(t) \cos \omega t \, dt = \pi A(\omega) \tag{3.26}$$

$$\int_{-\infty}^{+\infty} g(t) \sin \omega t \, dt = \pi B(\omega). \tag{3.27}$$

$A(\omega)$ and $B(\omega)$ are again related to $a(\omega)$ by (3.21).

There is a very powerful theorem in Fourier theory that is represented by the equation

$$\Delta\omega \, \Delta\tau > 2\pi. \tag{3.28}$$

Here $\Delta\tau$ is some measure of the duration of a wave train and $\Delta\omega$ the range of frequencies that must be used to construct it. It is possible to give more exact definitions of both quantities, but we will not do so here. The simplest cases are those where $\Delta\tau$ represents the actual length of a signal that has no disturbance either preceding or following it. The example just above of a single rectangular pulse has frequency components whose strength first reaches zero when $\sin(\omega\tau/2) = 0$, or $\omega\tau/2 = \pm\pi$. Then we could take $\Delta\omega$ to be about $2\pi/\tau$ or more, which would agree with (3.28). Problems 3.14 and 3.15 provide further illustrations.

Equation 3.28 is often called the uncertainty principle, for it says that it is impossible in principle to precisely define both a unique frequency and a definite time of occurrence for a wave: the shorter its duration, the broader the range of frequency components required to construct it, and only a wave lasting forever can have a perfectly defined frequency. This is a property of all waves—light, radio, water waves, and so on, as well as sound. It is perhaps most famous for its application to quantum waves, where it says that the precision in our knowledge of frequency (or energy) and of duration or lifetime cannot both be made arbitrarily small at the same time.

*3.6 RELATING THEORY AND MEASUREMENT

A truly complete Fourier analysis would determine both the magnitude and phase of the function $a(\omega)$, or equivalently the two real functions $A(\omega)$ and $B(\omega)$. All this information would be needed to carry out a Fourier synthesis that would reproduce the original waveform. Many electronic spectrum analyzers, however, determine *only* the absolute value of the complex function $a(\omega)$; this is equivalent to finding only $A^2 + B^2$ but neither A nor B individually. Generally we are satisfied with this partial information, since it does tell us how much energy is represented at each frequency. We must understand, however, that as long as we lack phase information we cannot actually reconstruct the original waveform (see Prob. 3.18).

If $g(t)$ represents the actual pressure history $p(t)$ of a sound wave falling on a microphone, we have two representations of its spectrum. In (3.8) we had an experimentalist's definition of the long-term average of p^2 as

$$p_{\text{rms}}^2 = \int_0^\infty S(f)\, df. \tag{3.29}$$

But a theorist using (3.23) and averaging over $0 < t < T$ would express this as

$$p_{\text{rms}}^2 = (2\pi/T) \int_{-\infty}^{+\infty} |a(\omega)|^2\, d\omega, \tag{3.30}$$

as you may show in Prob. 3.17. Since $a(-\omega)$ and $a(\omega)$ both have the same absolute value, the last integral may be replaced by twice the integral from 0 to ∞. Then, using $\omega = 2\pi f$, we must have the relation

$$S(f) = (8\pi^2/T)|a(2\pi f)|^2. \tag{3.31}$$

This still involves the very puzzling factor T. It appears because $S(f)$ was constructed to represent a finite average signal strength (total energy over a very long time T already divided by T), whereas $a(\omega)$ represents the entire signal over all time (the division by T remaining to be done). Let us illustrate that rather delicate point with some examples.

First take the finite-duration single pulse, $g(t) = V$ for $-\tau/2 < t < +\tau/2$ and zero otherwise. We found above that it has a finite amplitude function

$$a(\omega) = (V/\pi\omega) \sin \omega\tau/2.$$

The corresponding $S(f)$ is the limit as T goes to infinity of

$$(8\pi^2/T)(V/2\pi^2 f)^2 \sin^2 (\pi f\tau).$$

It is correct that this appears infinitesimal, for if the finite energy in the single pulse is averaged over an extremely long time it will appear that the average power is nearly zero.

For another example take a finite-duration sinusoid,

$$g(t) = V \sin \omega_0 t \quad \text{for} \quad -\tau/2 < t < +\tau/2.$$

Problem 3.14 shows that $a(\omega)$ has magnitude

$$(V\tau/4\pi)[(\sin y)/y - (\sin z)/z]$$

with

$$y = (\omega - \omega_0)\tau/2 \quad \text{and} \quad z = (\omega + \omega_0)\tau/2.$$

Note that $a(\omega)$ at $\omega = \omega_0$ grows without bound as τ becomes infinite, though $\int a(\omega)\, d\omega = 0$ and remains finite. In this case (3.31) gives

$$S(f) \simeq (V^2\tau^2/2T) \sin^2 y/y^2$$

since the term involving z is relatively small. If the averaging time T goes to infinity while τ remains finite, this becomes infinitesimal as did the preceding example. But if T and τ are set equal and both become large together, $S(f)$ actually grows with T in a narrow range around ω_0; but that range shrinks as T grows in such a way that $\int S(f)\, df = V^2/2$ and remains finite. This is exactly the mean-square average we would expect for a sinusoidal wave.

Finally, consider two important types of noise. **White noise** is defined to have equal strength per unit bandwidth at all frequencies: $S(f)$ = constant. If $g(t)$ is taken to be a sample of white noise with strength S_1 for $0 < t < T$ but zero otherwise, then $a(\omega)$ has the same magnitude $(1/2\pi)\sqrt{TS_1/2}$ for all ω. **Pink noise,** in order to have equal strength in each octave band, has

$$S(f) = K/f \quad \text{and} \quad |a(\omega)| = \sqrt{TK/4\pi\omega},$$

with K = constant (see Prob. 3.19). In both cases $a(\omega)$ grows with T because longer samples of noise must contain more energy.

The heart of this matter is that any practical measurement must be made within a finite time. Simple as it may make the theory to consider infinite wave trains, those are never the basis of our work with electronic spectrum analyzers. The effective averaging time T will be the same as the duration τ of the analyzed signal and can only be large, not infinite. Furthermore, the measurement cannot distinguish between τ as finite duration of the signal itself and τ as the longest portion of a waveform that can be stored at one time in a digital analyzer, or τ as the effective integration time of any other analyzer. Since by definition the analyzer cannot include data beyond this interval, the spectrum it produces must not depend on such data.

Heterodyne analyzers effectively assume that the missing data are identically zero. Thus even though theory tells us that the spectrum of a periodic wave exists only at a discrete harmonic series of frequencies, the electronic wave analyzer will never show us perfectly sharp spikes. Instead, each peak will be broadened to a bandwidth of about $1/\tau$. Thus the shorter the time in which we attempt to measure a spectrum, the fuzzier the peaks must become; indeed if we push τ too small, we may smear all the peaks together into an apparent continuum. You will best appreciate this point if you actually experiment with a continuous periodic signal (e.g., a 100-Hz square wave) and a heterodyne analyzer, trying to rush the analysis by forcing the sweep period below 1 s.

FFT analyzers calculate as if the data seen through the time window $0 < t < \tau$ were one cycle of an infinite repeating signal with period τ. This can give rise to bothersome effects, for it may introduce fictitious jumps into $g(t)$. (Imagine, for example, that τ happened to be 4.7 times the period of a sine-wave signal, as in Fig. 3.6.) In order to avoid false broadening of spectrum lines, such analyzers generally make available a modified algorithm (under a name such as Hanning or cosine weighting) which assigns less importance to the data near the beginning and end of the time window.

It is possible that a microphone could receive a total signal coming from several different sources, for which there is both a continuous (noise) spectrum

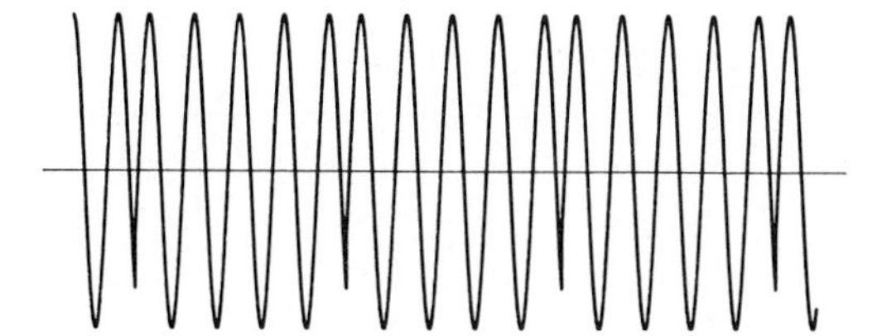

Figure 3.6 The signal effectively analyzed by an FFT when a sine wave can only be sampled in a time window of finite length. Fictitious discontinuities are introduced.

and also some discrete components superimposed on it. In such cases, whether the discrete components are actually recognized as such ("resolved" from the smooth continuum) may depend on how narrow a bandwidth is used in the analysis.

PROBLEMS

3.1. **(a)** If that portion of a signal within the 500-Hz octave band has rms pressure 0.4 Pa, what is the PBL for that band?
(b) If the energy in this octave band is divided equally among its three third-octave bands, what is the band level for each of them?

3.2. **(a)** If a certain signal has band levels 37, 50, 48, 43, 45, and 35 dB in the octave bands from 125 through 4000 Hz, respectively, and negligible strength outside that range, what is its SPL?
(b) What is the *A*-weighted sound level for this signal?

3.3. **(a)** If $S_p(f) = (3 \times 10^{-7}\ \text{Pa}^2/\text{Hz}^3)f^2$ for $f < 1000$ Hz, and $(3 \times 10^5\ \text{Pa}^2\text{Hz})/f^2$ for $f > 1000$ Hz, what is the pressure spectrum level?
(b) What is the pressure band level for the 500-Hz octave band?
(c) For the 2-kHz octave band?
(d) What is the overall SPL of this signal?

3.4. For the signal described in Prob. 3.2, estimate the average $S_p(f)$ and average PSL in the 500-Hz band alone.

3.5. **(a)** If the PSL of a signal is $87 - 20 \log f$ dB(*re* S_{ref}), what is its spectral density $S_p(f)$?
(b) What would be the band level of the 1000-Hz third-octave band?

3.6. For the example at the end of Sec. 3.2, use the approximation of Eq. 3.11 to show which octave band has the highest PBL. Can you do this elegantly, rather than merely grinding out PBL for all bands and then picking the largest?

3.7. Find the Fourier-series representation of a sawtooth wave (Fig. 3.7*a*) with maximum amplitude V and period T. This means you are to determine the coefficients A_n and B_n (or C_n and ϕ_n, or a_n) in terms of V and n, with Eqs. 3.16 and 3.17. [*Hint:* you must begin by writing an explicit mathematical expression for $g(t)$ that will have the property of rising linearly from $-V$ to $+V$ during time $\Delta t = T$, so that you can perform integrations with it.]

3.8. Find the Fourier series of a rectified sine wave, that is, $g(t) = |\sin 2\pi t/T|$. (See Fig. 3.7*b*.) Notice the relation of one of the amplitudes to Prob. 2.9.

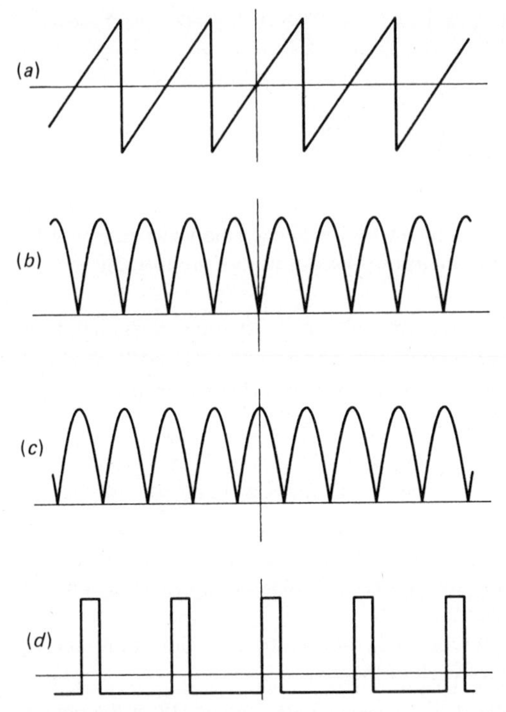

Figure 3.7 Signals used in Probs. 3.7 through 3.10.

3.9. Repeat Prob. 3.8, replacing sine by cosine (Fig. 3.7*c*). Discuss the relation between the two superficially different sets of amplitudes.

3.10. Show that the Fourier amplitudes of a pulse wave with duty factor β (Fig. 3.7*d*) are

$$C_n = 2V \sin \beta n\pi/(1 - \beta)n\pi.$$

[*Hints:* The algebra can be minimized by shifting the origin of time to represent the wave as an even function. It will be simpler still if you define $g(t)$ as $V_0/(1 - \beta)$ for $\Delta t = \beta T$ and zero otherwise, arguing that adding a steady dc component to the wave defined in Prob. 2.8 will not change any of the other amplitudes.] Verify that you recover the known answer for a square wave in the case $\beta = 1/2$. If $\beta = 1/p$ for some other integer p, which harmonics have zero amplitude?

3.11. For at least one of the preceding four problems, plot (preferably with the aid of a computer) a graph like Fig. 3.5 to show how addition of more terms in the series improves the approximation to the original function.

3.12. Consider any two components of different frequency in a signal: $p_1(t) = a \cos \omega t$, $p_2(t) = b \cos \gamma t$. Show that the long-term rms average of $p_1 + p_2$ is related to the two individual rms averages in accordance with Eq. 3.2.

3.13. Consider the integral described just above Eq. 3.24.

(a) Supposing that the two integrals may be done in either order, show why there should ultimately be no contribution from any ω not equal to γ.

(b) For $\omega = \gamma$ the integral may appear infinite, or at least indeterminate, but get around that as follows: Make an argument why $a(\omega)$ may be treated as a constant so that you need only consider the double integral of $\exp[j(\omega - \gamma)t]$. Do the integral over dt from $-T$ to $+T$ for some large but finite T, then do the integral over $d\omega$, and finally let T become infinite. Show that this produces the factor 2π in (3.24).

3.14. Consider a sine wave that is turned on only for a finite length of time: $g(t) = \sin \omega_0 t$ for $-\tau/2 < t < +\tau/2$, but zero otherwise (Fig. 3.8*a*). Considering ω_0 and τ as fixed, find the Fourier transform $a(\omega)$ for this wave. Sketch this $a(\omega)$, and adopt some reasonable definition $\Delta\omega$ for the range of ω in which $a(\omega)$ is large. Describe how $a(\omega)$ and $\Delta\omega$ change if τ is changed. Do your results support the uncertainty principle?

3.15. Consider a sinusoidal disturbance whose amplitude first grows and then decays exponentially:

$$g(t) = \sin \omega_0 t \exp(-|t|/\tau).$$

(See Fig. 3.8*b*.) Follow the same instructions as in Prob. 3.14.

3.16. Show a derivation of Eqs. 3.25 to 3.27. To do this, write out both $a(\omega)$ and $\exp(j\omega t)$ in terms of their real and imaginary parts. Use $a(-\omega) = a^*(\omega)$ to limit all integrals over $d\omega$ to the range $0 < \omega < \infty$.

3.17. If $p(t)$ is represented as a Fourier transform (Eq. 3.23), try to compute the average of p^2. In writing p twice, be sure you distinguish between the dummy variables ω_1 and ω_2 of the two integrations. The long-term average is then a third integral over time (say from $-T/2$ to $+T/2$), divided by T, in the limit as T goes to infinity. In the same imprecise spirit as in Prob. 3.13, change the order of integration to take dt first. Argue that the result is zero unless $\omega_1 + \omega_2 = 0$, and that a factor 2π will result when that condition is satisfied, thus justifying Eq. 3.30.

3.18. In order to demonstrate the effect of missing phase information, use for amplitudes the first four components of a sawtooth wave ($C_n = 2V/\pi n$ as shown in Prob. 3.7 for $n = 1, 2, 3, 4$ only) and plot the synthesized waveform that results for each of several arbitrary choices of the phases ϕ_n.

3.19. **(a)** Show that every octave band level for white noise is 3 dB higher than the preceding octave band level.

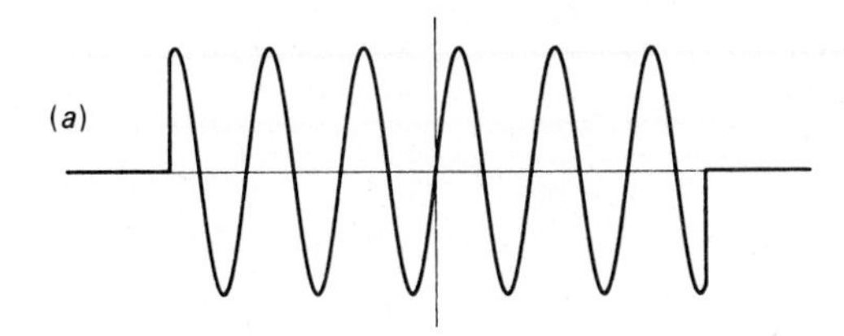

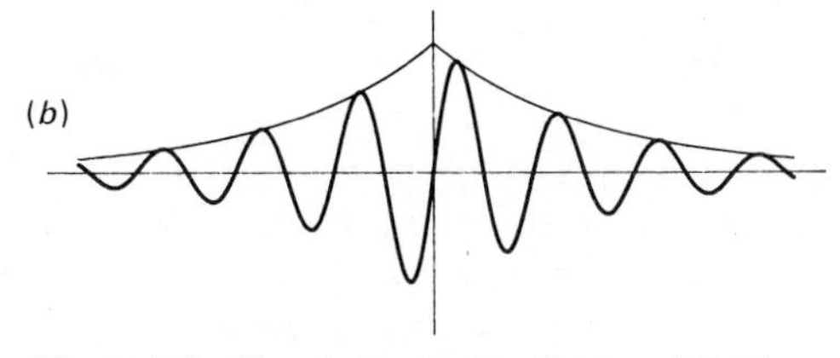

Figure 3.8 Signals for Probs. 3.14 and 3.15.

(b) Show that the assumption of equal band level in every octave band for pink noise requires $S(f) = K/f$. Use that to find the corresponding magnitude of $a(\omega)$.

LABORATORY EXERCISE

3.A. Get acquainted with the operation of whatever kind of spectrum analyzers you have available to you. Consult the manufacturer's instruction manuals, and try analyzing simple controllable sources in the lab. Once you are confident how to proceed, see what you can find out about a variety of real sound sources. Try in at least one case to show calibrated results with explicit numbers for the spectrum level.

Chapter 4

Some General Wave Phenomena

There are several more important things we can understand about sound purely on the basis of its being a wave phenomenon. Specifically, we make use of the superposition principle (introduced in Chap. 1) to derive results for various combinations of waves meeting at a common place and time. We can study these topics without knowing any details of how a sound wave actually works. All the developments in this chapter are equally applicable to radio, light, or any other kind of wave.

4.1 BEATS

Suppose that two sine waves with different frequencies happen to be arriving at the same point in space. What would our microphone detect there? As long as we limit ourselves to sounds of moderate strength, the superposition principle assures us that the actual pressure $p(t)$ at any moment is simply the sum of the two pressures each wave would have provided alone. Consider first a case where both amplitudes are the same, so that

$$p(t) = C(\cos \omega_1 t + \cos \omega_2 t). \qquad (4.1)$$

A standard trigonometric identity suggests another way of looking at this:

$$p(t) = 2C \cos [(\omega_1 - \omega_2)t/2] \cos [(\omega_1 + \omega_2)t/2]. \qquad (4.2)$$

The alternative expression (4.2) is most helpful in the case where $\Delta\omega = \omega_1 - \omega_2$ is small, illustrated in Fig. 4.1. The two waves remain nearly in phase, reinforcing each other, for a considerable time. When $\omega_1 t$ eventually gets near 180° ahead of $\omega_2 t$, the crests of one wave just cancel the troughs of the other

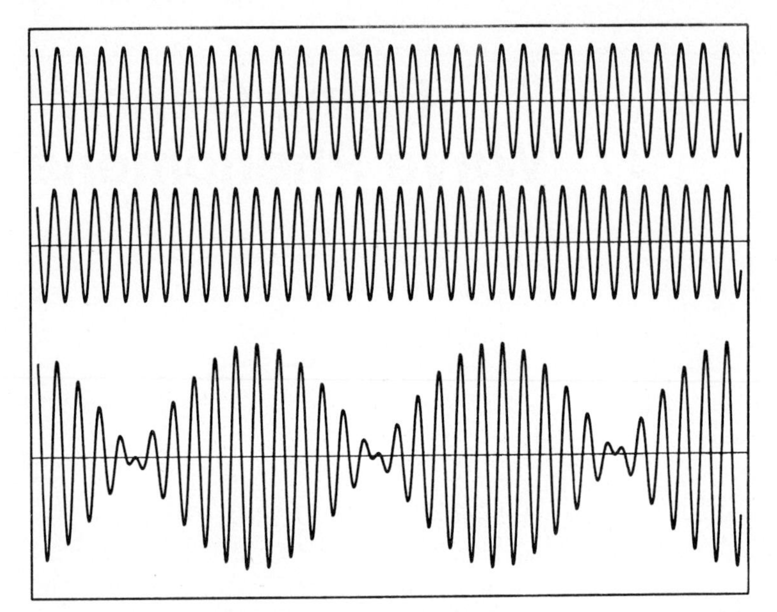

Figure 4.1 Two sine waves of equal amplitude whose frequencies differ by 10 percent, and their sum, illustrating the phenomenon of beats.

and there is practically no disturbance. After a long enough time for one wave to have gained a full cycle on the other, crests will again arrive together and the total will again become large. This suggests that we might want to regard this as a single wave with slowly changing amplitude:

$$p(t) = A(t) \cos \omega_{av} t, \tag{4.3}$$

where $\omega_{av} = (\omega_1 + \omega_2)/2$ and

$$A(t) = 2C \cos (\Delta\omega/2)t. \tag{4.4}$$

Positive and negative $A(t)$ (cosine of 0° or 180°) both represent large oscillations; so $(\Delta\omega/2)T_b = \pi$ gives the period T_b from one maximum to the next. Thus $T_b = 2\pi/\Delta\omega$, and the corresponding frequency is

$$\omega_b = \Delta\omega \quad \text{or} \quad f_b = f_1 - f_2. \tag{4.5}$$

This is called the **beat frequency,** since we hear the recurring surges as a series of rhythmic beats.

Beats may still be identified when the signals have unequal amplitudes (Fig. 4.2). In the general case with

$$p(t) = C \cos (\omega_1 t - \phi_1) + D \cos (\omega_2 t - \phi_2), \tag{4.6}$$

the trigonometric identities are less obvious. But by using phasors it is not difficult to show (Prob. B.3 in Appendix B), supposing C to be the larger of the two amplitudes, that we may write

$$p(t) = A(t) \cos [\omega_1 t - \phi(t)]. \tag{4.7}$$

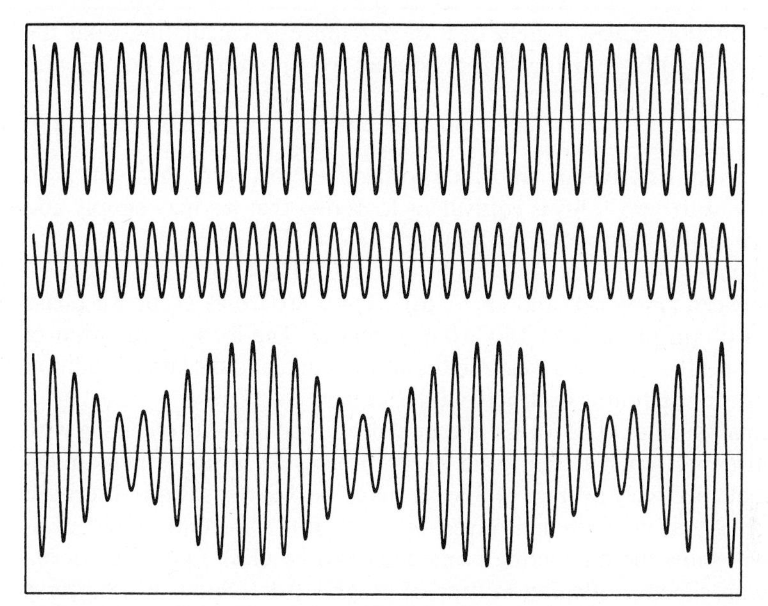

Figure 4.2 Addition of waves with slightly differing frequencies but unequal amplitudes, showing beats in which the minimum strength is not zero.

Here the slow variation $\phi(t)$ is uninteresting; much more important is

$$A^2(t) = C^2 + D^2 + 2CD \cos(\Delta\omega t - \phi_b), \tag{4.8}$$

which describes the amplitude of the slowly varying envelope in Fig. 4.2. The last expression shows that $A(t)$ oscillates at the beat frequency f_b between a maximum $C + D$ and a minimum $C - D$, justifying what is already strongly suggested in Fig. 4.2.

For instance, waves with amplitudes $C = 6$ Pa and $D = 4$ Pa (109.5 and 106 dB, respectively, when alone) and frequencies 256 and 261 Hz would combine to make 5 beats per second. The maximum beat strength would be $C + D = 10$ Pa (114 dB) and the minimum $C - D = 2$ Pa (100 dB).

4.2 COHERENCE

Two signals that have a definite fixed relative phase relation are called **coherent.** This could be the case for two signals both deriving from the same source, such as two speakers both being driven by the same signal generator or light from a single laser beam being split and recombined. If, on the other hand, there is no actual connection and so no physical reason for any particular phase relation, the sources are **incoherent.** This would be true of two violinists trying to play the same note together or two atoms emitting light of about the same color in an ordinary (nonlaser) source.

The distinction makes a great deal of difference in calculating what the combined signal will be. For incoherent sources we would follow Eq. 3.2 and Prob. 3.12 in saying that

$$p_{tot}^2 = p_1^2 + p_2^2, \tag{4.9}$$

where each p is an rms average and the prospective cross term involving $p_1 p_2$ has averaged out to zero. This is equivalent to saying that we may simply add energies or intensities for incoherent waves.

But for coherent sources (4.9) will be incorrect. The total strength may be anywhere between $(p_1 + p_2)^2$ and $(p_1 - p_2)^2$. Those extremes occur for phase differences of 0° (in phase) and 180° (out of phase). The former condition of maximum cooperation is called **constructive interference,** and the latter condition of maximum cancellation is **destructive interference.**

For signals such as (4.6) whose phase relation is definite yet changing with time, we must specify what time scale we are interested in. If we make measurements that effectively average (4.8) over many beat periods, we will seem to have a steady signal with strength given by (4.9). But sufficiently rapid measurements will show the fluctuating beats described by (4.8). Our ears, for example, have an effective integration time of roughly 0.1 s. So we will perceive fluctuations in loudness for beat frequencies much less than 10 Hz, but when f_b is much above 10 Hz the sound will seem to have constant strength.

In a similar way, we may solve a paradox presented by Eqs. 4.1 and 4.2. The first of these suggests that a Fourier analyzer would find two steady components at frequencies f_1 and f_2, while the second indicates only a single spectral peak whose height changes rhythmically. As you can easily verify in the lab, you may see either case depending on whether the effective integration time for the control setting chosen is longer or shorter than the beat period $T_b = 1/f_b$.

If three or more sine waves of slightly different frequencies are combined, every possible pair will gradually come in and out of phase with each other. The multitude of beats going on at several different beat rates simultaneously will have a somewhat irregular sound which is called **chorus effect** in music. It means that the sound of a dozen violins playing in unison is qualitatively different from the sound of a single instrument amplified.

In the limit of a great number of components all within a narrow band Δf in frequency, we have a narrow-band noise signal with a steady continuous long-term average spectrum. If our ears or other measuring apparatus have a shorter integration time than $1/\Delta f$, however, we notice irregular fluctuations in signal strength.

4.3 SPATIAL INTERFERENCE PATTERNS

If two coherent sources are sending sound into the same region of space, we may have constructive or destructive interference at each point. We must expect the nature of the combination to change from place to place, simply because of the time it takes each wave to get from its source to the observing point (Fig.

4.3). If L_1 is the distance from source S_1 to P and the speed of sound is c (340 m/s in air), the time required for a signal to cover this distance is L_1/c. Then at time t the point P is receiving a wave that left S_1 at time $t - L_1/c$. The arriving wave $p_1(t)$ must be proportional to what was radiated at that earlier time by S_1. Using the same argument about the second wave, and supposing both sources to be sinusoidal with the same frequency, we obtain

$$p(t) \propto C_1 \cos[\omega(t - L_1/c) - \phi_1] + C_2 \cos[\omega(t - L_2/c) - \phi_2]. \quad (4.10)$$

Thus the relative phase of the arriving waves is

$$\Delta\phi = -\omega(L_1 - L_2)/c - (\phi_1 - \phi_2). \quad (4.11)$$

If this is any integral multiple of 2π, crests arrive together to make constructive interference. In the simplest case where the two sources are in phase this reduces to

$$L_2 - L_1 = n2\pi c/\omega = n\lambda. \quad (4.12)$$

Clearly, any integral number of wavelengths plus an extra half is what would make a crest of one wave arrive with a trough of the other for destructive interference.

As the receiver moves around, L_1 and L_2 change, and so must the signal strength received. It is quite easy to hear this by moving your head around while listening to a signal of 3 or 4 kHz frequency, since this means a wavelength on the order of 10 cm. In an ordinary room it is not even necessary to send the

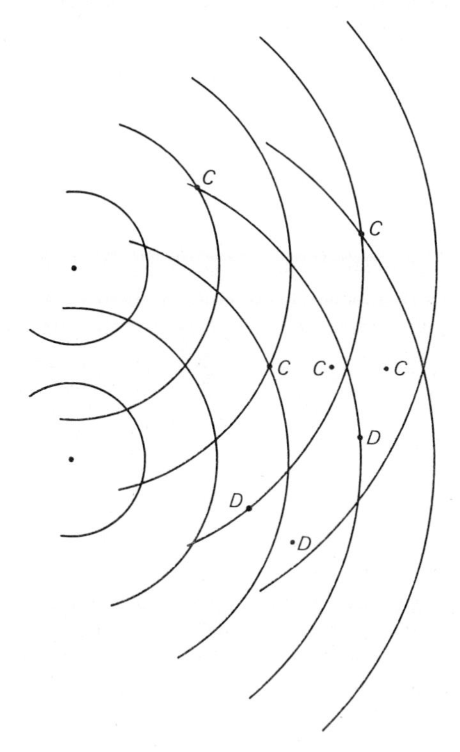

Figure 4.3 Snapshot of wave crests moving outward from two sources. Constructive interference occurs at points marked C, destructive at D. (Take a minute to be sure you are convinced that they are correctly marked!)

signal through two speakers, for reflections from the walls can combine with the original signal just as if they came from mirror-image speakers behind the walls.

One simple and interesting application for this theory is in understanding "stack" or column-array speakers, commonly used for sound reinforcement in popular music concerts. Consider N identical speakers all driven by the same signal so that all ϕ_m have the same value ϕ, and spaced evenly with center-to-center distance d (Fig. 4.4). Suppose for the moment that each speaker acts as a point source radiating more or less equally in all directions. (We learn under what conditions that is true in Chap. 11.) If point P is only at a moderate distance from the array, there will be two difficulties: (1) the trigonometry for comparing all the L's will be complicated, and (2) we will have to be concerned that the arriving signals have somewhat different strengths, since some have come farther from their sources. But let us consider only the case where every L is so much larger than Nd that the L's are nearly parallel. (In optics this is called the Fraunhofer approximation.) In this limit all N signals have practically the same amplitude (call it C), and

$$p(t) = \sum_{m=1}^{N} C \cos\left[\omega(t - L_m/c) - \phi\right]. \tag{4.13}$$

This is a case where phasor notation makes further progress much easier; if you are not already familiar with it, you should study Appendix B now. When each cosine in (4.13) is replaced by an imaginary exponential of the same argument, we can easily take outside the summation a factor that is common to all terms:

$$p(t) = \sum_{m=1}^{N} Ce^{j[\omega(t-L_m/c)-\phi]}$$

$$= Ce^{j[\omega(t-L_1/c)-\phi]} \sum_{m=1}^{N} e^{-j\omega(L_m-L_1)/c}. \tag{4.14}$$

The first exponential merely says there is an oscillation, and the magnitude of that factor is just unity. It is the summation that carries all the interesting information about how the amplitude changes from place to place, for the sum may have any magnitude between zero and N. Figure 4.5 shows that in the Fraunhofer approximation we have simply

$$L_m - L_1 = (m - 1)\, d \sin\theta. \tag{4.15}$$

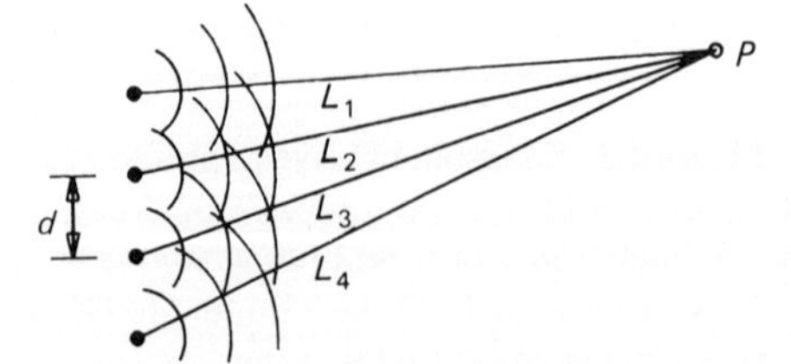

Figure 4.4 Identical signals from four point sources with mutual spacing d along a line travel distances L_i and reach point P with different phases.

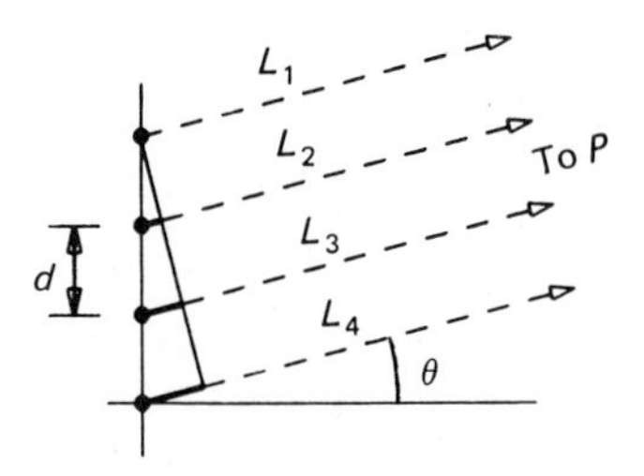

Figure 4.5 For an observing position P very far away in the direction θ, all paths are parallel and each L_i exceeds the one before by $d \sin \theta$. This is the Fraunhofer limit.

Thus the wave amplitude received in the direction θ is

$$p(\theta) = C \sum_{m=1}^{N} e^{-j[\omega(m-1)d/c] \sin \theta}. \tag{4.16}$$

If we define

$$z = e^{-(j\omega d/c) \sin \theta} \quad \text{and} \quad q = m - 1, \tag{4.17}$$

we can see that this is just the sum of a standard form known as a geometric series (Prob. 4.6):

$$\begin{aligned} p(\theta) &= C \sum_{q=0}^{N-1} z^q = (z^N - 1)/(z - 1) \\ &= C[e^{-j(N\omega d/c) \sin \theta} - 1]/[e^{-j(\omega d/c) \sin \theta} - 1]. \end{aligned} \tag{4.18}$$

But the numerator and denominator each may be further simplified because they have the form

$$\begin{aligned} e^{-jy} - 1 &= e^{-jy/2}[e^{-jy/2} - e^{+jy/2}] \\ &= e^{-jy/2} 2j \sin (y/2). \end{aligned} \tag{4.19}$$

Using again the fact that $e^{-jy/2}$ has magnitude unity, the amplitude finally simplifies to

$$p(\theta) = NC \frac{\sin [(N\omega d/2c) \sin \theta]}{N \sin [(\omega d/2c) \sin \theta]}. \tag{4.20}$$

The fraction in (4.20) simply approaches unity as θ goes to zero (Fig. 4.6). The smallest angle for which complete cancellation occurs must be at $\pi = (N\omega d/2c) \sin \theta$, or

$$\theta_1 = \arcsin (2\pi c/N\omega d) = \arcsin (\lambda/Nd). \tag{4.21}$$

This tells us that we could confine most of the radiated energy within a limited range of θ (where the audience is located) by having Nd somewhat greater than λ. At $f = 500$ Hz, for example, $\lambda = 0.7$ m and an angular range of $\pm 30°$ would be achieved for $Nd = 1.4$ m. If you prefer to specify instead the range of angles in which the level remains above a certain fraction of its maximum (for instance, within 6 dB of the highest level), that range will be somewhat less than θ_1. Figure 4.6 suggests that the 6-dB drop is reached when θ is roughly half of θ_1, and Prob. 4.7 asks you to show that in the limit when $N \gg 1$ it occurs at $0.61\ \theta_1$.

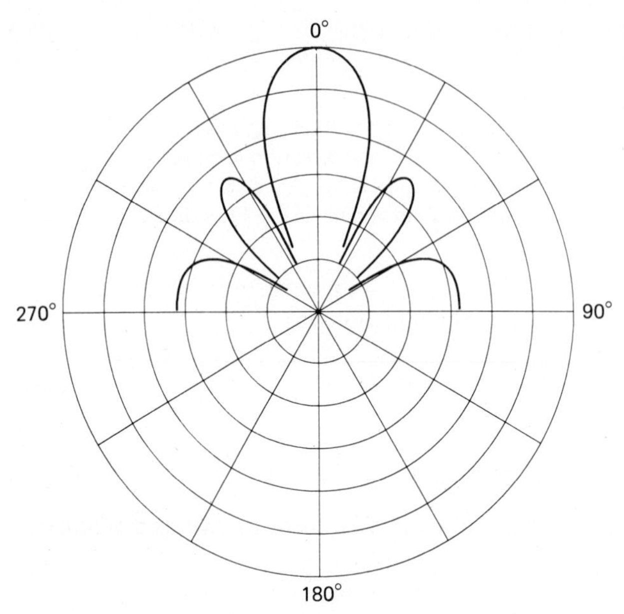

Figure 4.6 SPL(θ) for a speaker array, calculated from (4.20) for a case with $\lambda/d = 2.0$ and $N = 5$. Radial intervals are 5 dB.

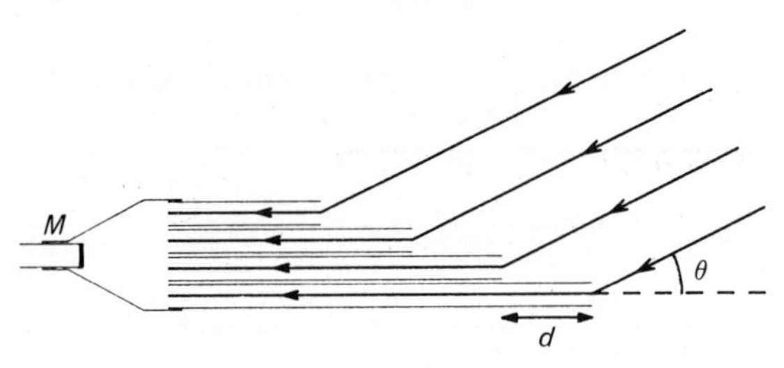

Figure 4.7 Paths followed by sound coming along a direction θ off axis of a shotgun microphone with successive openings separated by distance d.

An entirely similar argument can explain how a "shotgun" microphone achieves high discrimination among different directions of signal arrival. In this case signals all from the same source may follow several different paths in order to reach the detector. In the version most resembling a shotgun you may see only a row of openings along the single "barrel" of the device, which is aimed in the direction of desired maximum pickup. But the theory is more readily apparent when you think of it as a bundle of thin tubes, each longer than the one before by an amount d, leading to a small plenum where the signals mix and meet the actual microphone (Fig. 4.7). Analysis of the path lengths (left to Prob. 4.9) shows that the same source located in different directions will produce a microphone signal amplitude

$$p(\theta) = NC \frac{\sin\left[(N\omega d/c)\sin^2(\theta/2)\right]}{N\sin\left[(\omega d/c)\sin^2(\theta/2)\right]}, \tag{4.22}$$

where again C is the strength of a single component and the ratio becomes unity at $\theta = 0°$. The directional sensitivity pattern is qualitatively like Fig. 4.6, with an angular halfwidth where the numerator first becomes zero given by

$$\theta_1 = 2 \arcsin \sqrt{\lambda/2Nd}. \tag{4.23}$$

In this case θ_1 decreases in inverse proportion to the square root of frequency rather than frequency itself.

For example, a shotgun with a total length $Nd = 0.5$ m would be sensitive for $f = 3.4$ kHz ($\lambda = 0.1$ m) mainly for

$$\theta < 2 \arcsin \sqrt{0.1} = 37°,$$

but for $f = 340$ Hz ($\lambda = 1$ m) this would broaden to

$$\theta < 2 \arcsin (1.0) = 180°.$$

High directionality can be achieved for low frequencies only by making Nd longer.

4.4 DIFFRACTION

Suppose you aim a loudspeaker across a room at an open window. The energy passing through the window does not simply continue entirely in the original direction like a stream of bullets from a machine gun; it spreads toward the sides as well. This tendency to spread, common to all kinds of waves, is called **diffraction.**

Some simple predictions about diffraction can be made by supposing that the wave in the window opening is the same one that would be there if the wall were entirely absent. (This is only an approximation, since in reality the window frame does modify the air motion close to it.) Then we can use Huygens' idea that each bit of wave disturbance at one moment acts as a source of new waves, whose combined total tells us what the wave disturbance will be at a later time. In the present example, this means that having a wave passing through the window (with equal strength and equal phase everywhere in the plane of the window) is just the same as if the window were filled with an array of tiny loudspeakers all radiating the same signal.

To keep the problem simple, let us make it one-dimensional by supposing the window is so tall that we are concerned only with the horizontal diffraction caused by its finite width W. And let us again consider only distances L beyond the window much greater than W, so that the Fraunhofer approximation applies (Fig. 4.5). If we let x label the location of points in the plane of the window ($0 < x < W$), the difference in lengths between any path beginning at x and the reference path from one edge of the window is

$$L(x) - L(0) = x \sin \theta. \tag{4.24}$$

Writing the amplitude contributed by an infinitesimal portion dx as $C\,dx$, we can calculate the sum of all waves received as

$$p(\theta) = \int_0^W e^{-j(\omega/c)x \sin\theta} C\,dx. \tag{4.25}$$

This integral is easily carried out (Prob. 4.10) to show that the amplitude of sound received in direction θ will be

$$p(\theta) = CW \frac{\sin[(\pi W/\lambda)\sin\theta]}{(\pi W/\lambda)\sin\theta}. \tag{4.26}$$

This is quite similar to (4.20), with W taking over the role of Nd. In fact, (4.26) can also be derived simply by taking the limit of (4.20) as N becomes infinite and d becomes zero, but in such a way that the product Nd remains finite and becomes W. Equation 4.26 still has the property that low-frequency sounds will spread more readily than high.

Diffraction of a uniform wave passing through a circular opening has a similar qualitative dependence, with stronger spreading for wavelengths larger than the opening. This case is very closely related to radiation from a circular speaker diaphragm, which will be presented in Chap. 11.

Another important illustration of diffraction occurs for people living adjacent to busy freeways. A sound barrier is often erected to shield homes from the traffic noise, but the sound waves tend to penetrate into the shadow region behind the barrier. The opening through which the sound passes is bounded on only one side and extends all the way to the sky on the other. This means it is impossible to get far enough back from this opening to claim that all arriving waves are practically parallel. Then the Fraunhofer approximation is not justified, and exact calculations are quite difficult.

But let us try to understand the basis for one useful prediction about how strong the diffracted sound will be, without trying to get all the details correct. Consider an observer at height $y = 0$ receiving the diffracted portion of a horizontally traveling plane wave cut off by a barrier of height H a distance D away. Think of the curved wave fronts in Fig. 4.8 as cross sections of a cylindrical diffracted wave. Then the amplitude of a Huygens wavelet coming from height y must be (according to Sec. 1.3) inversely proportional to the square root of the distance $\sqrt{D^2 + y^2}$ it has traveled. Compared with the component that just grazed the top of the barrier, one coming from any other height $y > H$ has traveled an extra distance

$$\Delta L(y) = \sqrt{D^2 + y^2} - \sqrt{D^2 + H^2}, \tag{4.27}$$

and this determines their relative phase. So the amplitude received should be calculated from

$$\int_H^\infty \frac{\exp[-j(2\pi/\lambda)\,\Delta L(y)]}{\sqrt[4]{D^2 + y^2}}\,dy. \tag{4.28}$$

This integral is certainly too difficult for us to do in closed form. But a graph of the integrand (Fig. 4.9) suggests that the positive and negative contributions from successive cycles will largely cancel. That cancellation will be more

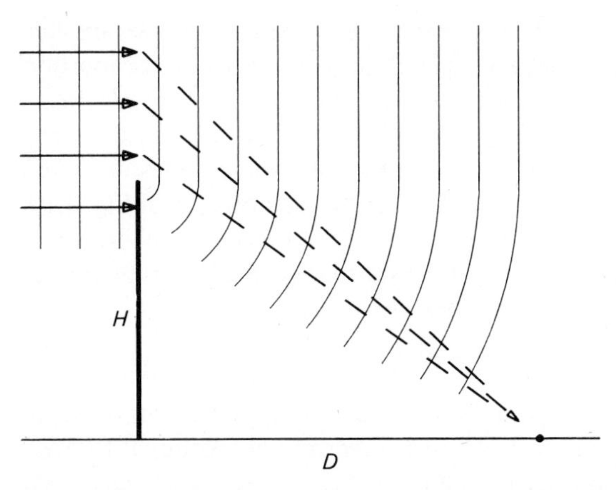

Figure 4.8 Plane waves diffracting over a barrier of height H to reach an observer a distance D behind it.

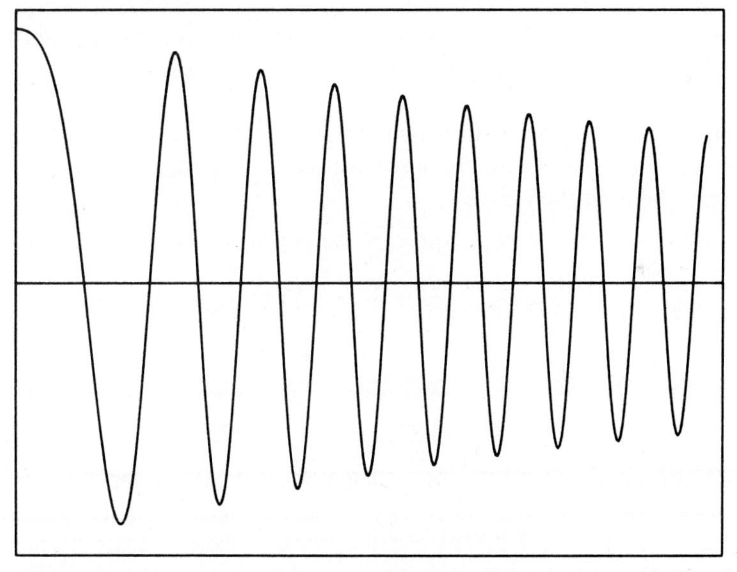

Figure 4.9 An example of the type of function being integrated in Eq. 4.28.

and more nearly exact for large values of y. But the very first quarter cycle has both larger amplitude and larger range of y; so its contribution can never be entirely erased. This first quarter cycle alone, then, would give us a (very crude) estimate for the entire integral. We must ask, then, for what value $y = y_1$ the phase of the integrand reaches 90°. The answer is that $\Delta L(y_1) = \lambda/4$, which can be solved to give

$$y_1 = \sqrt{H^2 + (\lambda/2)\sqrt{D^2 + H^2} + \lambda^2/16}. \tag{4.29}$$

For values of H in the range $\sqrt{\lambda D} \ll H \ll D$, this simplifies to

$$y_1 \simeq H + \lambda D/4H. \tag{4.30}$$

Since the dominant part of the integral is proportional to $(y_1 - H)$, we are able to see that for fixed D the amplitude received will drop off in inverse proportion to H. Then a doubling of H would produce a 6-dB decrease in received SPL.

The exact theory (discussed at length in Chap. 9 of the advanced text by Pierce) justifies this conclusion and extends it to the case where both source and receiver are at finite distances from the barrier. Using the notation of Fig. 4.10, the result is most conveniently expressed in terms of the Fresnel number, which is the number of half wavelengths by which the shortest path grazing the barrier exceeds the straight-line path through the barrier:

$$N_F = (D_1 + D_2 - S)/(\lambda/2). \tag{4.31}$$

It is accurate for values of $N_F > 2$ (and still useful though crude on down to $N_F \sim 0.2$) to say that the amplitude received is inversely proportional to $\sqrt{N_F}$. Then the **insertion loss** of the barrier is about

$$\mathrm{IL(dB)} = 16 + 10 \log N_F. \tag{4.32}$$

This tells how much the sound level is reduced compared with what it would have been in the absence of any barrier, and is useful in estimating the size and location of barriers needed to make specified reductions in noise.

For example, take a barrier with height $y = H = 4$ m at $x = 0$, a source located at $x = -6$ m, $y = 2$ m, and a receiver at $x = +8$ m, $y = 1$ m. The Pythagorean theorem then gives

$$D_1 = \sqrt{6^2 + 2^2} = 6.32 \text{ m},$$

$$D_2 = \sqrt{8^2 + 3^2} = 8.55 \text{ m},$$

and

$$S = \sqrt{14^2 + 1^2} = 14.04 \text{ m},$$

so that

$$D_1 + D_2 - S = 14.87 - 14.04 = 0.83 \text{ m}.$$

For sound of frequency 1.7 kHz (for which $\lambda/2$ is 0.1 m), we get

$$N_F = 0.83/0.1 = 8.3$$

and

$$\mathrm{IL} = 16 + 10 \log (8.3) = 25 \text{ dB}.$$

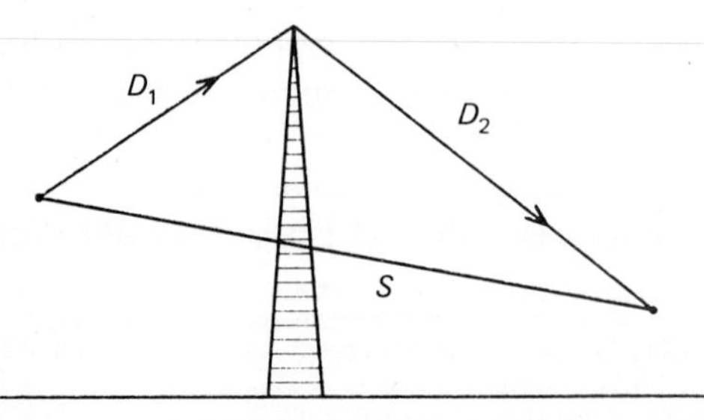

Figure 4.10 Definition of distances used to calculate Fresnel number for diffraction over a barrier.

But at 170 Hz, $\lambda/2$ is 1 m; so $N_F = 0.83$ and the predicted IL is only 15 dB. If it were required to achieve IL = 20 dB at 170 Hz, that would mean

$$10 \log N_F = 4 \quad \text{or} \quad N_F = 2.5,$$

so $D_1 + D_2 - S$ would have to be increased to 2.5 m. That requires that the barrier height H be raised to nearly 6 m, illustrating that low-frequency sounds are particularly difficult to block out.

4.5 REFRACTION

Another wave property familiar to most people from its occurrence in optics is **refraction.** This means *bending* of waves away from straight lines of travel. This bending is in one or another specific direction and must be distinguished from the *spreading* in any and all directions characteristic of diffraction. The physical reason for refraction is always a difference in wave speed from place to place so that different portions of a wave front do not move at the same rate.

In the case of light we most often think of refraction as occurring suddenly as the waves cross a sharp interface between air and glass at the surface of a lens, because the light travels more slowly in glass than in air. We do not ordinarily encounter the corresponding acoustic effect at audible frequencies, for the wavelengths are so long that the apparatus to do "acoustic optics" would have to be very large. But at high ultrasonic frequencies the development of the **acoustic microscope** has been of great practical interest. The device works as indicated in Fig. 4.11. Plane waves may be launched by a piezoelectric transducer into a little block of sapphire, which has a spherical hollow ground in its opposite face. To the extent that the waves travel much faster in the crystal than in the adjoining liquid, the central portion of each wave front is held back relative to the outer

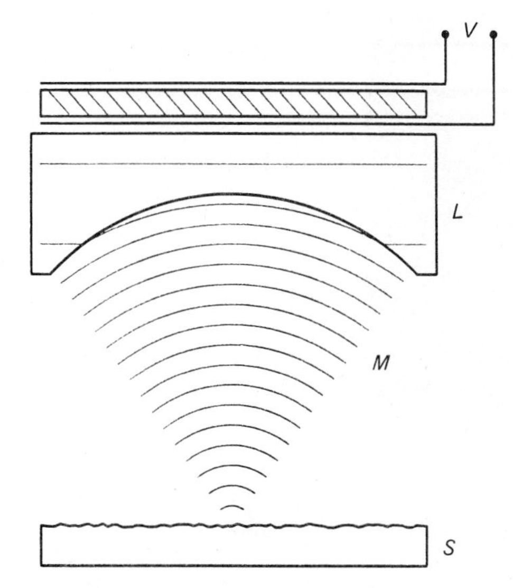

Figure 4.11 Essential elements of an acoustic microscope. Terminals V deliver input voltage to electrodes on opposite faces of a piezoelectric crystal, which launches a short train of waves into the lens L. When these waves enter the liquid medium M, in which their speed is slower, they are focused toward a point on the surface of the sample S. Reflected waves follow the same paths in reverse, and, by the time they arrive, the piezoelectric transducer has been switched to detector mode.

parts. All parts of the wavefront enter the liquid at nearly the same time, and nearly spherical waves are launched inward toward the focal point at the center of curvature. The strength of the reflection depends on the nature of the sample surface at the focal point. The reflected wave is gathered by the same lens and detected by the same transducer. In using a single transducer both to launch and to receive signals, this device resembles radar and sonar systems. Since information is gotten from only one point at a time, the sample must be moved in a raster pattern in the focal plane of the microscope while a picture is gradually accumulated in a computer memory.

Consider using signals with frequency 3 MHz in water, for instance: These would have wavelength $(1480 \text{ m/s})/(3 \times 10^6/\text{s}) = 500\ \mu\text{m}$ and, therefore, would offer at best a resolution of about half a millimeter in distinguishing small features on any object to be observed. Clearly the best resolution of very small objects results from using as high a signal frequency as possible. But in water one cannot go much higher because such signals are very strongly absorbed (Fig. 1.5). The next possibility is to use a different medium with c as small as possible and, even more important, with less absorption than water. An attractive choice turns out to be liquid helium, with which instruments have been built to operate up to $f = 8$ GHz and so with wavelengths as small as $0.03\ \mu\text{m}$. An example of information obtained with an acoustic microscope is shown in Fig. 4.12. You may read an interesting review of the current state of development of acoustic microscopes in an article by C. F. Quate in *Physics Today* (August 1985, pp. 34–42).

Refraction also occurs if the wave speed c changes only gradually with position, as it does in the atmosphere because the air temperature is not the same everywhere. On spatial scales of a few meters this is unimportant, but over large distances effects are cumulative. Figure 4.13 suggests how a sound speed that increases with height can cause sound energy to curve back down toward

Figure 4.12 An acoustic microscopic image of bipolar transistors on a silicon integrated circuit chip. The scale bar is 3 microns long. (Courtesy of C. F. Quate)

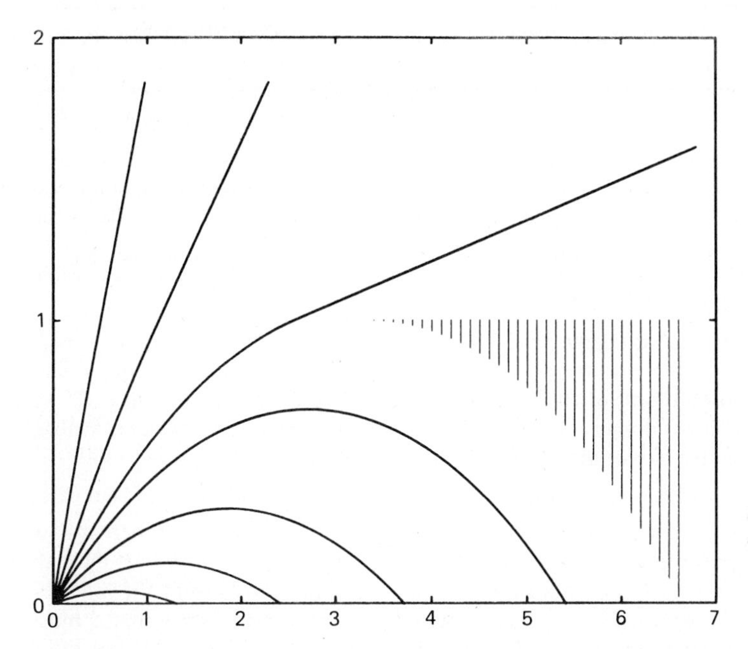

Figure 4.13 Sound rays leave the origin in various directions in a layered medium. The speed of sound was assumed here to increase 20 percent from $z = 0$ to $z = 1$ and then to remain constant for $z > 1$. This exaggerates the variations normally encountered in practice in order to clearly show bending of the rays and existence of a shadow zone (hatched) into which none of this signal can penetrate.

the ground. Such effects can make the sound levels measured a few kilometers away from an outdoor concert site depend drastically on weather conditions.

Nonuniform sound speed is also a very important factor in underwater acoustics because of temperature and salinity gradients that are always present in the ocean. It is not unusual to have a minimum in c at some depth on the order of a kilometer, with higher values both above and below that layer. By imagining an extension of Fig. 4.13 with its own mirror image below it, you can see some interesting possibilities. One is communication through *sound channels* in which trapped signals traveling horizontally would be stronger than if they had been able to spread in all directions. Another is the existence of *shadow zones,* where waves from a particular source never arrive, so that they might be good places for a submarine to hide.

PROBLEMS

4.1. If an oscilloscope trace shows a beating signal like that in Fig. 4.2, with maximum and minimum amplitudes 15 and 3 mV, what would be the amplitude for either component signal alone?

4.2. Consider two sine waves with frequencies 498 and 502 Hz, and suppose each by itself produces a SPL at a certain position of 75 and 70 dB, respectively.

(a) What is the long-term average p_{rms}, and corresponding SPL, when the two waves are added?

(b) What are the maximum and minimum SPL of the short-term beats?

(c) Relate your answers to the Slow and Fast responses of a sound-level meter.

4.3. Let p_1, I_1, and SPL_1 represent the strength of a single sound wave.

(a) Let N such waves, all of the same strength and mutually incoherent, be combined with each other. How is the intensity I_N of the total related to I_1? What is the relation of p_N to p_1, and SPL_N to SPL_1?

(b) If all N waves are mutually coherent, what are the greatest and smallest possible values for p_N, I_N, and SPL_N?

4.4. If four sine waves with frequencies 498.0, 499.2, 501.0, and 503.2 Hz are combined, how many different beat frequencies are there and what are they? How long would you have to wait for every beating pair to return to the same point in its cycle, so that the whole pattern would repeat? If you have some computer graphics available, it is informative to plot the signal.

4.5. Roughly how wide must a noise band be to sound smooth and steady to your ears?

4.6. In case you are not already familiar with summing a geometric series, here is how you may prove the result quoted in Eq. 4.18. Let $S = \Sigma\, x^q$ for $q = 0$ through $N - 1$. Observe that xS contains most of the same terms as S, so that if you subtracted $xS - S$ term by term most would cancel. Use this to show that S must equal $(x^N - 1)/(x - 1)$.

4.7. **(a)** If we defined beam width according to where the signal level is 6 dB below its maximum, instead of null, what would replace Eq. 4.21 in the case where N is quite large? (*Hint:* Show that you need to solve $\sin y = y/2$, and remember Appendix A.)

(b) By this criterion, what beam width would be available from a stack of five speakers at 15-cm spacing for 500-Hz signals? For 2 kHz?

4.8. For what value of $y = (N\pi d/\lambda) \sin\theta$ does the first side-lobe maximum occur in the pattern of Fig. 4.6, and what is the maximum signal level there **(a)** for very large N, **(b)** for $N = 5$? Describe it in terms of how many decibels this maximum is below the main lobe. [*Hint:* First orient yourself by giving an approximate answer based on pretending that the denominator of (4.20) is constant. Then show that the exact solution boils down to $N \tan(y/N) = \tan y$, and follow Appendix A.]

4.9. Show that the difference in path length for two adjacent tubes in the shotgun microphone (Fig. 4.7) is $d(1 - \cos\theta)$. Add the N phasors, taking hints from the derivation of Eq. 4.20, to find Eqs. 4.22 and 4.23. What would be the angular pattern width for $Nd = 80$ cm and $f = 400$ Hz? For $f = 4$ kHz?

4.10. Fill in the steps between Eqs. 4.25 and 4.26.

4.11. What is the smallest angle for which $p(\theta) = 0$ in Eq. 4.26? What is the narrowest window opening that will keep sound mostly confined in a narrow beam (say $\pm 20°$) for $f = 400$ Hz? For 4 kHz?

4.12. For what angles other than zero does $I(\theta)$ in (4.26) have maxima? How many decibels below the central maximum is the strongest of these? (*Hint:* This is equivalent to the limit where N becomes infinite in Prob. 4.8, and requires solving $\tan y = y$.)

4.13. Suppose you find the side lobes bothersome in the pattern of Fig. 4.6. Investigate whether they can be suppressed by sending unequal signal strengths to the five speakers. Do this by writing an expression for $p(\theta)$ when the amplitude at the two end speakers is some fraction β of that for the central three, and using a computer to generate graphs of this amplitude for several values of β between zero and one.

4.14. Find the Fresnel number N_F in terms of D and H for the geometry of Fig. 4.8. Show that for $H \ll D$ it reduces to $H^2/\lambda D$, so that our conclusions from (4.29) are consistent with those in the general case.

4.15. Suppose your bedroom window is at about the same height as a freeway and 10 m away. Estimate how much the noise will be reduced by having a solid fence halfway between which rises 2 m above the line of sight; do this for both $f = 2$ kHz and $f = 250$ Hz. If in either case the insertion loss is less than 20 dB, how high must you make the fence to achieve that much reduction?

LABORATORY EXERCISES

4.A. Study the interference pattern produced by a sinusoidal signal driving a pair of speakers. Verify that spacing between minima depends as it should on spacing d and frequency f. Do this either in an anechoic room or outdoors, to minimize confusion due to wall reflections.

4.B. Try to verify the prediction about doubling H for 6 dB reduction, following Eq. 4.29, using a narrow-band noise source. It is probably more convenient to do the experiment in a horizontal plane (around the corner of a building) instead of vertical, although this risks some complication due to ground reflections.

Chapter 5

Simple Harmonic Motion

It is now time to fill in some of the fundamental theory we need in understanding acoustic phenomena. Our first goals must be to gain a thorough understanding of the nature of vibration and to develop a variety of mathematical tools with which to describe it accurately. Sound sources, and sound waves themselves, involve complex vibrations of solid bodies and regions filled with gas or liquid. But it is helpful to start with the simplest oscillating system we can think of. This will provide us with important concepts that can aid our later study of more interesting and realistic systems.

In the following sections you will consider first the basic equation of motion of the harmonic oscillator, and the parameters in that equation that identify the systems' physical properties. Then the solutions of this equation will be considered under two main headings: decaying free oscillation, with only internal forces involved, and driven oscillation, in which a sinusoidal external force maintains the motion indefinitely. Real-life situations demand that we also consider the presence of decaying components in the motion when driving forces are turned on or off. Finally, we indicate very briefly how Fourier methods can extend these solutions to other types of driving forces.

5.1 THE HARMONIC OSCILLATOR

The standard model for simple oscillating motion is a mass on a spring. Practical lab exercises often involve suspending the mass below the spring so that it oscillates vertically. It is conceptually clearer, however, to think of the mass sliding on a horizontal plane as indicated in Fig. 5.1. This will help us avoid the false impression that gravity has any essential role in the motion.

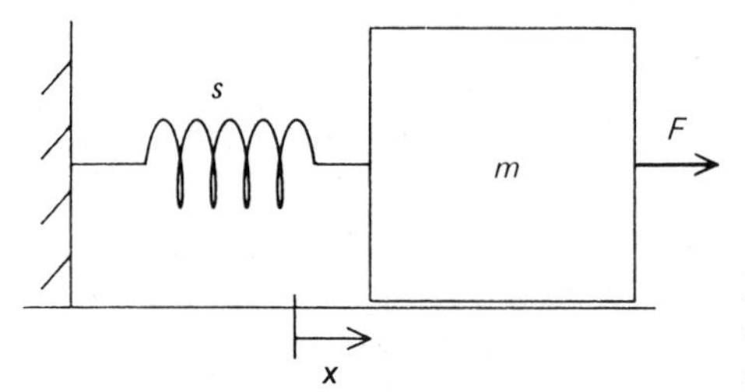

Figure 5.1 A mass m moving through a distance x while attached to a spring serves as a model simple harmonic oscillator.

There is one position of m for which the spring is relaxed and exerts no force on the mass. This is called the **equilibrium** position, and it is usually convenient to choose this as our origin of coordinates, $x = 0$. Whenever the spring is stretched, it pulls the mass back toward the left: $x > 0$ causes $F < 0$. But for $x < 0$ the compressed spring will push m toward the right, so that $F > 0$. This may be described as a **restoring force**—one that always tends to restore the system to its equilibrium configuration.

If you pull m out to some positive x and let it go, why doesn't it just return to $x = 0$ and stop there? Newton's first law of motion reminds us that any massive body in motion tends to continue in motion: only the action of a force toward the right can slow down the leftward motion, and the spring will not begin to do that until x is already negative. The mass is said to have **inertia,** the property of persisting in whatever state of motion it already has. In oscillating systems of every type, no matter how complicated, you may always identify something acting as a restoring force and something else playing the role of an inertia.

For an exact description of the motion we must look to Newton's second law, $F = ma$. Here F stands for the total of all forces acting on m to produce its acceleration a. You have undoubtedly already studied the case where m slides on a perfectly smooth surface and the only horizontal force is that of an idealized spring described by Hooke's law, $F = -s\,\Delta L$. Here s stands for **stiffness,** and we suppose it is a constant characteristic of the spring material and dimensions. ΔL is the change in length of the spring from its relaxed state. If we were to encounter a case where F is not strictly proportional to ΔL and thus not described by a constant s, we would call that a nonideal or nonlinear spring. Its motion would be much more difficult to analyze.

When we identify ΔL with x and a with the second derivative of $x(t)$, we have

$$ma = F \quad \text{or} \quad m\ddot{x} = -sx \quad \text{or} \quad \ddot{x} + (s/m)x = 0. \tag{5.1}$$

We are already familiar with two functions (sine and cosine) that have the property needed here, namely, a second derivative that is the negative of the original function. But you cannot take the sine of a time t; only a dimensionless number can serve as the argument of this function. Therefore, we must use the combination Ωt, where Ω has the units of a frequency. You can easily verify by substitution that either

$$x(t) = A \cos \Omega t + B \sin \Omega t \tag{5.2}$$

or
$$x(t) = C \cos(\Omega t - \phi) \tag{5.3}$$

will satisfy (5.1) if and only if

$$\boxed{\Omega = \sqrt{s/m}.} \tag{5.4}$$

For other types of vibrating systems, their analogs of stiffness and inertia will appear in a formula just like (5.4) to determine their natural frequencies.

As an example that we use several more times later, let a mass $m = 0.01$ kg on a spring with $s = 25$ N/m be called vibrator V. This system will have a natural oscillation frequency

$$\Omega = \sqrt{25/0.01} = 50 \text{ radians/second}$$

or
$$f = \Omega/2\pi = 8 \text{ Hz}.$$

Notice an important distinction here: s and m (and hence Ω) are *intrinsic* properties of the system and do not depend on how it is being used. Amplitude and phase (A and B, or C and ϕ), on the other hand, are *extrinsic:* the nature of the system itself does not determine what they will be. They are determined instead by outside influences, such as our grasping and pulling the spring or giving a kick to the mass. If such outside influences acted continually and arbitrarily, the motion $x(t)$ could be practically anything. But we should study first the simpler situation of **free oscillation:** once some motion is begun, the system is left alone to do whatever it will.

So external influences are often expressed only through **initial conditions.** If the position x and velocity v are specified at one time, then A and B (or C and ϕ) can be determined and the motion at all subsequent times is known. The reason we are considering two initial conditions and determining two constants of the motion is simply that we are solving a second-order differential equation. It is usually convenient (and has no physical significance) to let $t = 0$ be the time when the motion began. In that case, using a subscript to denote time zero, it follows from (5.2) and (5.3) that

$$A = x_0, \qquad B = v_0/\Omega \tag{5.5a}$$

and
$$C = \sqrt{A^2 + B^2}, \qquad \phi = \arctan(B/A). \tag{5.5b}$$

5.2 DAMPED HARMONIC OSCILLATION

Once set in motion, the simple oscillator studied so far would continue vibrating forever with the same amplitude C. But none of the real-life systems we are trying to model have that property. A tuning fork, for example, gradually loses energy in the form of the sound it gives off as well as in heating of the flexed metal. So after some minutes have gone by you can no longer detect any motion.

It would make our model more realistic, then, to suppose that some friction acted to slow down the moving mass. Though it is not literally a good representation of sliding friction, one type of resistance that will be the most useful analogy for acoustic and electric applications is a force proportional to velocity:

$$F_{fr} = -Rv. \tag{5.6}$$

We include a minus sign to indicate that this force always acts in a direction opposite to the motion. The constant R must have units N/(m/s), which is the same as kg/s. We call this **mechanical resistance,** anticipating that it will play the same role here that electrical resistance does in circuit theory.

When we include this new term, the equation of motion becomes

$$ma = -Rv - sx \quad \text{or} \quad m\ddot{x} = -R\dot{x} - sx. \tag{5.7}$$

Let us continue to let Ω^2 stand for s/m and introduce the additional definition

$$\boxed{\beta = R/2m.} \tag{5.8}$$

This **damping rate** β will have units 1/s, just like Ω. The differential equation we must solve is then

$$\ddot{x} + 2\beta\dot{x} + \Omega^2 x = 0. \tag{5.9}$$

You may already know definite mathematical techniques for finding an $x(t)$ that will obey (5.9). But our awareness of oscillating motions that gradually die away, like those of a tuning fork, can physically motivate the way we search for a solution. We would still expect sinusoidal functions to be involved somehow, though trial and error shows that the calculations are often easier if we write them as phasors. Probably we should use a phasor oscillation multiplied by a gradually decreasing amplitude. We might reasonably try an amplitude proportional to a negative power of t, but that would turn out to lead nowhere. Another guess that does end up working very well is an exponential decay:

$$x(t) = (Ce^{-\eta t}) \exp[j(\Omega_d t - \phi)]. \tag{5.10}$$

Here we have allowed for the possibility that the **damped oscillation frequency** Ω_d might differ from Ω. Substitution of this trial solution into (5.9) shows that it is correct if and only if

$$(-\eta + j\Omega_d)^2 + 2\beta(-\eta + j\Omega_d) + \Omega^2 = 0. \tag{5.11}$$

Requiring both real and imaginary parts of (5.11) to be true shows that

$$\eta = \beta \quad \text{and} \quad \Omega_d = \sqrt{\Omega^2 - \beta^2}. \tag{5.12}$$

This damped oscillation exists only if $\Omega > \beta$, that is, $sm > R^2/4$, but our concern in this book is always with cases where R is small enough to satisfy this condition.

Using again our example with $m = 0.01$ kg and $s = 25$ N/m, suppose that vibrator V also has $R = 0.2$ kg/s. Then

$$\beta = R/2m = 0.2/0.02 = 10 \text{ s}^{-1}$$

and
$$\Omega_d = \sqrt{(2500 - 100)} = 49 \text{ s}^{-1}.$$

Thus the damped oscillation frequency is reduced to

$$f_d = 49/2\pi = 7.8 \text{ Hz}.$$

A single cycle takes time

$$t_1 = 1/f_d = 1/7.8 = 0.128 \text{ s},$$

so the maximum excursion in each cycle is only a fraction

$$e^{-\beta t_1} = e^{-(10/\text{s})(0.128\text{s})} = 0.28$$

of that in the preceding cycle. This would be considered a very heavily damped oscillation, as illustrated in Fig. 5.2.

Nearly all of our applications will involve systems with little enough damping that they can complete many cycles before their amplitude is reduced drastically. In such cases, with $\beta \ll \Omega$, the approximation $\Omega_d \cong \Omega$ will be quite accurate.

Notice that R (like m and s) and β (like Ω) are intrinsic parameters of the system. Another very useful combination of these quantities is the **quality factor,** defined as

$$\boxed{Q = \sqrt{sm}/R = \Omega m/R = \Omega/2\beta.} \tag{5.13}$$

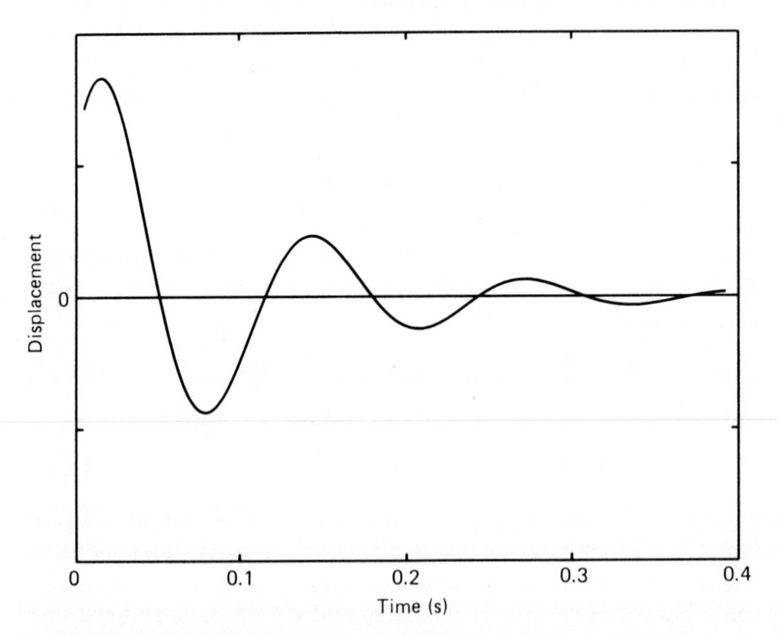

Figure 5.2 Damped oscillation of a system with damped natural frequency 7.8 Hz and damping rate $\beta = 10/\text{s}$, thus with $Q = 2.5$.

Our example above has $Q = 50/(2 \times 10) = 2.5$. Systems with Q of order unity are heavily damped, and the extent to which Q is much greater than 1 is a good indicator of how well the oscillator will "ring" at its natural frequency once disturbed. For large Q, the approximation $f = \Omega_d/2\pi \cong \Omega/2\pi$ can be used to write the time-dependent amplitude in terms of Q as

$$A(t)/A(0) = e^{-\beta t} = e^{-\Omega t/2Q} \cong e^{-\pi N/Q}, \tag{5.14}$$

where $N(t) = ft$ is the number of cycles completed in time t. This decay can also be expressed in decibels as

$$20 \log [A(t)/A(0)] = (20)(0.434)(-\pi N/Q) = -27.3N/Q, \tag{5.15}$$

where 0.434 is $\log_{10}(e)$. This emphasizes that Q is uniquely connected with the number of cycles regardless of the time scale on which they occur.

For example, any oscillator (mechanical, optical, acoustic, or otherwise) with $Q = 1000$ will after $N = 200$ cycles have its amplitude reduced by a factor

$$\exp(-0.2\pi) = 0.524.$$

This corresponds to

$$(-27.3)(0.2) = -5.5 \text{ dB}.$$

Or if we ask how long it takes for an amplitude to decay 30 dB, the answer is given by

$$-30 = -27.3N/Q$$

or

$$N = (30/27.3)Q = 1.1Q.$$

For $Q = 1000$ that is 1100 cycles, and if f were 500 Hz they would take $(1100)(2 \text{ ms}) = 2.2$ s.

Now that we know the form (5.10) of the general solution for the lightly damped free oscillator, we may take its real part:

$$\boxed{\begin{aligned} x(t) &= Ce^{-\beta t} \cos (\Omega_d t - \phi) \\ &= e^{-\beta t}(A \cos \Omega_d t + B \sin \Omega_d t). \end{aligned}} \tag{5.16}$$

Both forms involve two constants that can be determined only from specific initial conditions. Problem 5.8 asks you to prove that (5.16) requires

$$A = x_0, \qquad B = (v_0 + \beta x_0)/\Omega_d. \tag{5.17}$$

If you prefer C and ϕ, they are still given by (5.5b). Can you can make a physical explanation of why B is no longer simply v_0/Ω?

If vibrator V, with $\Omega_d = 49$/s and $\beta = 10$/s, is set in motion with initial velocity $v_0 = -2$ m/s from initial position $x_0 = 0.1$ m, what is its subsequent motion? That will be given by (5.16), either with

$$A = 0.1 \text{ m}, \qquad B = (-2 + 10 \times 0.1)/49 = -0.02 \text{ m}$$

or $$C = \sqrt{0.1^2 + 0.02^2} = 0.102 \text{ m}$$

and $$\phi = \arctan(-0.02/0.1) = -11.3°.$$

The other possibility for this arctangent, +169°, must be rejected because when used in (5.16) it would give the wrong signs for x_0 and v_0.

5.3 DRIVEN HARMONIC OSCILLATION

What is the response of a harmonic oscillator when it is acted on by a continuing external force? We postpone the general case until Sec. 5.5 and consider first the simple example of a sinusoidal applied force. So we ask what motion will take place when $-Rv$ and $-sx$ are joined by an additional force with amplitude F_0 and frequency ω,

$$F_{ext}(t) = F_0 \cos(\omega t - \phi_F). \tag{5.18}$$

You should think of ω as a parameter: it is constant for any one motion under consideration, but by letting ω take on different values we learn about a whole family of possible motions. Again the calculus will go much more easily if we use phasor notation in adding this term to Eq. 5.7:

$$m\ddot{x} = -R\dot{x} - sx + F_0 \exp[j(\omega t - \phi_F)] \tag{5.19}$$

or $$\ddot{x} + 2\beta\dot{x} + \Omega^2 x = (F_0/m) \exp[j(\omega t - \phi_F)]. \tag{5.20}$$

Let us ask first about the possibility of motions that are steady in the sense of constant-amplitude oscillation. Even though things may be more complicated when the force first begins to act, it is reasonable that after the force has been on long enough the mass may settle into a simple motion repeating with the same period as the driving force. To see if such a solution is possible, we modify (5.3) by assuming frequency ω rather than Ω:

$$\boxed{x(t) = D \exp[j(\omega t - \phi_x)].} \tag{5.21}$$

It is too much to expect that F and x would be exactly in phase with each other; so we must use subscripts to keep careful track of each phase lag separately. This trial solution will indeed satisfy (5.20) if and only if

$$(-\omega^2 + 2j\beta\omega + \Omega^2)De^{-j\phi_x} = (F_0/m)e^{-j\phi_F}. \tag{5.22}$$

Requiring both real and imaginary parts of (5.22) to hold gives

$$(-\omega^2 + \Omega^2)D = (F_0/m) \cos(\phi_x - \phi_F) \tag{5.23}$$

and $$2\beta\omega D = (F_0/m) \sin(\phi_x - \phi_F). \tag{5.24}$$

These are easily solved: squaring and adding the two equations gives

$$D = \frac{F_0/m}{\sqrt{(\Omega^2 - \omega^2)^2 + (2\beta\omega)^2}}, \tag{5.25}$$

while taking the ratio requires

$$\phi_{xF} = \phi_x - \phi_F = \operatorname{arccot} \frac{\Omega^2 - \omega^2}{2\beta\omega}. \tag{5.26}$$

Note the satisfying result that only the phase of x relative to F is physically relevant, and the solution is not really influenced by the choice of when $t = 0$. There is one pitfall in blindly punching a calculator to solve (5.26): there are always two different angles that both have the same cotangent. Looking back at (5.24), where the sine of the angle is required to be positive, determines that you must always choose the solution lying in the range $0° < \phi_{xF} < 180°$.

Of all driving frequencies ω, it is those close to Ω that will minimize the denominator of (5.25) and produce large amplitude D. Figure 5.3 illustrates how these amplitudes depend on ω for different values of β and the corresponding $Q = \Omega/2\beta$. The large-amplitude response when the driving force nearly matches the system's intrinsic natural frequency is called **resonance.** The less damping the system has, the higher its Q and the more marked this resonant response peak will be.

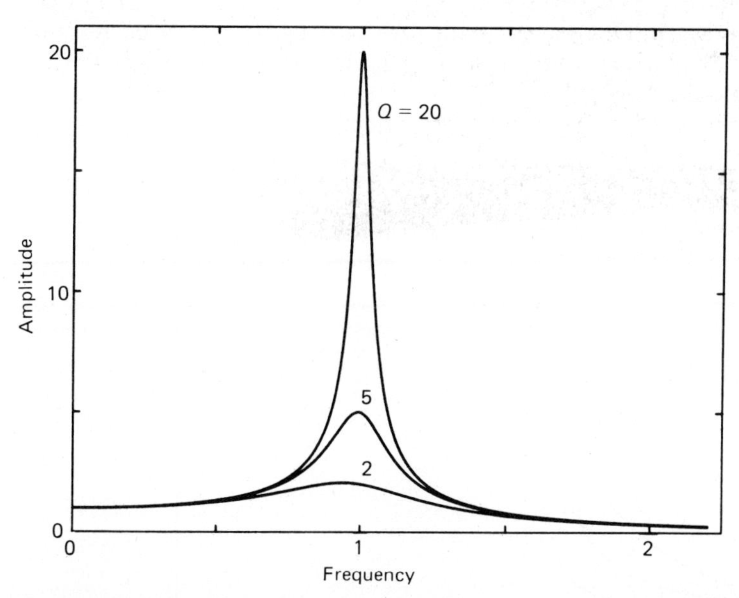

Figure 5.3 Steady-state displacement amplitude as a function of driving frequency for three different amounts of damping. Amplitude is measured in units of F_0/s and frequency in units of Ω.

Figure 5.4 shows the corresponding variation of phase response. For low driving frequencies, $\omega \ll \Omega$, ϕ_{xF} is close to zero; this says that the displacement x reaches maximum at nearly the same time as does the force. Such motions are slow enough that there is plenty of time for the mass to respond to the driving force. Since the restoring force of the spring is nearly in balance with the driving force at all times, this motion is **stiffness-limited**—notice that D is about equal to $F_0/m\Omega^2 = F_0/s$ and does not depend on R or m at all. If, on the other hand, $\omega \gg \Omega$, the motion is **mass-limited.** That is, $D = F_0/m\omega^2$ and the amplitude does not depend on R or s. Here the relative phase is 180°—greatest negative displacement occurs at the same time as greatest positive force. Finally, the height of the resonant peak at $\omega = \Omega$ is $D = F_0/m2\beta\omega = F_0/R\omega$, so that this motion is **resistance-limited.** Here $\phi_{xF} = 90°$ and the greatest positive displacement occurs a quarter cycle after the maximum positive force.

When we develop analogies with electric circuits in the next chapter, we find less emphasis placed on x and more on the velocity v, which is more like electric current. By differentiating (5.21) we obtain $v = j\omega x$, which says that phasor v is always 90° ahead of phasor x. Thus the phase lag ϕ_{xv} is always 90°, so that the velocity lags the force by $\phi_{vF} = \phi_{xF} - 90°$. Then the same graph will serve to show ϕ_{vF} if it is simply labeled to cover the range from −90° to +90°, as we have done with Fig. 5.4. The corresponding formula is

$$\boxed{\phi_{vF} = \arctan \frac{\omega^2 - \Omega^2}{2\beta\omega}.} \tag{5.27}$$

Let us use again our example with $m = 0.01$ kg, $s = 25$ N/m, $R = 0.2$ kg/s, $\Omega = 50$ rad/s, and $\beta = 10$/s. If we apply a force with amplitude $F_0 =$

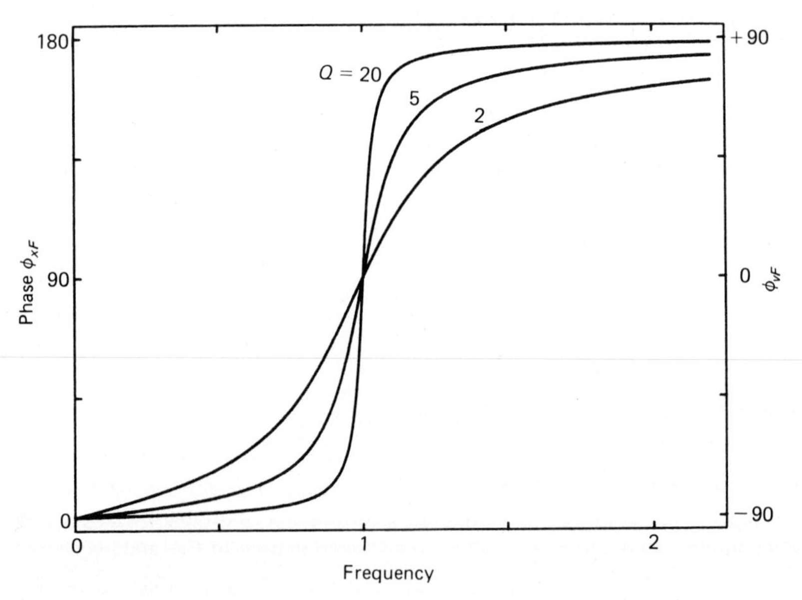

Figure 5.4 Phase angles in degrees for the same cases as in Fig. 5.3.

2N and frequency $\omega = 20$ rad/s (that is, $f = 3.2$ Hz), we find that the steady motion will have displacement amplitude

$$D = (2\text{N}/0.01\ \text{kg})/\sqrt{[(2500 - 400)^2 + (20 \times 20)^2]/\text{s}^4}$$
$$= (200\ \text{m})/\sqrt{4.57 \times 10^6} = 9\ \text{cm}$$

and velocity amplitude

$$v_{\max} = \omega D = (20/\text{s})(0.09\ \text{m}) = 1.8\ \text{m/s}.$$

Note how the argument that $\omega \ll \Omega$ so that the system should be stiffness-limited gives an estimate

$$D \simeq F_0/s = (2\ \text{N})/(25\ \text{N/m}) = 8\ \text{cm},$$

which is not far wrong. The displacement will lag behind the force by

$$\phi_{xF} = \text{arccot}\ (2100/400) = 11°,$$

but the velocity leads the force by

$$\phi_{vF} = \arctan\ (-2100/400) = -79°.$$

A force with the same amplitude but frequency close to 8 Hz would cause a resonant response, which could be as large as

$$D_{\max} \approx F_0/\omega R = (2\ \text{N})/(50/\text{s})(0.2\ \text{kg/s}) = 20\ \text{cm}.$$

5.4 TRANSIENTS

The preceding section avoided the question of what happens immediately after the driving force is turned on. It is not reasonable to suppose that the oscillator immediately obeys (5.21) when the force has not been acting long enough even to firmly establish what its frequency is. From experience with real oscillators we know it takes a while for the motion to settle into the steady state. The corresponding mathematical knowledge is contained in the statement that we could take any motion of the form (5.16) that is a solution of the "homogeneous" equation 5.9 and add it on to the "particular" solution (5.21). That is,

$$x(t) = x_h(t;\ A,\ B) + x_p(t;\ D,\ \phi)$$
$$= e^{-\beta t}(A \cos \Omega_d t + B \sin \Omega_d t) + D \cos (\omega t - \phi_x). \tag{5.28}$$

The sum will still be a solution of (5.20), but now it has two constants to allow for initial conditions, and so it is the most general possible solution. The amplitude D and phase ϕ of the steady-response term are still given by (5.25) and (5.26), but (5.17) must be generalized to

$$A = x_0 - D \cos \phi_x, \qquad B = (v_0 - \omega D \sin \phi_x + \beta A)/\Omega_d. \tag{5.29}$$

Be careful not to confuse this ϕ_x for the steady-state solution with that in (5.5*b*), which is an alternate representation of A and B in the transient term.

Since the transient $x_h(t)$ includes the factor $\exp(-\beta t)$, it eventually dies out and leaves $x_p(t)$ as the steady-state solution. We can now say specifically that we have to wait until $t \gg 1/\beta$ for this simplification.

Figure 5.5 shows an example with $F_0 \cos \omega t$ suddenly turned on at $t = 0$, which means $\phi_F = 0$ in (5.18). The oscillator was at rest at $x = 0$ until the force was applied. The driving frequency has been chosen as $\omega = 2\Omega$, so that the oscillator is finally forced to move at twice its preferred frequency as evidenced in the transient. Results are graphed for three different values of Q. In the last case with $Q = 20$, six cycles of Ω are enough to reduce the transient to about $1/e$ times its initial amplitude. If we use (5.4) and (5.13), we can find the essentials of this motion without specifying separate numerical values of m, R, or s. First write

$$D = (F_0\Omega^2/s)/\sqrt{(\Omega^2 - \omega^2)^2 + (\Omega\omega/Q)^2}$$

$$= (F_0/s)/\sqrt{(1 - \omega^2/\Omega^2)^2 + (\omega/Q\Omega)^2}$$

$$= (F_0/s)/\sqrt{(-3)^2 + (2/20)^2} = 0.333F_0/s$$

and

$$\phi_x = \text{arccot}\,[(1 - \omega^2/\Omega^2)/(\omega/Q\Omega)]$$

$$= \text{arccot}\,[(-3)/(2/20)] = \text{arccot}\,(-30) = 178.1°,$$

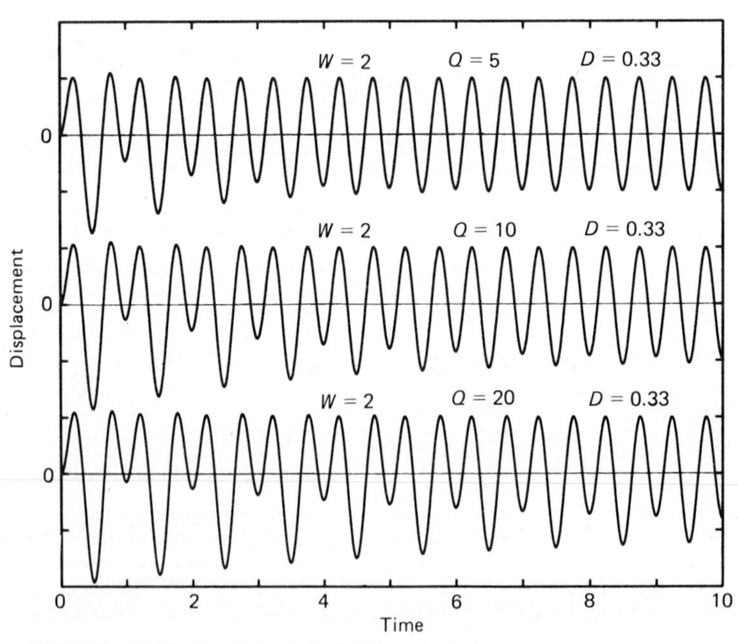

Figure 5.5 Starting transient for force $F_0 \cos \omega t$ suddenly applied at $t = 0$ when $\omega = 2\Omega$. Note how the transient persists longer for higher Q. The vertical axis is marked in units of the steady-state amplitude D, which is approximately $0.33F_0/s$ in all three cases. The horizontal axis is in units of the natural period $2\pi/\Omega$.

then $\quad A = 0 - D \cos \phi_x = +0.999D$

and $\quad B = 0 - (\omega/\Omega)D \sin \phi_x + A/2Q$

$$= -2D(0.033) + D/40 = -0.041D.$$

When x_0 and v_0 are zero, the factor D is common to every term in (5.28); so by using D as a unit we can plot Fig. 5.5 in such a way that each curve shown applies to every case that has the same values of ω/Ω and Q. Figure 5.6 shows other examples with different frequency ratios ω/Ω.

*5.5 FOURIER AND THE SIMPLE OSCILLATOR

Let us return to the question of steadily repeating but nonsinusoidal driving forces. We can give a general and elegant answer to that question if we recognize that we are dealing with a linear system. That is, its governing differential equation is linear, so that the superposition principle applies. This means that if $x_1(t)$ is the solution of Eq. 5.19 for $F_1(t)$, and $x_2(t)$ is the response to $F_2(t)$, then $x_1 + x_2$ correctly gives the motion for applied force $F_1 + F_2$. This allows us to use the

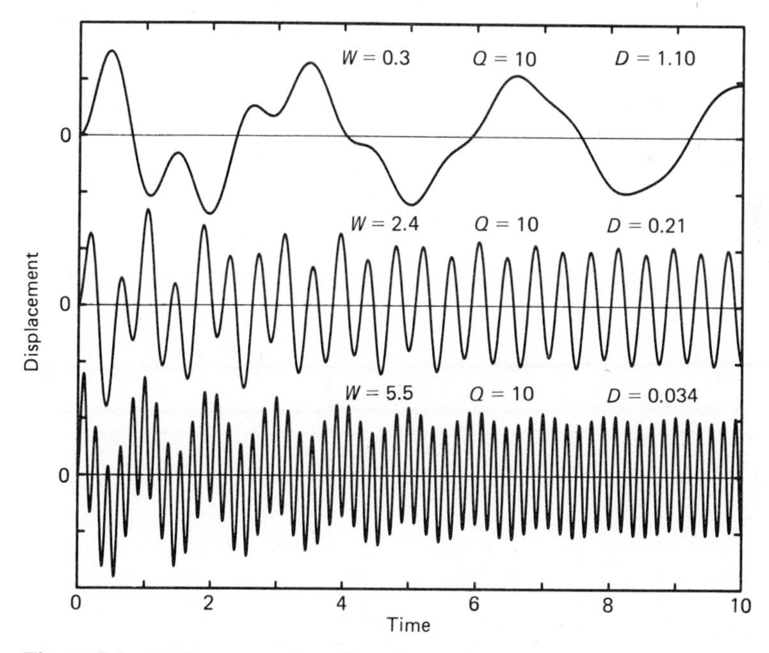

Figure 5.6 Further examples of starting transients, all with the same Q as the middle case in Fig. 5.5, but with relative driving frequencies ω/Ω = 0.3, 2.4, and 5.5. In each case transient vibrations at frequency Ω eventually give way to steady response at the driving frequency ω. Note how some of the regularity of Fig. 5.5 was due to the choice of a special case where ω/Ω was an integer. Vertical scales have been adjusted to show all steady-state amplitudes as equal, but as indicated by the values of D the amplitude is actually much smaller for higher frequencies.

techniques of Fourier analysis. Since we have learned in Sec. 5.3 what will happen for pure sinusoidal forces of all frequencies, every other possibility can be built up as a sum of Fourier components.

As a simple example, suppose the force is described by a square wave of amplitude F_0 and frequency ω. One way to analyze its effect is to say this will produce an infinite series of transients, one for each half cycle. This will be most useful if $\omega \ll \beta$, so that the most recent transient is much larger than any of the earlier ones.

A different approach, offering different insights, is to look at the Fourier components of the force. If $t = 0$ is chosen at one of the moments when the force changes from negative to positive, our example from Chap. 3 shows that we can regard this as the sum of many harmonic driving forces,

$$F_n(t) = (4F_0/n\pi) \sin n\omega t, \tag{5.30}$$

for all odd integers n. Each of these driving forces may be considered to generate its own separate response:

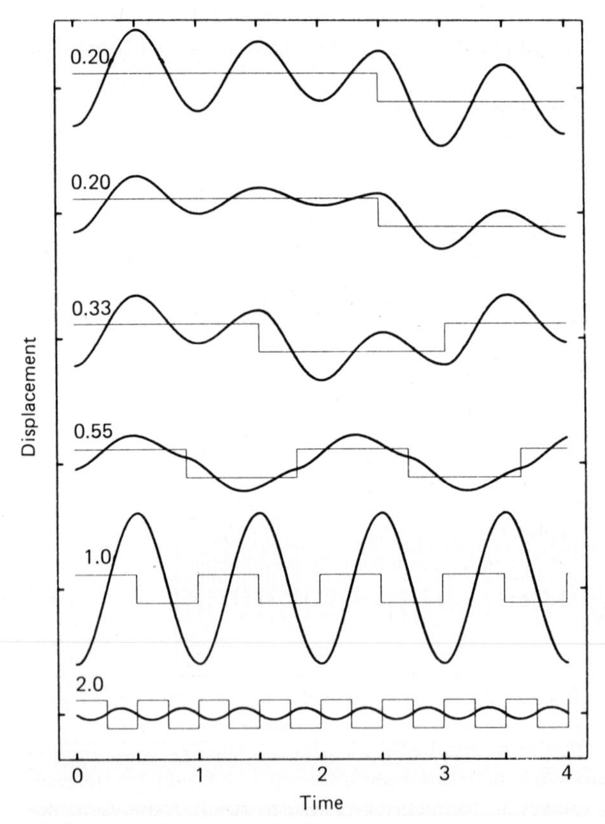

Figure 5.7 Steady response of an oscillator driven by square waves of several frequencies. Graphs are labeled with values of ω/Ω, and the time axis is marked in units of the natural frequency $2\pi/\Omega$. First trace at top has less damping, $Q = 7$, while all others have $Q = 3.3$. Note stronger response when harmonics $n = 1$, 3, or 5 in driving signal match Ω.

$$C_n = \frac{4F_0/n\pi m}{\sqrt{(\Omega^2 - n^2\omega^2)^2 + (2n\beta\omega)^2}} \tag{5.31}$$

$$\phi_n = \text{arccot}\,\frac{\Omega^2 - n^2\omega^2}{2n\beta\omega}\,. \tag{5.32}$$

Contemplation of (5.31) suggests several conclusions. For $\omega > \Omega$, the $n = 1$ term will dominate; we might think of the oscillating mass as a low-pass filter that rounds off the sharp corners of the input (force) to produce the output (displacement). For $\omega < \Omega$, that component for which n most nearly equals Ω/ω will have the largest amplitude; its resonance acts as a bandpass filter. If $\beta < \omega$, this resonance is sharp enough that the response will be dominated by the single component closest to resonance. But for $\beta > \omega$ the response curves are broad enough that several components are comparably important and the total response does not look sinusoidal. Figure 5.7 illustrates some possibilities.

Another interesting case is when $F(t)$ is highly erratic. Let us suppose that it has a white-noise spectrum with the same strength S_F in N^2/Hz at all frequencies, where we are making the same sort of description we did in Eq. 3.4. There is now no particular phase relation among components for either input force or output motion, and the output waveform is also random (Fig. 5.8). But

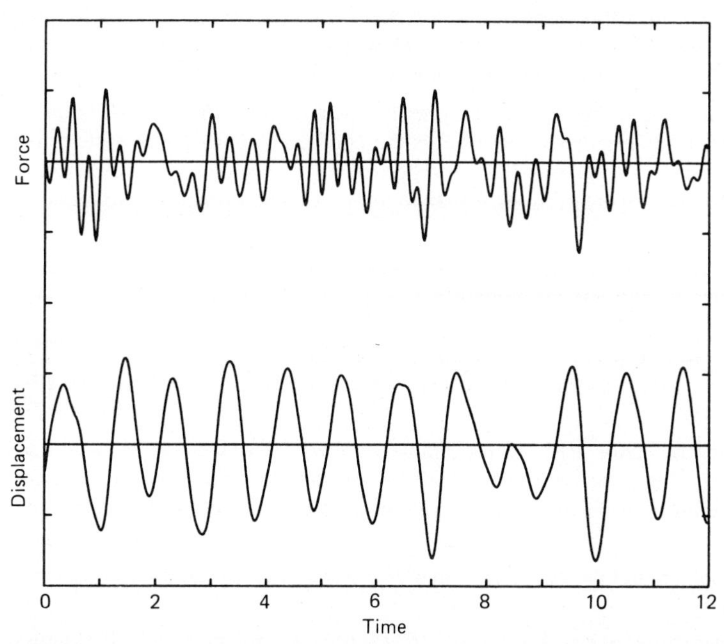

Figure 5.8 A random forcing signal with $S_F(f)$ = constant, simulated by 50 sinusoidal components with frequencies randomly chosen in the range between 0 and 4Ω and random phases, and the resulting displacement of an oscillator with $Q = 5$. Note the tendency for vibrations to occur at approximately the natural frequency Ω, whose periods are marked on the time axis.

(5.25) still gives the amplitudes correctly and tells us that the random motion $x(t)$ has a spectrum

$$S_x(f) = \frac{S_F/16\pi^4 m^2}{(f_0^2 - f^2)^2 + (\beta f/\pi)^2} \,. \tag{5.33}$$

Here we have written f_0 for $\Omega/2\pi$. This says that the spectrum of the motion has a peak at the oscillator's natural frequency f_0. If β is small, this peak may be sharp enough that the motion is nearly sinusoidal even though the driving force was completely random. These are lines along which one could develop a detailed theory of how a loudspeaker with pronounced natural resonances could impose its own coloring on whatever signal was sent through it.

PROBLEMS

5.1. If an oscillator has m = 5 g, s = 5000 N/m, and R = 0.02 kg/s, how much can you say about its free motion? How much about its driven motion? What can you not say without further information?

5.2. Let a massless spring have stiffness $s = 10^4$ N/m. How much mass would you add in order to get a natural frequency of 500 Hz? How much resistance could the system have and still retain half its original amplitude after vibrating for 1 s?

5.3. Describe how much the amplitude of a system with Q = 500 has decreased after completing 300 cycles of free oscillation. How many cycles will it take before the amplitude is down to 1 percent of its starting value?

5.4. If you observe that it takes about 40 cycles for the amplitude of a free oscillator to decay to $1/e$ of its original value, what is the Q of the system? If this system has m = 0.1 kg and s = 10 N/m, what is its effective R? (Notice that this is likely to be a much easier way of determining R than to follow its definition and try to measure F_{fr} and v.)

5.5. Sometimes the decay factor in Eq. 5.10 is written as $\exp(-t/\tau)$.
(a) Describe the physical meaning of τ.
(b) Express τ in terms of β.
(c) Express it in terms of the oscillation period T and Q.
(d) What is τ for the example in the text with m = 0.01 kg, s = 25 N/m, and R = 0.2 kg/s?

5.6. Consider the oscillator used as an example with m = 0.01 kg, s = 25 N/m, and R = 0.2 kg/s. Let it be released from rest at an initial displacement x_0 = 5 cm from its equilibrium position. To describe its subsequent free motion, determine C and ϕ.

5.7. Repeat Prob. 5.6, but changing the initial displacement to zero and initial velocity to −2.5 m/s.

5.8. Show the derivation of Eqs. 5.17 from 5.16. Then show similarly how (5.29) follows from (5.28).

5.9. Suppose you are asked to design an oscillator with the following properties. For an applied alternating force of 2 N amplitude, the greatest steady displacement amplitude is to be 1 mm, and this is to occur for driving frequency

5 kHz. But the response must not exceed 10 μm at 4 or 6 kHz. What will you choose for the mass, stiffness, and resistance of your system? What is its Q? (*Hint:* Extract as much information as possible about 5 kHz first before worrying about 4 or 6.)

5.10. If $F \sin \omega t$ (rather than cosine) is suddenly turned on at $t = 0$, the resulting transient may be entirely different. Illustrate this by taking $\phi_F = 90°$ for a system with $\omega = 500/\mathrm{s}$, $\Omega = 250/\mathrm{s}$, $\beta = 12.5/\mathrm{s}$, $F/m = 5000$ N/kg. Calculate D, ϕ, A, and B, plot $x(t)$, and compare with Fig. 5.5*b*.

5.11. You suddenly turn on a system whose natural frequency is 5 Hz and Q is 200. How long will you have to wait for the transient to decay 20 dB below its initial level? How much would you need to multiply the R of this system in order to reduce that waiting time to 5 s? What effect will that have on the steady-state amplitude for driving frequency close to 5 Hz? What effect for driving frequency much different from 5 Hz?

5.12. Show that the velocity amplitude of a steadily driven oscillator may be written as

$$v = \omega D = \frac{QF_0/\sqrt{ms}}{\sqrt{1 + Q^2(\omega/\Omega - \Omega/\omega)^2}}.$$

5.13. Consider an oscillator with $Q = 3.3$ driven by a square-wave force with frequency $\omega = 0.55\Omega$. Calculate the amplitudes and phases of enough Fourier components of the response to show that their graphical superposition reproduces the fourth curve in Fig. 5.7.

LABORATORY EXERCISE

5.A. Write a program for a computer with graphic output that will embody the general case of a driven oscillator and its transient. That is, the completed program should accept values of the parameters m, s, R, x_0, v_0, F_0, and ϕ_F at run time and produce a plot of the corresponding $x(t)$. A good program will concern itself with scaling the output so that the interesting features of the graph nicely occupy the available paper area, with appropriate labeling of the axes.

Chapter 6

Circuit Analogies and Impedance

We can develop a deeper understanding of what we have said about the harmonic oscillator by drawing parallels with simple electric circuits. These analogies can help guide our study of more complex vibrating systems, and we will eventually extend them to apply in acoustic problems as well. We will specifically focus on the concept of impedance as a way of summarizing how any system interacts with its surroundings. It will help in appreciating the usefulness of these analogies to begin by looking at the energy involved in simple harmonic motion.

6.1 ENERGY AND POWER IN SIMPLE HARMONIC MOTION

A simple harmonic oscillator with mass m and displacement $x(t)$ must have kinetic energy

$$\boxed{E_k(t) = m\dot{x}^2/2.} \tag{6.1}$$

The rate at which this kinetic energy changes is then

$$dE_k/dt = m\dot{x}\ddot{x}. \tag{6.2}$$

This suggests that the energy change under the action of any external force $F(t)$ could be interpreted by multiplying Eq. 5.19 by $\dot{x}$:

$$m\dot{x}\ddot{x} = -R\dot{x}^2 - sx\dot{x} + F\dot{x}. \tag{6.3}$$

By writing $sx\dot{x}$ as $d(sx^2/2)/dt$, we may recognize it as the rate of change of the elastic potential energy stored in the spring,

$$\boxed{E_p(t) = sx^2/2.} \tag{6.4}$$

Let us move that term to the left side of (6.3) to obtain

$$d(E_k + E_p)/dt = -R\dot{x}^2 + (F\,dx)/dt \tag{6.5}$$

If we regard $E = E_k + E_p$ as the total mechanical energy belonging to the oscillator, this says there are two reasons for that energy to change. One is the rate at which work $F\,dx$ is being done upon it by outside forces. The other is the irreversible conversion of useful energy into waste heat by the resistive term. Let us look more closely at three cases.

1. $R = 0$ and $F = 0$: For the undamped free oscillator we have $dE/dt = 0$ and $E = E_k + E_p =$ constant. If $x(t) = C\cos(\Omega t)$,

$$E_k = (m\Omega^2C^2/2)\sin^2(\Omega t) = (m\Omega^2C^2/4)[1 - \cos(2\Omega t)] \tag{6.6}$$

and
$$E_p = (sC^2/2)\cos^2(\Omega t) = (sC^2/4)[1 + \cos(2\Omega t)]. \tag{6.7}$$

Since $m\Omega^2 = s$, E_k and E_p both oscillate forever between the same maximum value $E = sC^2/2$ and zero.

2. $R \neq 0$ and $F = 0$: A lightly damped free oscillator can be described by

$$x(t) = Ce^{-\beta t}\cos\Omega_d t,$$

$$dx/dt = Ce^{-\beta t}(-\beta\cos\Omega_d t - \Omega_d\sin\Omega_d t). \tag{5.16}$$

Thus it will gradually lose energy at a rate calculated most easily from (6.5):

$$dE/dt = -RC^2\exp(-2\beta t)(\beta^2\cos^2\Omega_d t + \Omega_d^2\sin^2\Omega_d t + \beta\Omega_d\sin 2\Omega_d t). \tag{6.8}$$

If we average out the rapid fluctuations of the individual cycles and limit our considerations to $\beta \ll \Omega$, for which the average energy $\bar{E} = (sC^2/2)e^{-2\beta t}$ changes only slowly, this becomes

$$(d\bar{E}/dt) = -(RC^2\Omega^2/2)\exp(-2\beta t) = -(R\Omega^2/s)\bar{E}. \tag{6.9}$$

Since $(R\Omega^2/s) = (R/m) = 2\beta$, this merely confirms that the energy will die away according to $\bar{E}(t) = E_0\exp(-2\beta t)$.

Consider an example with $m = 0.02$ kg, $s = 400$ N/m, and initial amplitude $C = 3$ mm. Its natural frequency is

$$\Omega = \sqrt{s/m} = \sqrt{400/0.02} = 141/\text{s} \quad \text{or} \quad f = \Omega/2\pi = 23\text{ Hz}.$$

If undamped, it will retain its initial energy

$$E_0 = sC^2/2 = (200\text{ N/m})(9 \times 10^{-6}\text{ m}^2) = 1.8\text{ mJ}$$

indefinitely. But if it has resistance $R = 0.05$ kg/s, the energy will be lost with an exponential time constant

$$1/2\beta = m/R = (0.02\text{ kg})/(0.05\text{ kg/s}) = 0.4\text{ s}.$$

The initial rate of energy loss at $t = 0$ is

$$2\beta E_0 = (2.5/\text{s})(1.8 \text{ mJ}) = 4.5 \text{ mW}.$$

3. $R \neq 0$ and $F \neq 0$: Let us suppose any transients have died out and we are concerned only with the steady-state driven oscillator. Though large amounts of energy will be traded between kinetic and potential forms during each cycle, the fact that

$$x(t) = D \cos(\omega t - \phi_x) \tag{5.21}$$

with constant amplitude D ensures that the average total $\bar{E} = sD^2/2$ remains steady. Therefore, the continual dissipation by friction must be exactly balanced by power input to the system from the external driving force to produce a zero total when (6.5) is averaged over time. The average dissipation rate is

$$-R(\dot{x}^2)_{\text{av}} = -R\omega^2 D^2/2, \tag{6.10}$$

while the instantaneous input power is

$$F(t)\, v(t) = -F_0 \cos(\omega t - \phi_F)\omega D \sin(\omega t - \phi_x). \tag{6.11}$$

It is easy to see in the last equation that whenever $\phi_F \simeq \phi_x$ (as occurs for $\omega \ll \Omega$) the long-term average input power is nearly zero. That is, whatever work is done upon the system in one quarter cycle is returned to the outside world nearly intact in the following quarter cycle. Similarly, for high driving frequencies and phase differences near 180°, very little net power is transferred. This is compatible with (6.10) because these are the circumstances in which the amplitude D is very small.

But near resonance, when D is large and (6.10) says considerable dissipation is taking place, (6.11) must justify that there is net power flowing in from the external force. Since this is the case where $\phi_x - \phi_F$ is near 90°, the cosine and sine are always opposite in sign; that is, F does positive work on the oscillator in every quarter cycle. The identity $\cos A \sin B = [\sin(A + B) - \sin(A - B)]/2$ can be used to show that, in general, balancing (6.10) with the time average of (6.11) requires that the power flow is

$$P = R\omega^2 D^2/2 = (F_0\omega D/2) \sin \phi_{xF}$$

$$= (F_0\omega D/2) \cos \phi_{vF}. \tag{6.12a}$$

This agrees exactly with our derivation of how D depends on ω at Eq. 5.24. It is interesting to use (5.24) again to eliminate D and write

$$P = (F_0^2/2R) \sin^2 \phi_{xF} = (F_0^2/2R) \cos^2 \phi_{vF}. \tag{6.12b}$$

This emphasizes that whenever velocity v is out of phase with force F, as it is for ω far from resonance, there cannot be much net power delivered. We can also use (5.25) to eliminate both D and ϕ, which enables us to write the average power input P to the oscillator in all cases as

$$\boxed{P = \frac{R\omega^2 F_0^2/2m^2}{(\Omega^2 - \omega^2)^2 + (2\beta\omega)^2}\,.} \tag{6.13}$$

Equation 6.13 again illustrates the resonance phenomenon that was shown in Fig. 5.3. As you may show in Prob. 6.5, for fixed F_0 the function $P(\omega)$ has its greatest value when $\omega = \Omega$, and that maximum is

$$P_{max} = F_0^2/4\beta m = F_0^2/2R. \tag{6.14}$$

The width of the resonance may be characterized by

$$\boxed{\Delta\omega = 2\beta = \Omega/Q.} \tag{6.15}$$

Here $\Delta\omega$ is the **half-power bandwidth,** defined as the range of frequencies within which P is not less than half its maximum value, that is, within 3 dB of the maximum. This provides a precise measure for describing how less damping (higher Q) means sharper, more selective, frequency response for the driven system.

Consider again the example above with $m = 0.02$ kg, $s = 400$ N/m, and $R = 0.05$ kg/s. Let this system be driven at resonance ($f = 23$ Hz) by a force with amplitude $F_0 = 2$ N, for which the steady-state displacement is

$$\begin{aligned} D &= (F_0/m)/(2\beta\omega) = F_0/R\omega \\ &= (2\text{ N})/\{(0.05\text{ kg/s})(141/\text{s})\} = 0.28\text{ m.} \end{aligned}$$

The power delivered then must be

$$P = F_0^2/2R = (2\text{ N})^2/(0.1\text{ kg/s}) = 40\text{ W.}$$

But if the driving frequency is reduced slightly to 22 Hz so that $\omega = (2\pi)(22) = 138.3/\text{s}$ and $\Omega = 141.4/\text{s}$, the power dissipation is reduced to

$$\begin{aligned} P &= (R\omega^2 F_0^2/2m^2)/[(\Omega^2 - \omega^2)^2 + (2\beta\omega)^2] \\ &= \frac{(0.05\text{ kg/s})(138/\text{s})^2(2\text{ N})^2/(0.0008\text{ kg}^2)}{(141^2 - 138^2)^2/\text{s}^4 + (2.5 \times 138)^2/\text{s}^4} \\ &= 5.8\text{ W.} \end{aligned}$$

This can be understood in terms of the relative phases: whereas $\phi_{vF} = 0$ for $\omega = \Omega$, it changes for $\omega = 138/\text{s}$ to

$$\phi_{vF} = \arctan\,[(138^2 - 141^2)/(2.5 \times 138)] = -68°.$$

Then the factor $\cos^2 \phi_{vF}$ in (6.12*b*) is reduced to 0.144 and P can also be found as (40 W)(0.144) = 5.8 W.

6.2 CIRCUITS AND IMPEDANCE

Much of what we have done above should remind you of the analysis of driving voltages, currents, and power flows in electric circuits. Specifically, we could recognize analogies for each term in Eq. 6.3: $-Rv^2$ is like the rate of energy dissipation $-Ri^2$ when current i flows through resistance R, and Fv is like the expression $\mathcal{E} i$ for the power delivered when an emf $\mathcal{E}$ drives current into a circuit. A little further thought shows that current i flowing into a capacitor C that already has charge q in it must be storing electric potential energy there at the rate $iV_C = \dot{q}(q/C)$, which is like $\dot{x}xs$. And the rate at which magnetic energy builds up in a coil with inductance L is $iV_L = i(L\, di/dt) = \dot{q}(L\ddot{q})$, just like $\dot{x}m\ddot{x}$.

This leads us to propose the formal analogy summarized in Table 6.1, in which subscripts m and e are needed to distinguish the mechanical and electrical cases. The seat of a mechanical resistance R_m is sometimes called a **dashpot.** Notice that it is the inverse of the spring stiffness, called **compliance,** that corresponds to capacitance. Let us consider the hypothesis that the "mechanical circuit" of Fig. 6.1*a* must behave in every way exactly like the electric circuit of Fig. 6.1*b*. Do not allow the superficial resemblance of the looping symbols to trick you into mistaking the spring as being analogous to the inductor! Neither should you jump to the conclusion that Fig. 6.1*a* "looks like a parallel circuit." What is important is that the velocity of the mass must be exactly the same as that of the spring and dashpot ends to which it is attached. That is like saying the same current must flow through L, C_e, and R_e, so that the simple harmonic oscillator must be thought of as acting like a simple series circuit.

TABLE 6.1 ANALOGIES BETWEEN MECHANICAL AND ELECTRICAL SYSTEMS*

Mechanical quantity		Electrical quantity	
Displacement	x	Charge	q
Velocity	v	Current	i
Mass	m	Inductance	L
Resistance	R_m	Resistance	R_e
Compliance	$C_m = 1/s$	Capacitance	C_e
Applied force	F	emf	$\mathcal{E}$
Energy and power:			
$mv^2/2 = E_k$		$Li^2/2 = U_B$	
$sx^2/2 = E_p$		$q^2/2C = U_E$	
$R_m v^2 = P_R$		$R_e i^2 = P_R$	
$Fv = P$		$\mathcal{E} i = P$	
Impedance and reactance:			
$Z_m = R_m + jX_m$		$Z_e = R_e + jX_e$	
$j\omega m = X_m$		$j\omega L = X_L$	
$s/j\omega = X_s$		$1/j\omega C = X_C$	

* Symbols Z, R, X, C play identical roles in both cases, and subscripts e and m are used wherever there might be doubt as to which is intended.

(a)

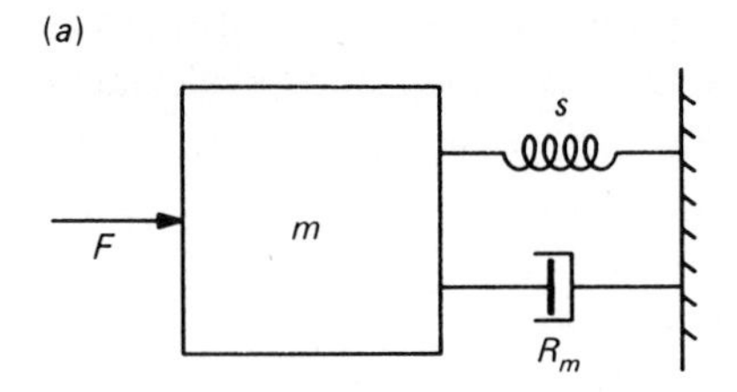

(b)

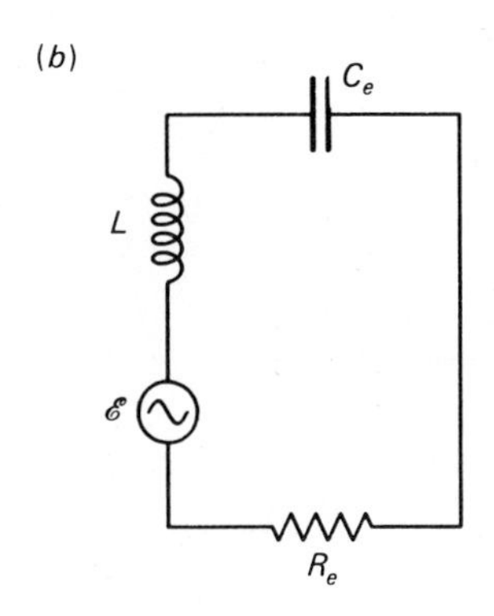

Figure 6.1 (*a*) A simple harmonic oscillator, including mechanical resistance represented as a dashpot or shock absorber. (*b*) An analogous electric circuit.

If you have already studied resonance in a series circuit, much of the preceding discussion should seem quite familiar. From the analogies you could have predicted that the natural frequency $\Omega = 1/\sqrt{LC}$ becomes $\sqrt{s/m}$, the damping rate $\beta = R_e/2L$ changes to $R_m/2m$, the quality factor $Q = \Omega L/R_e$ to $\Omega m/R_m$, the maximum power $P = \mathscr{E}^2/2R_e$ to $F^2/2R_m$, the bandwidth $\Delta\omega$ is simply unchanged when expressed as Ω/Q, etc. A particularly helpful aspect of the analogy, which we have not used thus far, is the impedance. When dealing with sinusoidal signals in electric circuits, it is useful to define the electrical impedance

$$Z_e = R_e + jX_e = V/i, \tag{6.16}$$

which for the simple series circuit is just $R_e + j\omega L + 1/j\omega C_e$. This makes it possible to pretend for many purposes that all circuit elements are simply resistors, and $V = iZ$ will keep track of both the magnitudes and the correct phase relations between the voltage drop V across any element and the current i flowing through it.

Let us then define **mechanical impedance** by

$$\boxed{Z_m = R_m + jX_m = F/v,} \tag{6.17}$$

where the imaginary part X_m is called **mechanical reactance.** The ratio F/v says that impedance means how much force must be applied to produce unit velocity, so that we should think of impedance as being a measure of "how hard it is to make the system move." We propose that all the results for the steady state of the driven harmonic oscillator would follow from simply using

$$Z_m = R_m + j\omega m + s/j\omega \quad \text{and} \quad v = F/Z_m. \tag{6.18}$$

Remember always that Z is not just a number but a complex-valued function of frequency, as illustrated in Figs. 6.2 and 6.3. Once this impedance is calculated, then by using $v = \dot{x} = j\omega x$ with $v = F/Z$ we can get

$$x = F/j\omega Z_m = -jZ_m^* F/\omega |Z_m|^2. \tag{6.19}$$

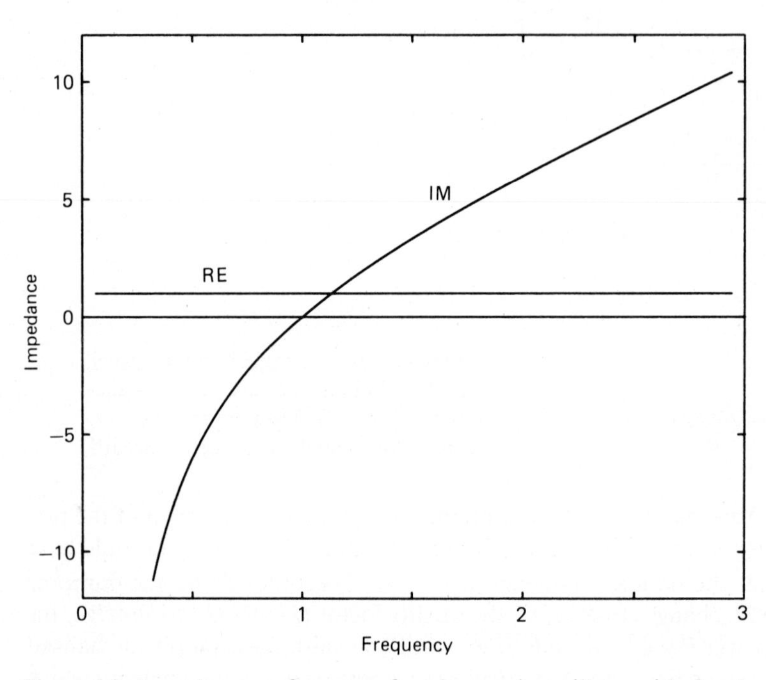

Figure 6.2 Impedance vs. frequency for a harmonic oscillator with $Q = 4$, using R_m and Ω as units for the vertical and horizontal axes. The real part R_m, or resistance, is constant. The imaginary part X_m, or reactance, is $\omega m - s/\omega$.

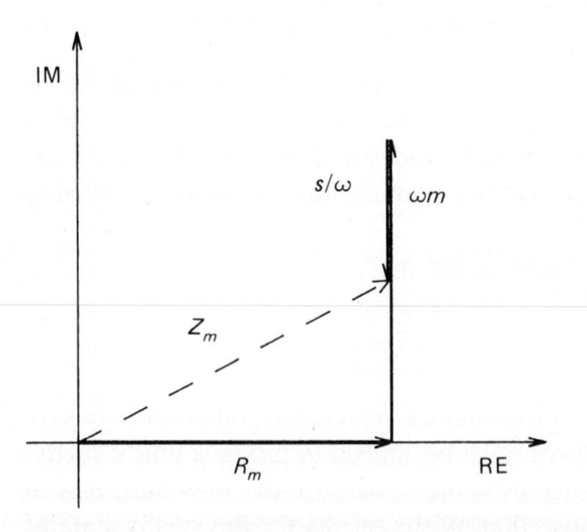

Figure 6.3 An alternative way of representing the impedance of a harmonic oscillator. Three components R_m, $j\omega m$, and $-js/\omega$ are added in the complex plane to get the total impedance Z_m. As ω increases from zero toward infinity, the tip of Z_m traces out the vertical line from $R - j\infty$ to $R + j\infty$, crossing the real axis when $\omega = \Omega$.

The magnitude and phase of this complex equation just reproduce Eqs. 5.25 and 5.26, as you may show in Prob. 6.8.

Let us illustrate this with the same system used above. For $\omega = 138.3/\text{s}$, $m = 0.02$ kg, $s = 400$ N/m, and $R_m = 0.05$ kg/s we have

$$Z_m = 0.05 + j(138.3)(0.02) + 400/j138.3$$
$$= 0.05 + j(2.77 - 2.89) = 0.05 - j0.12 \text{ kg/s}.$$

This can also be expressed in terms of magnitude and phase angle:

$$Z_m = \sqrt{0.05^2 + 0.12^2} \exp[j \arctan(-0.12/0.05)]$$
$$= 0.13 \text{ kg/s at } -68°.$$

The velocity amplitude then is

$$v = F/Z = (2 \text{ N})/(0.13e^{-j68°} \text{ kg/s})$$
$$= 15\, e^{j68°} \text{ m/s}$$

and the displacement

$$x = v/j\omega = (15e^{j68°} \text{ m/s})/(j138/\text{s})$$
$$= 0.11e^{-j22°} \text{ m}.$$

The angles of these phasors tell us that the velocity leads the force by 68° but the displacement lags the force by 22°. Equations 5.25 and 5.26 give the same result, predicting $D = 0.11$ m and $\phi_{xF} = +22°$.

The average power delivered to a circuit is given by

$$\boxed{P = (Fv)_{\text{av}} = (1/2)\,\text{Re}\{F^*v\};} \tag{6.20}$$

Appendix B will remind you why the factor 1/2 appears when we treat F and v as phasors. Using $v = F/Z$, this becomes

$$P = (F^2/2)\,\text{Re}\{1/Z\} = (F^2/2)\,\text{Re}\{Z^*\}/|Z^2| = F^2R/2(R^2 + X^2), \tag{6.21}$$

which reproduces (6.13). This suggests that we might sometimes find it an advantage to use the **admittance**

$$\boxed{Y = 1/Z = G + jB.} \tag{6.22}$$

(Though we do not emphasize it below, you may encounter the terminology **conductance** for the real part G and **susceptance** for the imaginary part B in electric-circuit theory.) The admittance of the series circuit is shown in Figs. 6.4 and 6.5. In these terms we would have

$$P = (F^2/2)\,\text{Re}\{Y\} = F^2G/2, \tag{6.23}$$

which suggests that the curve for G in Fig. 6.4 shows directly how the power absorbed varies with frequency. In (6.21) and (6.23), each occurrence of the

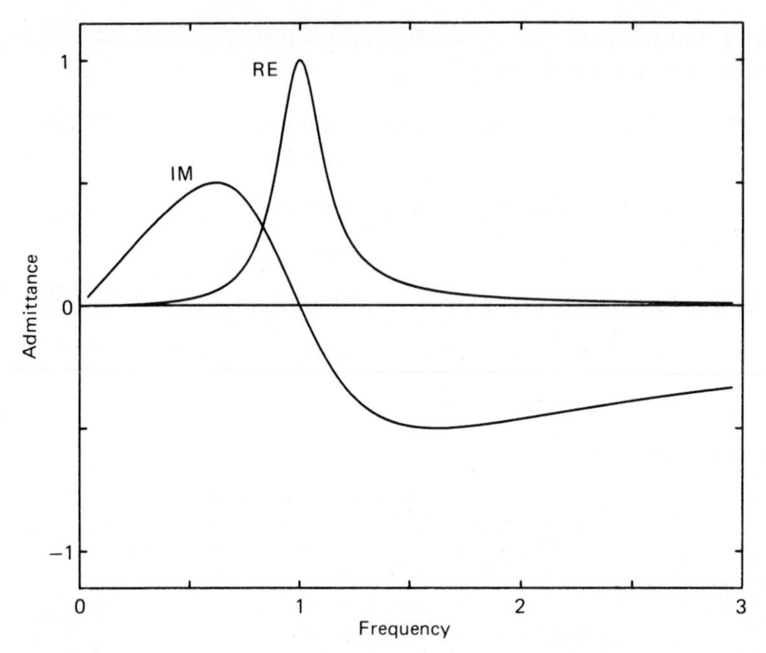

Figure 6.4 Admittance $Y = G + jB$ of a harmonic oscillator as a function of frequency. $1/R_m$ and Ω serve as units. Since $Y = 1/Z$, this graph contains exactly the same information as does Fig. 6.2.

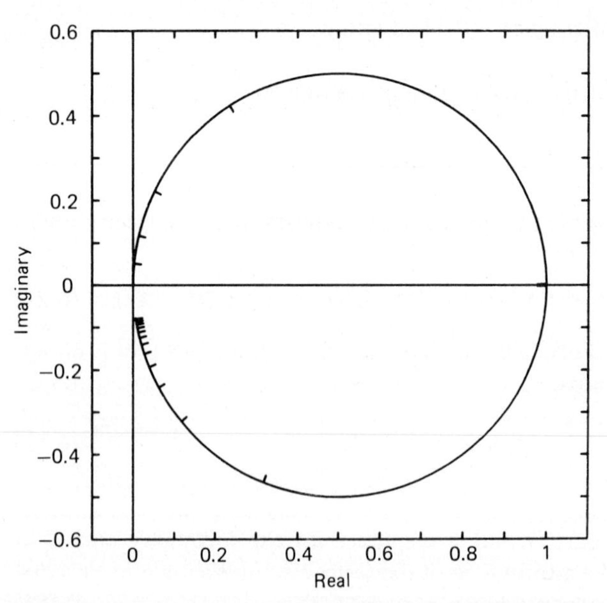

Figure 6.5 Alternative representation of admittance for the harmonic oscillator. As frequency varies from 0 to ∞, Y starts from zero, traces the circle clockwise, crosses the real axis when $\omega = \Omega$, and returns to the origin. Both admittance axes are in units $1/R_m$, and tick marks show multiples of 0.2Ω.

factor $F^2/2$ should be interpreted either in terms of the maximum amplitude F_0 as $F_0^2/2$ or in terms of the rms average force as F_{rms}^2 without the half.

Continuing the example above, the system driven at $\omega = 138/\text{s}$ has admittance

$$Y_m = 1/Z_m = 1/(0.13e^{-j68^\circ}\ \text{kg/s})$$

$$= 7.7e^{+j68^\circ}\ \text{s/kg} = 2.9 + j7.1\ \text{s/kg}.$$

Then the power dissipation for $F_0 = 2$ N is again found to be

$$P = F_0^2G/2 = (2\ \text{N})^2(2.9\ \text{s/kg})/2 = 5.8\ \text{W}.$$

6.3 POWER SOURCES

Devices for driving electric circuits are often built to approximate one or the other of two limiting ideal behaviors. An ideal voltage source always provides exactly the same electromotive force regardless of how much current is drawn. An ideal current source always maintains the same current flow, no matter how much (or how little) emf it must generate to send that current through the attached circuit.

We can make a similar distinction in the mechanical case between two types of external drivers. A **force-driven** system means one assumed always to experience the same amplitude of force, regardless of how much motion that force produces or of the frequency at which it is applied. An ideal force driver is analogous to an ideal voltage source but is seldom easy to achieve in practice. It can be approximated sometimes by using magnetic forces or by transmitting forces through lightweight elastic bands.

A **displacement-driven** system, on the other hand, is more like a circuit with an ideal current source. (Strictly speaking, the ideal current source corresponds to a system with prescribed velocity rather than displacement.) The displacement-driven system is automatically provided with however much force is needed to always produce the same amplitude of motion. This will be the case when the system is rigidly fastened to a very massive shaker that is practically oblivious to its load. With such a system one might even choose not to view the applied force as causing a displacement but rather to think of the prescribed displacement bringing about a force of reaction from the system upon the driver. Newton's third law, then, says that the driver supplied an equal and opposite force upon the system.

When dealing with a force-driven system, it may be most convenient to write the power delivered to it as we did in (6.23). In thinking about how P changes as the driving frequency varies, we would emphasize that $F^2/2$ is constant and $\text{Re}\{Y\}$ has a marked peak (associated with a zero crossing of $\text{Im}\{Y\}$) at the system's resonant frequency. And at a fixed frequency, the system would absorb *less* power if its impedance were increased, since that would cut down on its amplitude of motion.

But when considering a displacement-driven system, we might choose to replace F by Zv in (6.20), so that instead of (6.21) we would have

$$P = (v^2/2)\,\mathrm{Re}\{Z\} = (\omega^2 x^2/2)R, \tag{6.24}$$

with constant x. Here it is not the power that varies radically with frequency in the vicinity of a resonance, but the way that power is delivered: at resonance, the power arrives as a roughly steady flow into the system, while off resonance the net power represents only a small leakage from a much larger amount of energy being shared reactively by the system and driver. And at a fixed frequency, the system would absorb *more* power if its impedance were increased, since that would give the driver the opportunity to exert a greater force through the same distance.

We can point out an acoustic analogy while on the subject. Because of the great difference in mass density between air and solid objects, a microphone receiving a sound signal is often well approximated as a force-driven (actually pressure-driven) system. But the air in a room being moved by a loudspeaker is usually better thought of as displacement-driven, since the air presents little enough load that the speaker motion is very little affected by it.

6.4 MEASURING MOTION AND IMPEDANCE

For complex systems where theoretical analysis is quite difficult, we might like to take an empirical approach to the determination of their properties. Let us briefly describe here ways in which we might simply measure the quantities we need to know, either in the laboratory or in the field. This is partly a matter of knowing what types of transducers are available to convert any desired measurement into electrical form, after which it may conveniently be amplified, measured, or analyzed.

First, consider the problem of measuring how much vibration is occurring in a solid body. Instead of a microphone, what we need is an electromechanical transducer. These exist both in forms that actually touch the vibrating body and in others that do not touch it. We have already commented in Chap. 2 about contact transducers as close analogs of microphones, and in particular about how the bending of a small needle held against the moving object can be detected either magnetically (moving-coil, moving-magnet, or moving-armature) or piezoelectrically.

A variant of the piezoelectric pickup that is very widely used is the **accelerometer.** Instead of communicating vibration from the moving body through a needle to be detected in a nearby rigidly mounted device, we can let the entire detector be mounted on the moving body and ride along with it. Figure 6.6 suggests schematically how a mass m inside the accelerometer can be forced to move along with the housing if that force is transmitted to it through one or more piezoelectric crystals providing stiffness s. For the intended range of application, the frequency of the vibrations to be measured lies below $\sqrt{s/m}$, so that the motion of m is stiffness-controlled. Then the small mass m is forced to

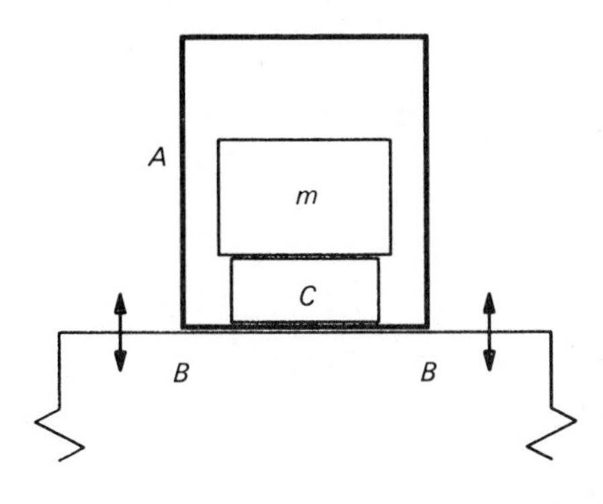

Figure 6.6 The basic idea of an accelerometer. A large vibrating body B carries the accelerometer A along with it. This motion is transmitted to the small mass m through the crystal C, in which the piezoelectric effect then generates a signal proportional to the acceleration. The two horizontal surfaces of C would have a conductive coating to serve as electrodes. Other types may derive their signal from bending instead of compression of the crystal.

move with the same acceleration as the vibrating body; but that force can only be supplied by a slight distortion of the crystal, so the piezoelectric effect gives an output signal directly proportional to that acceleration.

Noncontact transducers come in both electric and magnetic types. In the electric case the transducer is rigidly mounted and forms one plate of a capacitor, while the vibrating body (or a small metal foil attached to it) is the other plate; so this is analogous to the electrostatic microphone. In the magnetic case, the transducer is a small permanent magnet encircled by a coil of wire, so that motion of nearby ferromagnetic material alters the magnetic field and thereby induces emf in the coil. Again, if the vibrating body itself happens to be made of nonmagnetic material, a small disk of iron can be attached to it. Notice that the voltage output of the capacitive pickup (or of a crystal contact transducer) is proportional to displacement, and that of the magnetic pickup (with or without contact) proportional to velocity, neither of which is like the accelerometer. Since the maximum amplitudes are related by $a = \omega v = \omega^2 x$, it appears that the relative signal strength favors use of the displacement pickup at low frequencies but of the accelerometer at high frequencies.

Four-terminal devices are available commercially to measure simultaneously the alternating force applied and the resulting vibration. If one of the outputs is held constant by a feedback circuit during a frequency sweep, the other provides a signal proportional to the impedance (or, with roles reversed, to the admittance); so it is called an **impedance head.** Graphs of the magnitude of the impedance vs. frequency can then be quickly generated, or with phase-sensitive amplification the resistance and reactance could be determined separately.

*6.5 ANALYSIS OF COUPLED OSCILLATORS

The concepts introduced in this chapter offer tools for theoretical predictions about systems with more than one moving part. You can find out a great deal about the steady-state operation of ac circuits involving several loops just by using Kirchhoff's laws and the complex impedances of all the individual circuit elements. So also we can deduce much about the steady response of complex mechanical systems to sinusoidal driving forces without having to start from the differential equations of motion.

Let us illustrate this by considering two masses connected with a spring as shown in Fig. 6.7*a*. In constructing the analogous circuit you must resist the temptation to suppose that the two masses are "obviously" in series. You must argue instead that whatever velocity m_1 has may be passed on as *either* compression of the spring *or* motion of m_2. Since velocity is analogous to current, that is like current arriving at a point where it has the choice to go through either of two branches. Hence we must consider the reactance of the spring to be in parallel with whatever lies beyond it, and the equivalent circuit is that of Fig. 6.7*b*.

We want to know the **input impedance,** meaning the impedance seen at the point of application of the force. Now when an impedance $j\omega m_1$ is put in series with the parallel combination of $s/j\omega$ and $j\omega m_2$, the total impedance is

$$Z_m = j\omega m_1 + (j\omega/s + 1/j\omega m_2)^{-1}$$

$$= j\omega[m_1 + m_2/(1 - \omega^2 m_2/s)]. \tag{6.25}$$

Several features of this system's behavior can be seen in this equation. First, Z_m is purely imaginary for all frequencies; so the impedance is entirely reactive. That is, no net power will flow into the steadily driven system since it has no resistance where that power could be dissipated. Corresponding to this lack of resistance is the prediction that there is a resonant frequency

$$\omega_2 = \sqrt{(m_1 + m_2)s/m_1 m_2} \tag{6.26}$$

at which $Z_m = 0$ and $Y_m = \infty$. The consequent infinite response to a finite driving force can be avoided only by allowing for some damping. There is also an antiresonant frequency

$$\omega_1 = \sqrt{s/m_2} \tag{6.27}$$

(a)

(b)

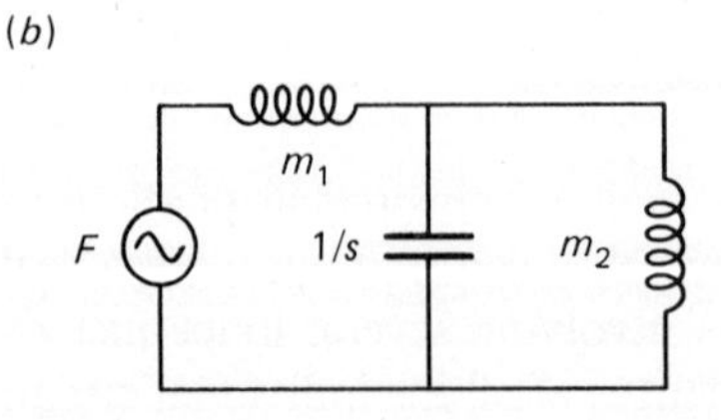

Figure 6.7 A two-mass system and its analogous circuit.

for which $Z_m = \infty$ and $Y_m = 0$, and finite force cannot produce any steady displacement of m_1 (although m_2 does move). In the low-frequency limit,

$$\omega \ll \omega_1: \qquad Z_m \simeq j\omega(m_1 + m_2). \tag{6.28}$$

This says, as it should, that for very low frequencies the two masses will nearly move together as a single unit. In the high-frequency limit,

$$\omega \gg \omega_2: \qquad Z_m \simeq j\omega m_1. \tag{6.29}$$

This too is expected on physical grounds, as the spring will tend to absorb high-frequency displacements and not pass on any motion to m_2. Further details are called for in Prob. 6.11.

In order to study a case that always has finite response, let us add another spring and a dashpot to create the system of Fig. 6.8. Now m_2, s_2, and R must all have the same velocity, so are placed in series in the equivalent circuit. The input impedance must be $j\omega m_1$ in series with the parallel combination of (1) $s_1/j\omega$ and (2) the series combination of $j\omega m_2$, $s_2/j\omega$, and R. That is,

$$Z_{\text{in}} = j\omega m_1 + \frac{(s_1/j\omega)[R + j(\omega m_2 - s_2/\omega)]}{(s_1/j\omega) + [R + j(\omega m_2 - s_2/\omega)]}. \tag{6.30}$$

A word of caution: Remember that applying the force at some other point will mean a different input impedance. As seen from a point between s_1 and m_2, for instance, the parallel combination of m_1 and s_1 would appear to be in series with the other three elements. This same issue is present in electric circuits, where the impedance encountered by a voltage source changes if it is moved to a different position in a circuit.

(a)

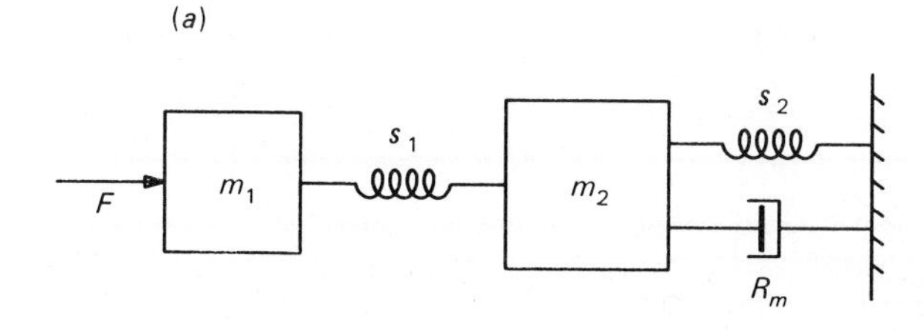

(b)

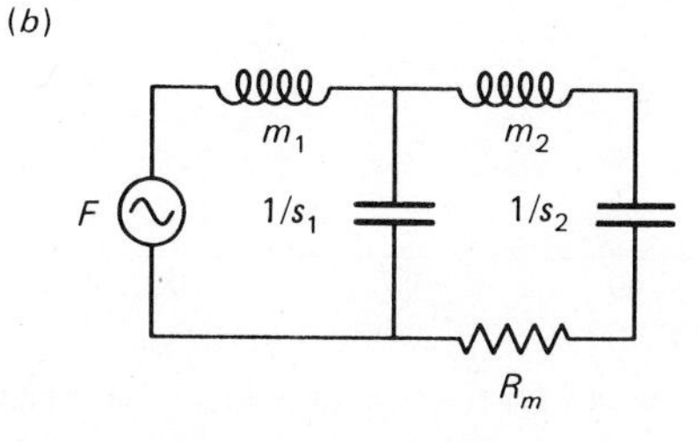

Figure 6.8 Another complex vibrator and its analogous circuit.

For any particular choice of parameters, Z_{in} is a function of ω, but (6.30) is too complicated to see easily what that function is like. Calculation with a short computer program and graphical presentation of the results can provide more insight. In order to give a concrete example, take the values $m_1 = 2$ kg, $m_2 = 3$ kg, $s_1 = 6$ kN/m, $s_2 = 3$ kN/m, and $R = 20$ kg/s. The impedance is then as shown in Figs. 6.9 and 6.10. Alternatively, we could plot the admittance Y_{in} as in Figs. 6.11 or 6.12. Either way, though perhaps more so with Y, there is a strong suggestion of repetitions of resonant behavior like that of Figs. 6.2 through 6.5 at two distinct frequencies.

In all lightly damped systems, the presence of distinct minima in the magnitude of Z (or corresponding maxima of Y) will serve as clues that there are resonances. Though resonance is strictly defined in terms of maximum power transfer, it is often more convenient to look at the zero crossings of the imaginary part of either Z or Y, since they occur very close to the resonant frequencies. For example, we could ask that the imaginary part of Z in (6.30) be zero, and (after rather long and dreary algebra) find the explicit formula

$$\omega^2 = (W \pm \sqrt{W^2 - 4m_1m_2s_1s_2})/2m_1m_2,$$

where W stands for $m_1s_1 + m_1s_2 + m_2s_1$. You can best understand what this means by convincing yourself that it produces sensible results in various limiting cases when each parameter becomes very large or very small. (For instance, if s_1 is very large we get a resonance at $\omega^2 = s_2/(m_1 + m_2)$, as we

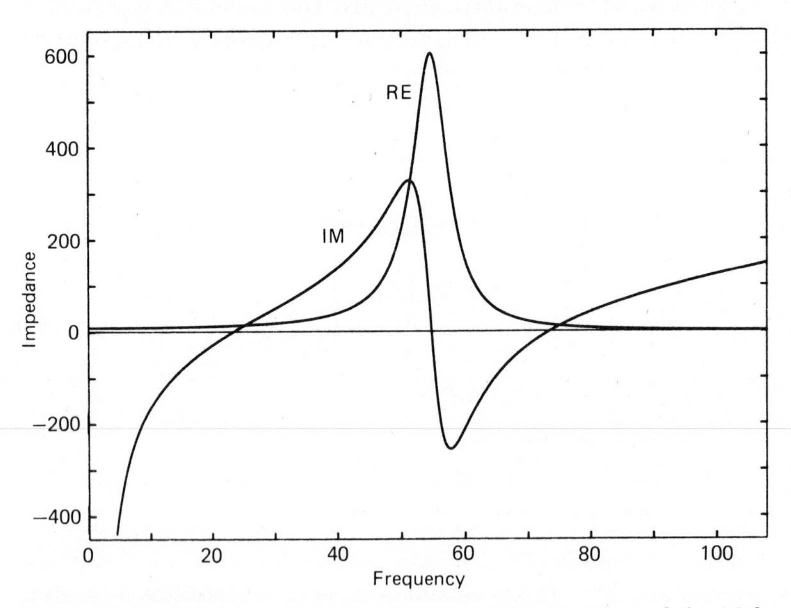

Figure 6.9 Impedance as a function of frequency for the system of Fig. 6.8 for parameter values used in example.

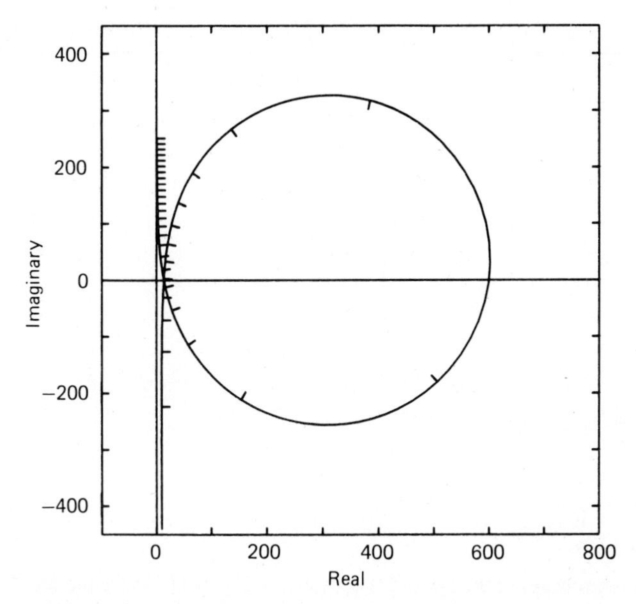

Figure 6.10 Alternative representation of the impedance given in Fig. 6.9. As ω increases from zero to infinity, Z traces a path from $-j\infty$ to $+j\infty$ like Fig. 6.3, with the addition of a clockwise loop. Tick marks are at 4-Hz intervals in frequency. The two upward crossings of the real axis occur at the resonant frequencies.

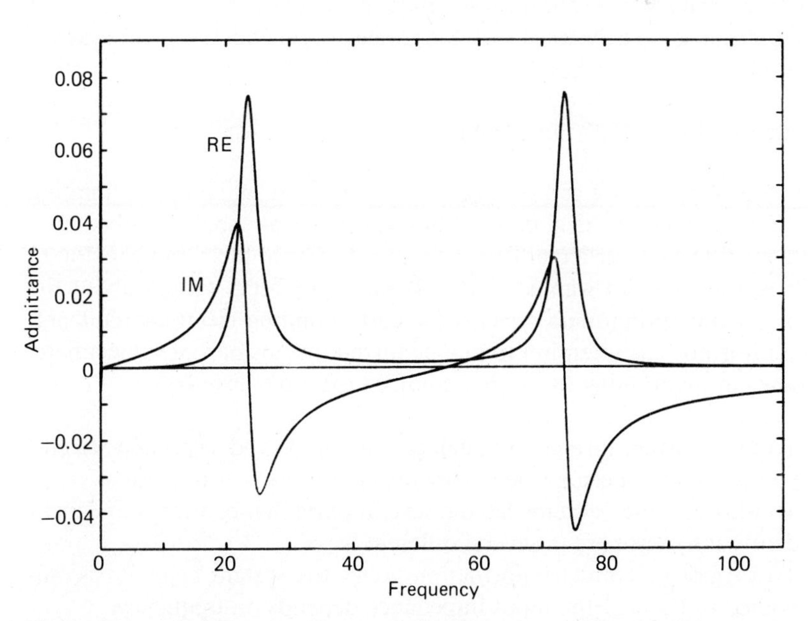

Figure 6.11 Admittance vs. frequency for the system of Fig. 6.8. Resonances appear as peaks of the real part of Y.

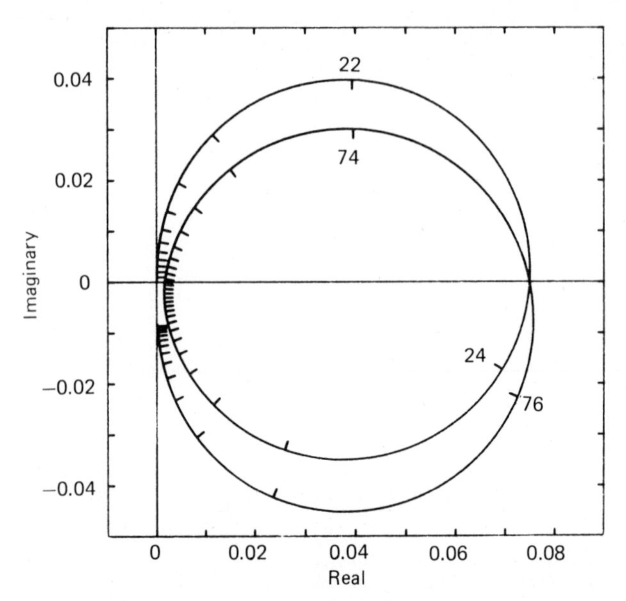

Figure 6.12 Alternative representation of the admittance shown in Fig. 6.11. As frequency goes from zero to infinity, Y traces two clockwise loops in the complex plane, beginning and ending at the origin. The two downward axis crossings at the right occur at the resonant frequencies.

would expect when the two masses are forced to move together.) For the numerical values used above, resonances are predicted for $\omega = 23.4$ and 75.2 rad/s, in good agreement with the graphs. In general, the low-frequency resonance always involves the two masses moving in phase with each other (though not necessarily equal amounts), and the high-frequency resonance has them moving 180° out of phase.

This sort of procedure can be used to study all kinds of complex vibrating systems and to deduce practical designs for such things as shock absorbers that will effectively filter out particular frequencies and prevent their transmission to places where they are not wanted. Entire books have been written about the analysis of general spring/mass systems (or corresponding electric-circuit networks), and it is not our intention to pursue further details of that subject here. Our purposes in mentioning this rather complex example have been:

1. To show further how circuit analogies are made and, especially, to emphasize how to decide when elements are in series or in parallel.
2. To illustrate that systems having several parts that can move independently may resonate in several different ways.
3. To suggest that much information about the system behavior is embodied in the way the input impedance depends on frequency.
4. To lay the groundwork for taking a similar approach to the analysis of sound vibrations in air in later chapters.

PROBLEMS

6.1. **(a)** If an undamped system with natural frequency 5 Hz and mass 20 g vibrates freely with energy 0.1 J, what is its displacement amplitude? What is its velocity amplitude?

(b) If a resistance of 0.01 kg/s is introduced, how rapidly is energy being lost when there is 0.1 J present? How long will it take for that to be reduced to 0.01 J, that is, a 10-dB loss?

6.2. For the system used as an example in Chap. 5, with $m = 0.01$ kg, $s = 25$ N/m, and $R = 0.2$ kg/s, what is the mechanical impedance at frequencies **(a)** $f = 2$ Hz, **(b)** $f = 8$ Hz, and **(c)** $f = 50$ Hz? How does it behave as f approaches zero or infinity?

6.3. Repeat Prob. 6.2 for admittance instead of impedance.

6.4. **(a)** A simple oscillator has $m = 5$ g, $s = 5000$ N/m, and $R = 0.5$ kg/s, and is driven by a sinusoidal force with amplitude 20 N and frequency 140 Hz. What is the impedance of the oscillator at this frequency? What is the amplitude and phase of the motion, and at what rate is power being absorbed?

(b) Repeat with the driving frequency changed to 160 Hz, and to 200 Hz.

6.5. Consider the power P as a function of frequency ω, as given in (6.13).

(a) Find the maximum of this function.

(b) Using the approximation $\beta \ll \Omega$, find the frequencies for which P equals half its maximum value, and verify (6.15).

6.6. Suppose we had defined resonant bandwidth as extending to all frequencies where response is within 10 dB of maximum, rather than 3 dB. In the case of light damping ($\beta \ll \Omega$), show that this would introduce a factor of 3 into (6.15).

6.7. **(a)** Consider again the example discussed at the end of Sec. 5.3, with $m = 0.01$ kg, $s = 25$ N/m, $R = 0.2$ kg/s, and $F = 2$ N. What is the maximum power and the half-power bandwidth?

(b) If you wanted that bandwidth to equal 0.01 Ω, what new value of R would be required? What would be the new maximum power?

6.8. **(a)** Show that (6.18) and (6.19) lead to (5.25) and (5.26).

(b) Show that (6.13) follows from (6.21).

6.9. How many different ways can we relate the quality factor Q to other important system properties? In Chap. 5 we started with the definition $\sqrt{sm}/R$, which was immediately equivalent to $\Omega m/R$ or $\Omega/2\beta$. In Prob. 5.5 we showed that $\tau = 1/\beta$ is the $1/e$ damping time scale; so Q can also be written as $\Omega\tau/2$. (5.15) shows that Q equals the number of cycles for the amplitude to decay by 27.3 dB. Equation 6.15 says that Q can be written in terms of the half-power bandwidth as $Q = \Omega/\Delta\omega$. Let us add two more interesting expressions to that list, again using the approximation $\beta \ll \Omega$ if needed.

(a) If ΔE is the amount of energy lost in one cycle of free damped oscillation, show that $Q = 2\pi E/\Delta E$.

(b) Show that $Q = (\Omega/2R)dX/d\omega$ if the derivative is evaluated at $\omega = \Omega$. This last expression is particularly interesting since it could be used to define a Q for each of the multiple resonances of a complex system, if Ω, R, and $dX/d\omega$ are all assigned values corresponding to one of the frequencies at which X passes through zero.

6.10. Show from (6.18) that the admittance phasor always lies on the circle of diameter $1/R$ shown in Fig. 6.5. Show that the half-power frequencies correspond to the top and bottom of the circle (in the approximation where $\beta \ll \Omega$).

6.11. Let the numerical values for the system of Fig. 6.6 be $m_1 = 2$ kg, $m_2 = 3$ kg, and $s = 6000$ N/m. Sketch graphs, including proper units, indicating the dependence of **(a)** input impedance Z_m, **(b)** velocity v_1, and **(c)** velocity v_2 upon frequency.

6.12. What is the input impedance at $f = 10$ Hz of the system shown in Fig. 6.8 if $m_1 = 2$ kg, $m_2 = 3$ kg, $s_1 = 6000$ N/m, $s_2 = 3000$ N/m, and $R = 20$ kg/s? If the driving force amplitude is 50 N, what is the amplitude and phase of the resulting motion? How much power is dissipated?

6.13. Verify that the resonant frequencies of the system used as an example in Sec. 6.5 are 3.7 and 11.8 Hz. What is the input impedance, the amplitude, and the power for each of those frequencies? Before calculating any numbers, can you say qualitatively how those will compare with the answers in the preceding problem?

LABORATORY EXERCISES

6.A. If any vibration transducers are available in your laboratory, learn how they work. Use them to measure the motion of some vibrating machinery, perhaps in your school's machine shop. In particular, study the signals with a spectrum analyzer if possible.

6.B. If an impedance head is available to you, use it to measure input impedance as a function of frequency for one or more small objects, such as a machine cover plate or a guitar string. Does $Z(f)$ indicate any resonances, and can you explain their nature? Check the manufacturer's calibration of the device [the sensitivities of the force transducer and accelerometer in V/N and V/(m/s^2)] by using a pure mass as a test object.

Chapter 7

Transverse Waves on a String

When we studied the simple harmonic oscillator, our central task was to find the single function $x(t)$ that would describe the entire history of the motion. More information was required to deal with two masses, each of which could move independently. In fact, for any finite number N of masses connected by various springs, we would need to determine $x_1(t)$, $x_2(t)$, . . . , $x_N(t)$. But now we want to move on to systems having infinitely many different parts, each of which may move in a different way. The patterns of motion we see in such a continuous medium, for instance, a bowl of quivering jelly, are called waves.

A flexible string under tension provides the easiest example for visualizing how waves work and developing physical concepts and techniques for their study. The vibrating string is interesting both for its own sake (as a source of sound on a guitar or violin) and as a model for the motion of other systems (such as air confined in a tube). Here we study both free and driven motion of a string, and the energy associated with them. The procedures we use will apply in our later study of other kinds of waves.

7.1 THE EQUATION OF MOTION

Let us consider a long, uniform string whose total mass is m_s and length is L. What is important for the local behavior of each part is the **linear mass density** $\mu = m_s/L$. This inertia needs a restoring force working against it to generate vibrations, and we suppose that is due to a **tension** T_0 applied to the ends of the string. We use the idealization of a perfectly flexible string, which does not have any stiffness of its own to resist being bent. Suppose we choose an x axis parallel to the undisturbed straight string, so that x serves as a label identifying various points on the string. This makes x an independent variable analogous to the subscripts 1, 2, . . . , N in the case of N coupled masses.

We need to distinguish x from the dependent variable telling how much any part of the string has moved. Let us call that displacement ξ, and consider for now only the case of **transverse waves** where ξ is perpendicular to the x axis. (It is enough for our present purposes to treat the case where the string remains always within one plane.) Since this displacement will differ both from point to point on the string and from time to time, we must study the function $\xi(x, t)$. Our predictions about ξ must all be based on insisting that every piece of the string obey Newton's second law of motion, $F = ma$. Let us see what this means for a typical piece lying between x and $x + \Delta x$, as shown in Fig. 7.1.

Suppose we limit our study to cases where gravity is unimportant, as it is for such light strings as those on a guitar. Then the only forces exerted on the small element of string are the tensions at the two points where it meets its neighboring elements. When the string is at rest, the tensions at x and at $x + \Delta x$ are precisely equal in magnitude and opposite in direction, making zero total force. Even for a string section with constant slope there would still be no net force or acceleration. It is only where the string has curvature (Fig. 7.1) that the tension acts in slightly different directions at x and $x + \Delta x$, thus pulling that segment in such a direction as to try to straighten out the curvature.

Now the tension at each point must be parallel to the slope of the string at that point, which is

$$\tan \delta(x) = \partial\xi/\partial x. \tag{7.1}$$

We must ask about both horizontal and vertical components of these vector forces and consider the possibility that T might not be the same throughout the string. The net force along the x axis is

$$F_x = T(x + \Delta x) \cos \delta(x + \Delta x) - T(x) \cos \delta(x). \tag{7.2}$$

The possibility of nonzero F_x leads to the consideration of longitudinal motion, which we want to defer until Chap. 9. In order to study purely transverse motion, we decree for now that $T \cos \delta = T_0$ is constant throughout the string so that $F_x = 0$.

The vertical component of force is

$$F_\xi = T(x + \Delta x) \sin \delta(x + \Delta x) - T(x) \sin \delta(x). \tag{7.3}$$

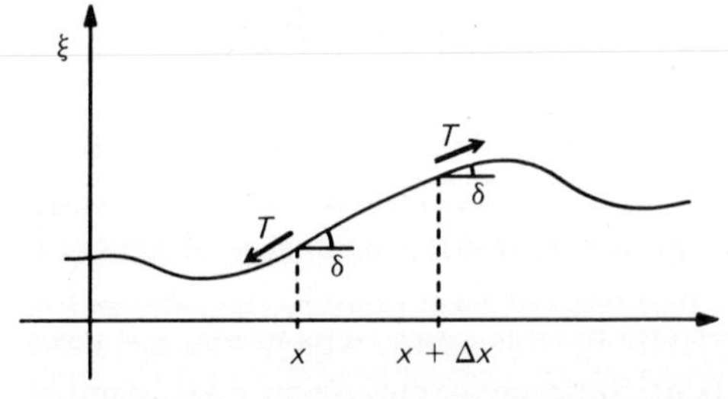

Figure 7.1 Transverse displacement $\xi(x, t)$ as a function of position x along a string shows its actual shape at time t. Slope angle δ also depends on x and t, and determines the direction in which tension T acts on each piece of string.

But each $T(x)$ may be replaced by $T_0/\cos\delta$, so that we may use (7.1) to write

$$F_\xi = T_0\, \partial\xi(x + \Delta x)/\partial x - T_0\, \partial\xi(x)/\partial x$$
$$\simeq T_0\, \Delta x\, \partial^2\xi/\partial x^2. \quad (7.4)$$

The last approximation will become exact in the limit as Δx approaches zero. If we also limit ourselves to small-amplitude waves for which $\delta \ll 1$ and $\cos\delta \simeq 1$ everywhere, there will be no need to distinguish T from T_0. Now the mass of this piece of string is $\Delta m = \mu\Delta x$, and its acceleration is $\partial^2\xi/\partial t^2$, so Newton's law gives

$$\mu\, \Delta x\, \partial^2\xi/\partial t^2 = T\, \Delta x\, \partial^2\xi/\partial x^2. \quad (7.5)$$

Upon canceling the Δx on both sides, we obtain an equation that no longer depends on the particular size of Δx as long as it is sufficiently small.

We may notice that T/μ has units N/(kg/m) = (kg-m/s^2)/(kg/m) = m^2/s^2, that is, the square of a velocity. It is very common to define a new constant c by

$$\boxed{c^2 = T/\mu,} \quad (7.6)$$

anticipating that we will be able to interpret c as the speed with which transverse waves travel along the string. In these terms, (7.5) can be written finally as

$$\boxed{\partial^2\xi/\partial t^2 = c^2\, \partial^2\xi/\partial x^2.} \quad (7.7)$$

7.2 TRAVELING WAVES

From our experience with ropes or Slinky springs, we might suspect that one possible type of solution to (7.7) would be a wave traveling steadily along the string with constant profile. So let us examine as a trial solution

$$\xi(x, t) = f(ct \pm x), \quad (7.8)$$

where $f(w)$ is any reasonably well-behaved function of a single variable w. You may easily verify by substitution that this is indeed a solution of (7.7).

To interpret (7.8), consider one particular point on a wave (a crest, perhaps) represented by a specific value of ξ. We could be assured of finding that same ξ if we would look at other combinations of location and time for which $(ct \pm x)$ always has the same value. But setting

$$0 = \Delta(ct \pm x) = c\Delta t \pm \Delta x \quad (7.9)$$

means choosing $\Delta x/\Delta t = \mp c$. That is, whatever feature we are looking at must appear to move along with speed c as time goes by. The plus sign in (7.8) must correspond to waves traveling toward the left, and minus toward the right.

Since (7.7) is a linear differential equation, any sum of two solutions is also a solution. Thus in general the string could support waves $f(ct - x)$ and $g(ct + x)$ at the same time. An important physical reason for such a situation could be the presence of waves reflected from the ends of the string; in that case there would have to be a relation between f and g. In order to see what that relation is, consider a **fixed end** where the string is firmly anchored to a solid wall. We may for convenience let that point be $x = 0$, with the string located at $x > 0$, and impose the **boundary condition**

$$\boxed{\xi(0, t) = 0 \quad \text{(fixed).}} \tag{7.10}$$

Now $f(ct - x) + g(ct + x)$ is restricted to $f(ct) + g(ct) = 0$, or

$$f(w) = -g(w), \tag{7.11}$$

so that the combination can be written as

$$\xi(x, t) = g(ct + x) - g(ct - x). \tag{7.12}$$

This has the interesting interpretation (Fig. 7.2) that the incoming wave $g(ct + x)$ produces a reflection $-g(ct - x)$ that has exactly the same amplitude and shape except for being upside down and traveling in the opposite direction. It is sometimes convenient even to pretend that there is no boundary and that someone far off to the left has used advance knowledge to generate an incoming wave $-g(ct - x)$ for $x < 0$. This phantom wave is regarded as continuing smoothly across $x = 0$, while the original $g(ct + x)$ continues onward toward $-\infty$.

The opposite extreme from the fixed end is a **free end,** which we imagine as a massless ring in the end of the rope sliding up and down without friction on a pole. (This seems quite artificial for the string, but the corresponding type of boundary condition for sound waves will turn out to be very important.) Here

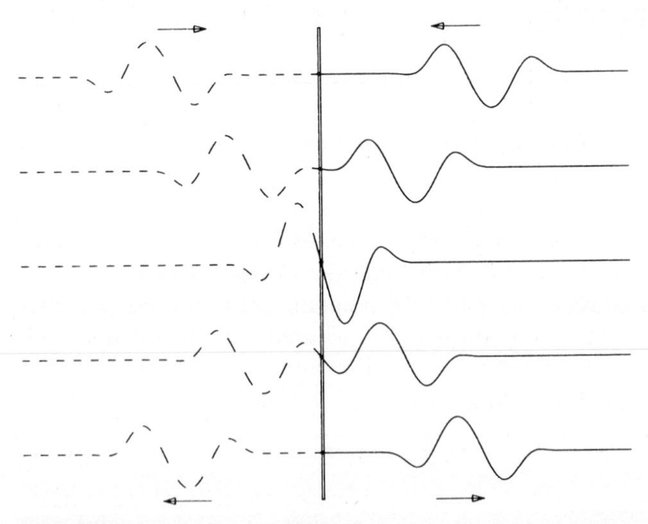

Figure 7.2 Time sequence, from top to bottom, of string shapes as a wave pulse approaches a solid wall from the right and is reflected. A fictitious extension of the string to the left may be thought of as bringing in a reflected wave such that the sum of the two waves always leaves zero displacement at the wall.

tension on the last piece of string (and thus the string itself) must be horizontal, because the mounting has been assumed incapable of exerting any vertical force on the string. Thus a free end (located again for our convenience at $x = 0$) must have

$$\frac{\partial \xi}{\partial x}(0, t) = 0 \quad \text{(free).} \tag{7.13}$$

This will be satisfied by $f(ct - x) + g(ct + x)$ only if $-df/dw + dg/dw = 0$, or

$$\xi(x, t) = g(ct + x) + g(ct - x). \tag{7.14}$$

Thus reflections from a free end again retain the original wave shape, but now without being inverted (Fig. 7.3).

In both these cases the reflected wave clearly carries just as much energy as did the incident wave, and that is because neither fixed nor free end allows the string to do any work upon the support (no displacement in one case, no force in the other). Any other case of a partially yielding support would mean some energy lost to that support, and so a reflected wave carrying only some fraction of the incident energy. In general the reflected wave would have not only reduced amplitude but also modified shape. We will wait until our study of sound waves to consider such cases of partial reflection.

7.3 STANDING WAVES AND NORMAL MODES

Consider now a string of finite length L. Describing all motions of this string in terms of traveling waves remains possible in principle. But because of repeated

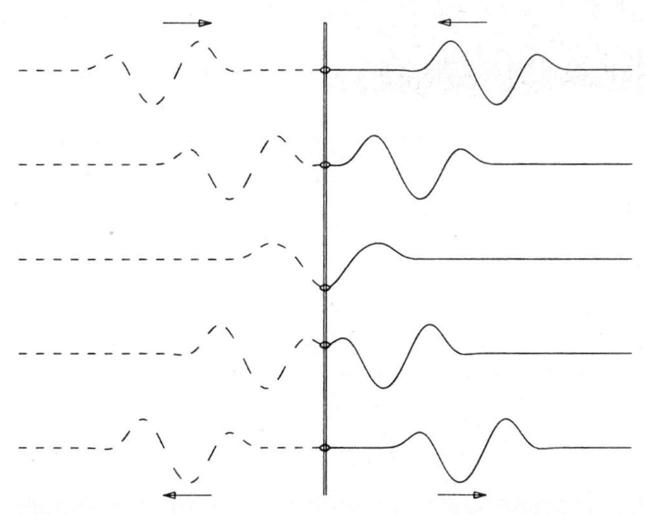

Figure 7.3 Reflection of a wave pulse as in Fig. 7.2, but with the string end free instead of fixed.

reflections between the two ends, that is usually not the most helpful description. We find it more convenient to study **standing waves.** We can see what this means by taking the example of two sinusoidal waves with equal amplitudes moving in opposite directions:

$$\xi(x, t) = A \sin k(ct + x) - A \sin k(ct - x), \tag{7.15}$$

where $k = 2\pi/\lambda$. Using $ck = \omega$ and familiar trigonometric identities, we can write this as

$$\xi(x, t) = 2A \cos \omega t \sin kx. \tag{7.16}$$

This expression has two important properties illustrated in Fig. 7.4, neither of which would be true for one traveling-wave term alone. First, there are some values of x where ξ remains zero at all times; whenever the crest of one traveling-wave component arrives there, it is always canceled by a trough of the other. Such points are called **nodes** of the standing wave. Second, there are some values of t for which ξ is zero simultaneously at all locations. That does not mean there is no wave, however, because at that moment the velocity $\partial\xi/\partial t$ has its greatest value.

Let us ask more generally about the possibility of motions satisfying (7.7) in which *all parts of the system oscillate in unison with simple harmonic motion of the same frequency*. Such a motion is called a **normal mode** of the system and may be thought of as a "natural" motion for which the system is particularly well suited. A trial solution representing this motion is

$$\xi(x, t) = \cos (\omega t - \phi)\, h(x), \tag{7.17}$$

and substitution shows that it will satisfy (7.7) if and only if the unknown function h obeys

$$-\omega^2 h = c^2 d^2h/dx^2. \tag{7.18}$$

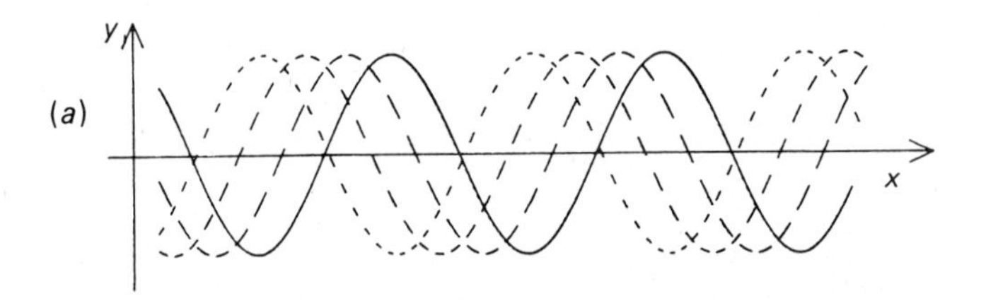

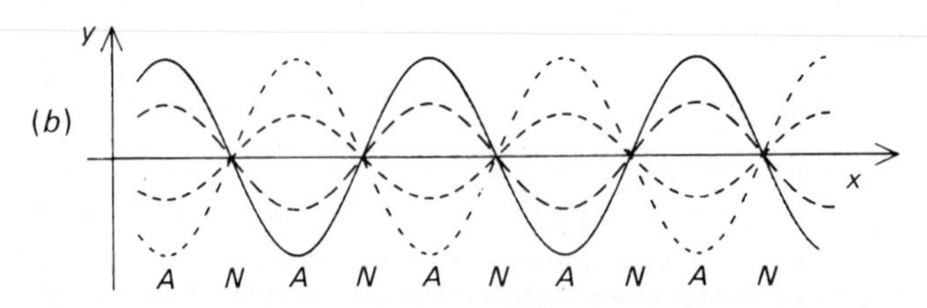

Figure 7.4 Sinusoidal traveling wave (a) contrasted with standing wave (b). Only the latter has nodes (N) where ξ remains zero at all times.

This would be satisfied throughout the interior of the string by any combination

$$h(x) = C \sin(\omega x/c) + D \cos(\omega x/c), \tag{7.19}$$

with arbitrary constants C and D.

But we must limit ourselves to solutions that meet the proper boundary conditions. Suppose specifically that both ends of the string are fixed; that is,

$$\xi(0, t) = 0 \quad \text{and} \quad \xi(L, t) = 0. \tag{7.20}$$

The only way to meet the first condition is to set $D = 0$. Then the second condition allows the amplitude C to be anything as long as we require that $\omega L/c$ be an integral multiple of π. Thus only the discrete set of frequencies

$$\boxed{\Omega_n = n\pi c/L \quad \text{or} \quad f_n = nc/2L} \tag{7.21}$$

is compatible with the normal-mode hypothesis. The corresponding wavelengths are

$$\lambda_n = c/f_n = 2L/n; \tag{7.22}$$

This should remind you of the argument that only if an integral number of half wavelengths matches L can you "come out right" with a node at both ends.

As an example, take a guitar string of length $L = 0.7$ m and mass $m_s =$ 4.2 g, under tension $T = 240$ N. Its linear mass density is

$$\mu = m_s/L = (0.0042 \text{ kg})/(0.7 \text{ m}) = 0.006 \text{ kg/m},$$

so the transverse wave speed on the string is

$$c = \sqrt{T/\mu} = \sqrt{240/0.006} = 200 \text{ m/s}.$$

The allowed frequencies are all integral multiples of

$$f_1 = c/2L = (200 \text{ m/s})/(1.4 \text{ m}) = 143 \text{ Hz},$$

and the corresponding wavelengths are $(1.4 \text{ m})/n$.

Normal-mode frequencies that form a harmonic series as in (7.21) are a very special feature of one-dimensional systems whose properties are uniform everywhere along their length. A nonuniform mass distribution on a string will have a nonharmonic list of mode frequencies, as will any essentially two- or three-dimensional system.

Again the superposition principle tells us that any sum of these sinusoidal standing waves is also a solution of both the equation of motion and the boundary conditions. So we can represent very general motions of a string fixed at both ends by

$$\boxed{\xi(x, t) = \sum_{n=1}^{\infty} C_n \cos(\Omega_n t - \phi_n) \sin(n\pi x/L).} \tag{7.23}$$

In fact, the techniques of Fourier analysis can show that every possible free vibration of the string is included in the range of motions describable in this way. In particular, motions that your intuition would visualize more readily as traveling waves are included in the scope of this machinery.

A standard problem in string motion is to determine the amplitudes C_n and phases ϕ_n when **initial conditions** have been specified. We are free to label the starting time as $t = 0$. Again we are looking at solutions of a second-order differential equation. So just as we needed both initial position and velocity for the simple oscillator in Chap. 5, here again we must have initial position and velocity for every point on the string. That is,

$$\xi(x, 0) = u(x) \quad \text{and} \quad \frac{\partial \xi}{\partial t}(x, 0) = v(x) \tag{7.24}$$

must both be specified. Evaluating (7.23) and its derivative at $t = 0$ gives

$$\sum C_n \cos \phi_n \sin (n\pi x/L) = u(x) \tag{7.25}$$

and

$$\sum \Omega_n C_n \sin \phi_n \sin (n\pi x/L) = v(x). \tag{7.26}$$

We can extract the desired information in exactly the same way we did in Sec. 3.4, for it is immaterial to the mathematics whether x or t is the independent variable. For any integer value of p, multiplying (7.25) and (7.26) by $\sin (p\pi x/L)$ and integrating over $0 < x < L$ gives

$$C_p \cos \phi_p = (2/L) \int_0^L \sin (p\pi x/L) u(x)\, dx \tag{7.27}$$

and

$$C_p \sin \phi_p = (2/L\Omega_p) \int_0^L \sin (p\pi x/L) v(x)\, dx. \tag{7.28}$$

Thus we have an explicit recipe for determining the amplitude and phase of each component mode whenever the complete initial conditions $u(x)$ and $v(x)$ are given.

As an example consider the motion of a guitar string after it is plucked at $t = 0$. Let the point of contact $x = x_c$ be a fraction β of the string length away from one end. Then at $x = \beta L$ it is pulled aside a small distance d into a triangular shape and released from rest. This means that $v(x) = 0$ and $u(x) = xd/\beta L$ for $0 < x < \beta L$ but $(L - x)d/(1 - \beta)L$ for $\beta L < x < L$. Carrying out the integrations shows that $\phi_p = 0$ for all p in this case, and

$$C_p = 2d \sin (\beta p\pi)/\beta(1 - \beta)p^2\pi^2. \tag{7.29}$$

As indicated in Fig. 7.5, this shows two important features in the spectrum. First, there is a general decrease of amplitudes in proportion to p^{-2}, as we could have anticipated from the mere fact that we used a function $u(x)$ with a discontinuous first derivative. Second, any mode for which $\beta p\pi$ is itself a multiple of π will have zero amplitude. For $\beta = 1/5$, for instance, $n = 5, 10, 15, \ldots$ would be "missing modes." This is precisely because the point of application of force is

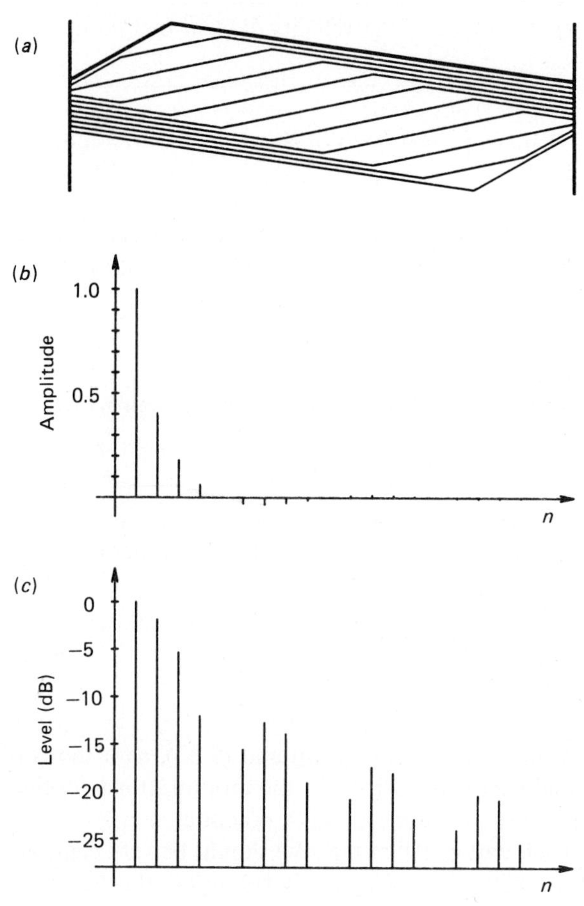

Figure 7.5 (*a*) Dark line: Shape of a plucked string at the moment it is released, if $\beta = 0.2$. Lighter lines: Subsequent shapes during the first half cycle of motion, displaced vertically for better visibility. The next half cycle repeats the same sequence of shapes in reverse order. (*b*) Fourier amplitudes corresponding to this motion, expressed on a linear scale as multiples of $C_1 = 0.744d$. (*c*) Spectrum made more easily visible with a logarithmic scale, using (7.37).

at a node of those modes; they cannot be excited by such a force any more than you can make a door move by pushing on its hinges. This is an extreme case of the more general property also embodied in (7.29), that the more motion a mode would involve at a given point, the better it will respond to forces acting at that point.

7.4 ENERGY IN STRING VIBRATIONS

As with all vibrating systems, we must have both kinetic and potential energy in the string. The kinetic energy of each small piece with mass $\Delta m = \mu \Delta x$ must simply be half this mass times the square of the velocity $\partial\xi/\partial t$ at that point. The total then is

$$E_k(t) = (\mu/2) \int_0^L [\partial\xi(x, t)/\partial t]^2 \, dx. \tag{7.30}$$

The potential energy exists because the entire string has been stretched like a spring. Instead of L, the length of the string when a wave is present is

$$L + \Delta L = \int_0^L \sqrt{dx^2 + d\xi^2}. \tag{7.31}$$

For small-amplitude waves we may use the leading term in a series approximation for the integrand:

$$dx\sqrt{1 + (\partial\xi/\partial x)^2} \simeq dx[1 + (\partial\xi/\partial x)^2/2]. \tag{7.32}$$

The resulting expression for ΔL can be multiplied by the tension T to find out how much work had to be done to accomplish that stretching; but that is simply the potential energy

$$E_p(t) = (T/2) \int_0^L [\partial\xi(x, t)/\partial x]^2 \, dx. \tag{7.33}$$

There is a very satisfying parallelism between (7.30) and (7.33), as if the two forms of energy play equal and symmetrical roles in the motion. It is tempting to take apart the integral in (7.33) and suppose that a potential energy density $(T/2)(\partial\xi/\partial x)^2$ is localized in each part of the string. We should be very cautious about such a step, however, for potential energy really belongs to the system as a whole and not in any unique way to the parts individually.

It is of particular interest to write expressions for $E_k(t)$ and $E_p(t)$ when string motion is described in terms of normal modes. Differentiating (7.23) gives for the velocity

$$\partial\xi/\partial t = \sum_{n=1}^{\infty} - \Omega_n C_n \sin(\Omega_n t - \phi_n) \sin(n\pi x/L). \tag{7.34}$$

Squaring the infinite series produces a double summation involving products $\sin(m\pi x/L) \sin(n\pi x/L)$. When these are integrated from 0 to L as prescribed in (7.30), the result is always zero unless $m = n$, and in that case the integral of $\sin^2(n\pi x/L)$ is just $L/2$. So we get simply

$$E_k(t) = (\mu L/4) \sum_{n=1}^{\infty} \Omega_n^2 C_n^2 \sin^2(\Omega_n t - \phi_n). \tag{7.35}$$

Using the same procedure with $\partial\xi/\partial x$, we get from (7.33)

$$E_p(t) = (TL/4) \sum_{n=1}^{\infty} (n\pi/L)^2 C_n^2 \cos^2(\Omega_n t - \phi_n). \tag{7.36}$$

These suggest that each normal mode may be considered to have its own energy E_n, independently of what the other modes are doing. If we recall that $T = c^2\mu$ and $\Omega_n = n\pi c/L$, we find that

$$\boxed{E_n = (\mu L/4)\Omega_n^2 C_n^2 = (\pi^2 T/4L)n^2 C_n^2.} \tag{7.37}$$

The mode kinetic and potential energies each have a time average equal to half of E_n, illustrating a property called **equipartition of energy.**

As an illustration, take again the example of the plucked string with $\beta = 0.2$ and suppose it was pulled aside 3 mm. Substitution of (7.29) into (7.37) gives

$$E_n = Td^2 \sin^2 (\beta n\pi)/L\beta^2(1 - \beta)^2 n^2\pi^2.$$

If $T = 240$ N, $L = 0.7$ m, and $d = 0.03$ m, the factor $Td^2/L\pi^2$ is 0.31 J. Using $\beta = 0.2$ then gives the first four mode energies as 4.2, 2.7, 1.2, and 0.26 J. Because of the factor n^{-2}, each doubling of frequency tends to reduce the mode energy by a factor of 4; so we might describe the spectrum of Fig. 7.5 as dropping off at the rate of 6 dB/octave.

It is also interesting to ask about the flow of energy along the string. The transfer of energy from any piece to its neighbor must be represented by work done by one upon the other. Referring to Fig. 7.1, we can see that the left half of the rope, pulling downward with a force component $-T \sin \delta$, would do negative work $\Delta W = -T \sin \delta \, \Delta\xi$ upon the right end during a positive displacement $\Delta\xi$. As before, we replace $T \sin \delta$ by $T(\partial\xi/\partial x)$ to find that the power flow in the direction of positive x is

$$\boxed{P = dW/dt = -T\frac{\partial\xi}{\partial x}\frac{\partial\xi}{\partial t}.} \tag{7.38}$$

In the case of a lone traveling wave $\xi = g(ct \pm x)$, this says

$$P(x, t) = \mp cTg'^2. \tag{7.39}$$

This nicely suggests an energy density Tg'^2 traveling with speed c in the same direction as the wave crests, in agreement with the sum of (7.30) and (7.33).

To illustrate this, take $g(w) = A \sin (w/b)$ with amplitude $A = 2$ mm, $w = ct + x$, and $b = 5$ cm, on a string with $T = 240$ N, $c = 200$ m/s, and infinite length. The energy density is

$$Tg'^2 = T(A/b)^2 \cos^2 (ct + x) = (0.38 \text{ J/m}) \cos^2 (ct + x),$$

and its spatial average is 0.19 J/m. The power flow is toward the left,

$$P = -cTg'^2 = -(76 \text{ W}) \cos^2 (ct + x),$$

and the average power passing any point is 38 W.

In the case of a standing wave like

$$\xi(x, t) = C_n \cos \Omega_n t \sin k_n x, \tag{7.40}$$

the power flow is

$$P(x, t) = (T/4c)\Omega_n^2 C_n^2 \sin 2\Omega_n t \sin 2k_n x. \tag{7.41}$$

This says that at any given time t energy is moving toward the right in some places and toward the left in others. It also says that for any given position x the energy flow oscillates as time goes on, but with zero average flow; this reinforces the feeling that a standing wave means a sloshing back and forth without any long-term progress in either direction. Note also that where $\sin 2k_n x$ is zero (which means at both the nodes and the antinodes of the normal mode) no energy ever passes in either direction.

7.5 DRIVING THE STRING

Just as we did with the simple harmonic oscillator, we would like to understand forced as well as free motions of the string. We would expect, since the string has its own natural mode frequencies, that we may encounter resonance effects when we apply driving forces that oscillate at those particular frequencies. What will be more complicated than before is that the string offers many different points at which the driving force might be applied.

Consider first a semi-infinite string that extends all the way to $x = +\infty$, with a free end at $x = 0$ where a transverse external force is applied. The very last bit of string at the end has both this force and the tension from the rest of the string acting on it (Fig. 7.6). If the total force were finite, and acting on an infinitesimal mass $\mu \Delta x$, it would produce infinite acceleration. That means the end of the string always adjusts itself very, very quickly in such a way as to keep that total force zero:

$$F_{\text{ext}}(t) + T \frac{\partial \xi}{\partial x}\bigg|_{x=0} = 0. \tag{7.42}$$

If there is only an outgoing traveling wave $\xi = f(ct - x)$, this boundary condition requires $F(t) - Tf'(ct) = 0$, or

$$F(t) = Tf'(ct). \tag{7.43}$$

But the string velocity at the end is

$$v_{\text{in}} = \partial \xi / \partial t = cf'(ct), \tag{7.44}$$

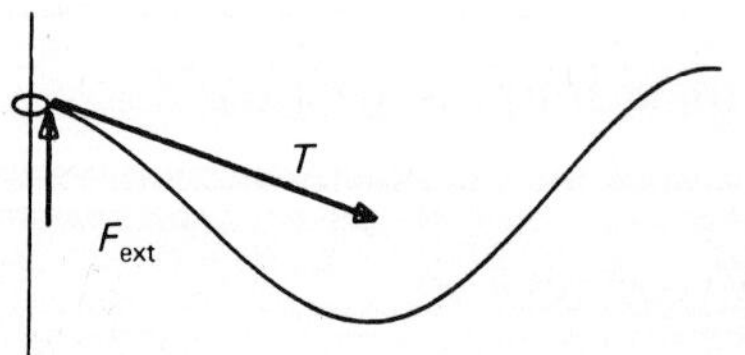

Figure 7.6 Applied force F_{ext} and tension T acting at the driven end of a string.

so it is most convenient to characterize the string's reaction by its mechanical **input impedance**

$$\boxed{Z_{in} = F/v_{in} = T/c = \mu c = \sqrt{T\mu}.} \qquad (7.45)$$

As before, this says how hard it is to produce motion in the system. We find here that both greater mass and greater tension increase this difficulty, just as we would expect. We also see that for the infinite string the impedance is a constant that does not depend on the frequency or waveform of the applied force. Furthermore, this impedance is real (that is, a pure resistance with no reactive component), which we interpret as meaning that all power put into the system is lost forever to the driver.

Things become much more interesting if the string has a fixed end a distance L away. Now reflected waves come back to the driving point and contribute to the force balance there. To keep things tractable, let us consider now only a sinusoidal driving force $F_0 \exp(j\omega t)$ that has been in operation long enough that all transients have been damped out. Then the driven system must be oscillating at the driving frequency, in a standing-wave pattern with a node at the fixed end $x = L$:

$$\xi(x, t) = C \exp(j\omega t) \sin [k(L - x)], \qquad (7.46)$$

where k must equal ω/c to satisfy (7.7). But this will satisfy the boundary condition (7.42) at $x = 0$ only if $F_0 - TCk \cos (kL) = 0$, or

$$C = F_0/[Tk \cos (kL)]. \qquad (7.47)$$

The velocity at the input point is $\partial\xi/\partial t$ evaluated at $x = 0$:

$$v_{in}(t) = j\omega C \exp(j\omega t) \sin (kL), \qquad (7.48)$$

so the input impedance is

$$\begin{aligned} Z_{in} &= F/v_{in} = F_0/[j\omega \sin kL(F_0/Tk \cos kL)] \\ &= (Tk/j\omega) \cot kL = -j(T/c) \cot \omega L/c. \end{aligned} \qquad (7.49)$$

This input impedance is plotted in Fig. 7.7. The factor j indicates that it is a pure reactance, which means that all energy fed into the string is temporarily stored there and then given back to the driver. The external force does work upon the string during one quarter of each cycle, and the string does an equal amount of work upon the driver in the following quarter cycle. For those frequencies where Z_{in} is positive imaginary, we may refer to it as masslike, just as an electric circuit with positive reactance may be called predominantly inductive. Similarly, the string is in some respects springlike whenever its reactance is negative, just as in a capacitance-dominated circuit.

For example, let a string with $L = 0.4$ m, $T = 500$ N, and $c = 800$ m/s be fixed at one end and driven at the other by a force with amplitude $F_0 = 10$ N and frequency $f = 900$ Hz. Then

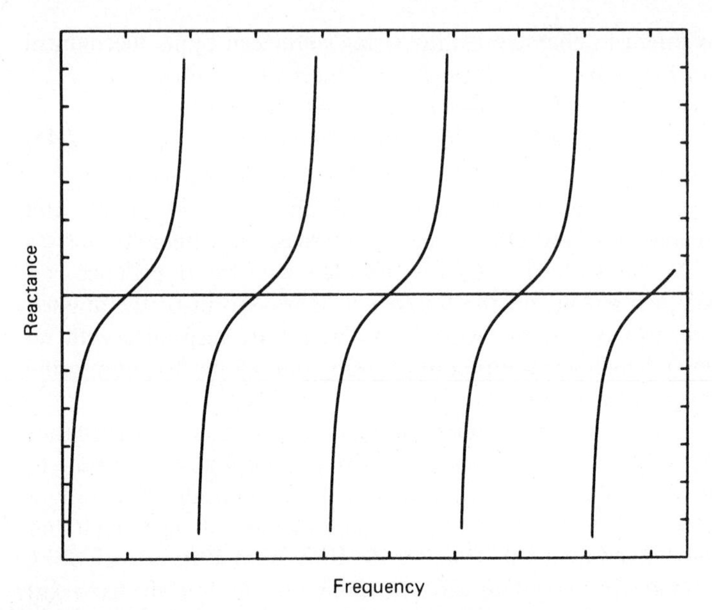

Figure 7.7 The (purely reactive) input impedance of a string driven at $x = 0$ and fixed at $x = L$. Zero crossings are resonances and occur when the driving frequency in hertz is any odd multiple of $c/4L$. Vertical axis is marked in units of T/c.

$$\omega = 2\pi f = (6.28)(900) = 5660/\text{s},$$

$$k = \omega/c = (5660/\text{s})/(800\ \text{m/s}) = 7.07/\text{m},$$

$$kL = (7.07/\text{m})(0.4\ \text{m}) = 2.83\ \text{rad} = 162°,$$

and

$$Z_{in} = -j[(500\ \text{N})/(800\ \text{m/s})] \cot(162°)$$

$$= +j1.92\ \text{kg/s}.$$

The velocity amplitude at the driving point is then

$$v_{in} = (10\ \text{N})/(1.92\ \text{kg/s}) = 5.2\ \text{m/s},$$

and the corresponding displacement is

$$\xi_{in} = v_{in}/\omega = (5.2\ \text{m/s})/(5660/\text{s}) = 0.92\ \text{mm}.$$

But the maximum string displacement occurs a quarter wavelength (0.22 m) from the fixed end, not at the driving point, and is

$$C = (10\ \text{N})/[(500\ \text{N})(7.07/\text{m}) \cos(162°)] = 3.0\ \text{mm}.$$

The zero impedance whenever ω is an odd multiple of $\pi c/2L$ represents resonance, for those are the natural mode frequencies of the string (see Prob. 7.10). Extremely large amplitudes of motion can be produced at resonance by steady application of a small driving force.

The infinite impedance whenever ω is a multiple of $\pi c/L$ could be described as antiresonance: no matter how great a force you apply, you cannot produce any appreciable steady motion at $x = 0$ because the reflected waves always maintain a node in the standing-wave pattern there. That does not mean, however, that the rest of the string does not move; indeed, (7.47) shows that a motion with amplitude F_0/Tk must have been established during the transient in order that the applied force F may always find itself balanced by the vertical component of string tension.

> In the last example above, changing the driving frequency to 500 Hz would produce zero impedance and infinite amplitude, since this is a resonant frequency of a system that was assumed to contain no resistive elements. But driving frequency 1000 Hz corresponds to infinite impedance, no finite motion at the driving point, and motion with amplitude
>
> $$F_0/Tk = (10\text{ N})/(500\text{ N})(7.85/\text{m}) = 2.5\text{ mm}$$
>
> at its antinode.

We would often be more concerned with applying a driving force to some interior point on the string rather than at an end (Fig. 7.8). We need not repeat a detailed analysis of the boundary conditions if we realize that the motion of the two string segments can be analyzed with the aid of circuit analogies. Each segment is being driven at a free end, and both must have the same velocity at the point where they meet. Therefore, the proper analogy is a series circuit, in which the same current passes through both elements. The impedance seen at the input point must be

$$Z_{\text{in}} = Z_1 + Z_2 = -j(T/c)[\cot(\omega L_1/c) + \cot(\omega L_2/c)], \tag{7.50}$$

where $L_1 = \beta L$ and $L_2 = (1 - \beta)L$. This expression will be zero when $\cos(\omega L_1/c)\sin(\omega L_2/c) + \sin(\omega L_1/c)\cos(\omega L_2/c) = 0$, but that is merely $\sin[\omega(L_1 + L_2)/c] = 0$, or $\omega L/c = n\pi$. It is reassuring to find resonances predicted at exactly the normal mode frequencies of the undriven system.

Figure 7.9 shows a few examples of the steady-state standing waves that may be set up in the string at different driving frequencies. The maximum am-

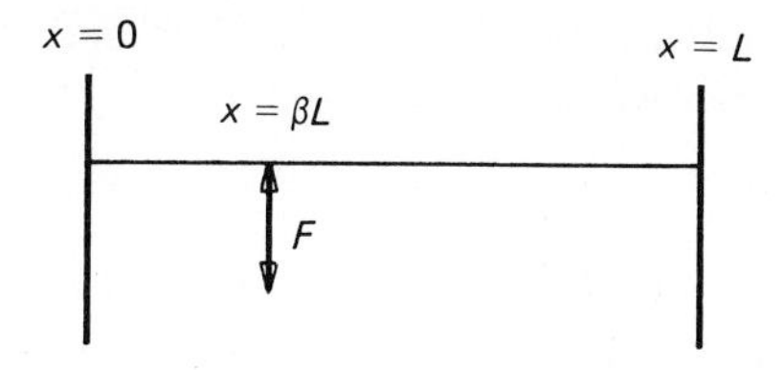

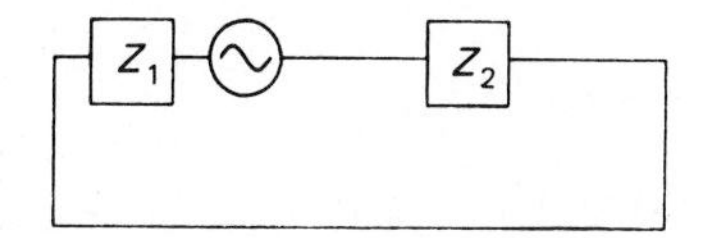

Figure 7.8 String of finite length driven at an interior point, and analogous circuit.

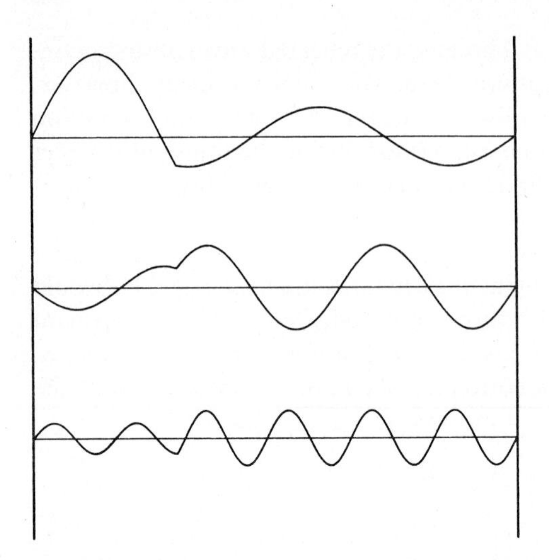

Figure 7.9 Standing waves produced on the string of Fig. 7.8 in cases where $\beta = 0.3$ and $f/f_1 = 3.7$, 5.5, and 11.7. Note that the string slope at the driving point may have either sign according to whether the input impedance is masslike or springlike. All three cases are pictured at a moment when the applied force is downward. In two cases this force is in phase with the input displacement.

plitude of the standing wave in each part of the string must be such that the velocities of the two driven ends both match that of the driver:

$$j\omega C_1 \sin kL_1 = F/Z_{\text{in}} = j\omega C_2 \sin kL_2. \tag{7.51}$$

Consider, for example, a rope with $T = 2000$ N, $\mu = 0.1$ kg/m, $L = 2$ m, and thus $c = \sqrt{T/\mu} = 141$ m/s. Its normal-mode frequencies are

$$f_n = nc/2L = n(141 \text{ m/s})/(4 \text{ m}) = n(35 \text{ Hz}),$$

and its characteristic impedance is

$$T/c = (2000 \text{ N})/(141 \text{ m/s}) = 14.1 \text{ kg/s}.$$

Let the rope be driven 78 cm from the left end, so that $L_1 = 0.78$ m and $L_2 = 1.22$ m. For driving frequency 50 Hz, we have

$$\omega = (6.28)(50) = 314/\text{s}$$

$$k = \omega/c = (314/\text{s})/(141 \text{ m/s}) = 2.23/\text{m}$$

and

$$kL_1 = (2.23)(0.78) = 1.73 \text{ rad} = 99°$$

$$kL_2 = (2.23)(1.22) = 2.72 \text{ rad} = 157°.$$

So the input impedance is

$$Z_{\text{in}} = -j(14.1 \text{ kg/s})(\cot 1.73 + \cot 2.72)$$

$$= -j(14.1)(-0.16 - 2.25) = +j34 \text{ kg/s}.$$

If the driving force has amplitude 68 N, the velocity at the driving point will have amplitude

$$v_{in} = F_0/Z_{in} = (68 \text{ N})/(34 \text{ kg/s}) = 2 \text{ m/s};$$

it lags 90° behind the force because the reactance is masslike. The displacement amplitude at the driving point is

$$\xi_{in} = v_{in}/\omega = (2 \text{ m/s})/(314/\text{s}) = 6.4 \text{ mm}.$$

But the standing-wave amplitudes on the left and right segments are

$$C_1 = v/\omega \sin kL_1 = 0.0064/\sin 1.73 = 6.5 \text{ mm}$$

and $$C_2 = v/\omega \sin kL_2 = 0.0064/\sin 2.72 = 15.6 \text{ mm}.$$

7.6 REALISTIC DRIVING CONDITIONS

The preceding discussion of resonant string response is rather artificial in that it admits no way for power to be dissipated within the finite-length string. In reality, of course, there must be irreversible transfers of energy out of the forms we are considering. This could represent the generation of heat because the rope is being flexed, or it might also account for energy delivered to slightly yielding supports. We can make some qualitative remarks about what happens with real strings without attempting to analyze those energy-loss mechanisms in detail.

The presence of dissipation must be represented by some nonzero resistive component R in the input impedance. Then the zero crossings of the reactance no longer represent infinitely large resonant response. Instead, the velocity amplitude will be limited to F/R in the force-driven case, and in the displacement-driven case the minimum required force will be $vR = \omega\xi R$ instead of zero. Similarly, R makes it possible to have a nonzero steady input power, and at the resonant frequencies this will simply be $F^2/2R$ or $\omega^2\xi^2R/2$. Each mode may have a somewhat different value of R. Graphs of impedance or admittance will resemble those at the end of Chap. 6, but with many resonances instead of only two. As long as the resistance is not very large, it produces practically no shift in the locations of the resonances on the frequency axis; that is, (7.21) remains valid.

PROBLEMS

7.1. What is the speed of transverse waves on a string with tension 800 N and linear density 0.005 kg/m? If this is a steel string (ordinary density 8×10^3 kg/m^3), what is its diameter?

7.2. If a wave approaching the origin from the right has the form $A \cos(\omega t + kx)$, what is the function $g(w)$ that would appear in (7.12) or (7.14)? What then is $\xi(x, t)$ in each case, including the reflected wave? What in particular is $\xi(0, t)$ in each case?

7.3. If the string of Prob. 7.1 has length 0.4 m, what are its normal-mode frequencies? If it is pulled aside 5 mm and released at a point 0.1 m from one end, what are the amplitudes of the normal modes in the subsequent motion?

7.4. Carry out the integration needed to prove (7.29).

7.5. If a string has initial displacement $u(x) = 0$ and velocity $v(x) = 2xV/L$ for $0 < x < L/2$ and $2(L - x)V/L$ for $L/2 < x < L$, what are the amplitudes and phases of the resulting normal modes?

7.6. Use (7.23) to argue that every free motion of the string is a periodic motion. What is the period of repetition, that is, the shortest time after which every term simultaneously returns to its initial value?

7.7. Show from Eqs. 7.30, 7.33, and 7.7 that any free motion of a string has constant total energy, as long as each end is either fixed or free. (*Hint:* You will examine dE/dt, and an integration by parts will help show that one term cancels another.)

7.8. What is the initial energy of the string in Prob. 7.3? If the amplitude decays as $\exp(-\gamma t)$ with $\gamma = 5/\text{s}$, what is the initial rate of power dissipation (smoothed over several cycles)?

7.9. What is the initial kinetic energy of the string described in Prob. 7.5? How much energy is associated with each mode in the subsequent motion? How many decibels per octave does this spectrum fall off? Can you show that the sum of all the E_n equals the proper total?

7.10. If a string is free at $x = 0$, what boundary condition must be satisfied there by $\xi(x, t)$? If it is fixed at $x = L$, what boundary condition must be satisfied there? What limitations do these place on the hypothetical standing waves $\xi(x, t) = \cos(\omega t - \phi)[C \sin(\omega x/c) + D \cos(\omega x/c)]$? And so what are frequencies of the normal modes for such a string?

7.11. In the limit as ω becomes very small, show that the impedance of a string (driven at $x = 0$, fixed at $x = L$) resembles that of a spring rather than that of a mass. What is the effective stiffness s? Show that this is the same stiffness you would have predicted by considering a small *static* displacement of the string end.

7.12. Let the string of Probs. 7.1 and 7.3 be driven at a point 0.13 m from one end. What is its input impedance for driving frequency 900 Hz? If the driving-force amplitude is 10 N, what is the velocity amplitude at the driving point? What is the maximum displacement amplitude at the antinodes on each segment of the string? How will these answers differ if the driving frequency is changed to 1100 Hz?

7.13. Suppose the free motion of mode number 2 of the string in the preceding problem decays to 0.1 times its original amplitude in 0.1 s. Viewing this mode as a simple harmonic oscillator, what is its Q? Argue from (7.35) that the "effective mass" of each mode can be taken as $m_s/2$, and use this to get a value for the effective resistance R_m. Just for the sake of getting a crude estimate, suppose we ignore the distinction between that R_m and an input resistance at a particular driving point. Then how would the answers to Prob. 7.12 differ for driving frequency 1000 Hz?

7.14. The theory presented in this chapter applies to small-amplitude oscillations only. Suppose you take as a criterion for smallness that the angle δ should never exceed 5°. What limit does that place on the amplitude C_n for a pure-mode motion on a string of length 1 m? Give explicit answers for $n = 1, 2, 3, 10,$ and 100.

7.15. The theory of Sec. 7.5 was developed for force-driven strings. Recall Sec. 6.3, and state analogous results for a displacement-driven string. That is, realizing that (7.49) is unchanged, what force must accompany a given input displacement for a string being driven **(a)** at one end or **(b)** at an interior point? **(c)** Describe what happens at the resonant and antiresonant frequencies (for zero resistance).

LABORATORY EXERCISE

7.A. Devise some means of measuring the motion of one point on a guitar string. You might consider a phonograph cartridge, or an optical system, or running the string itself between the poles of a strong magnet and connecting the string ends to an oscilloscope. Study the output both on an ordinary scope and on a spectrum analyzer for the damped free vibrations created when you pluck at various positions. Choose a means of applying an ongoing sinusoidal force to the string, such as feeding an alternating current through it while it is located in a magnetic field, and study the motion produced as a function of the driving frequency.

*Chapter 8

Waves: Further Examples

Before going on with a detailed theory of sound waves in fluids, it may be helpful to look briefly at a few more examples of waves in solid materials. It is easier in some ways to understand these, since we can see or feel the vibrations directly. In this optional chapter we give only a brief introduction to this topic; our purpose is primarily to point out how much of the ideas of Chap. 7 can be carried over into the study of waves in two and three dimensions, and to see what additional concepts may be needed.

8.1 TRANSVERSE WAVES ON A MEMBRANE

We can see two-dimensional waves in their simplest form by considering a thin, flexible membrane. This can serve as an idealized model of what happens on a drumhead or on a microphone diaphragm. Restoring force is provided by tension applied at a boundary, analogous to that on the ends of the one-dimensional string. But there is this difference: Where a finite amount of force was concentrated at a point of application for the string, now that force is spread along the line forming the membrane boundary. Therefore the appropriate measure of this tension is not force, but force per unit length. We emphasize the distinction by using the symbol $\mathcal{T}$ (with units N/m) instead of T.

We consider only the simple case where the boundary curve lies within a plane. Then the equilibrium position of the membrane itself is also in that same plane, and we use coordinates x, y to label individual locations on the membrane. At each such point the presence of transverse vibration means there is some displacement in the third dimension, which we will call $\xi(x, y, t)$. Our task is to find functions ξ that are compatible with the relevant physical laws.

Just as with the string, we must insist that every infinitesimal piece of the membrane obey Newton's $F = ma$. We picture in Fig. 8.1 a small rectangular piece with length Δx and width Δy. Since we assume perfect flexibility, the tension that acts all along the edges of this piece is everywhere tangent to the membrane itself. It is physically possible that the applied tension could be greater in one direction than in another, and we would have to allow for that if we wanted to study motions of an improperly adjusted drumhead. But we consider here only the case where the tension is applied uniformly everywhere along the membrane boundary.

Just as for the string, let us limit our study to transverse motion, so that the force components parallel to the xy plane remain precisely in balance. Let us also consider only small-amplitude waves that create slopes $\partial\xi/\partial x$ and $\partial\xi/\partial y$ that always remain very small compared with unity. Then we will be able again to ignore any distinction between $\mathcal{T}_0$ applied at the boundary and $\mathcal{T}$ at any interior point, as well as to freely exchange $\sin\delta$ for $\tan\delta = \partial\xi/\partial x$ or $\sin\epsilon$ for $\tan\epsilon = \partial\xi/\partial y$.

What is the total vertical force acting on the rectangle in Fig. 8.1? For any section of membrane that remains planar (even if tilted), the forces are all exactly balanced and it experiences no acceleration. Like the string, the membrane must have curvature in order to experience nonzero net forces. But now it may have different curvature in different directions. To find the proper expression of that effect, take first the side of the rectangle closest to you. Every bit of length dx has force $\mathcal{T}dx$ acting nearly in the $-y$ direction but with a small downward component

$$-\mathcal{T}dx\sin\epsilon(x, y) \simeq -\mathcal{T}dx\,\partial\xi(x, y)/\partial y. \tag{8.1}$$

The total force along this side is the integral of this expression from x_0 to $x_0 + \Delta x$. We plan in the end to make Δx very small; so we can use Taylor-series expansion to express the integrand entirely in terms of the derivatives of ξ evaluated at the single point x_0, y_0:

$$\xi_y(x, y) = \xi_y(x_0, y_0) + (x - x_0)\xi_{yx}(x_0, y_0) + (y - y_0)\xi_{yy}(x_0, y_0) + \cdots. \tag{8.2}$$

Here the subscripts on ξ denote partial derivatives, not vector components! If

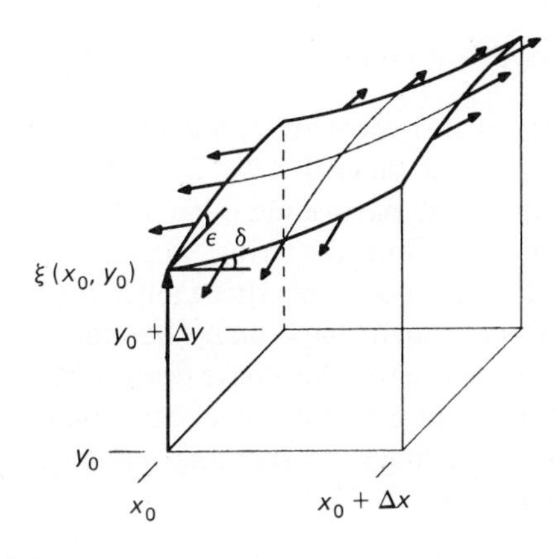

Figure 8.1 A small portion of membrane displaced a distance $\xi(x, y)$ above its equilibrium position in the xy plane. Tension acts along tangents at angles $\delta(x, y)$ parallel to the xz plane and $\epsilon(x, y)$ parallel to the yz plane.

we discard all terms of third or higher order in the small quantities Δx and Δy, the integral of (8.1) along the line where $y = y_0$ becomes simply

$$-\mathcal{T}[\xi_y(x_0, y_0)\,\Delta x + \xi_{yx}(x_0, y_0)(\Delta x)^2/2]. \tag{8.3}$$

The vertical force along the back side of the rectangle differs from this in two ways: positive ξ_y now gives upward force instead of downward, and $y - y_0$ is always Δy instead of zero. Thus this force is

$$+\mathcal{T}[\xi_y\,\Delta x + \xi_{yx}(\Delta x)^2/2 + \xi_{yy}\,\Delta x\,\Delta y] \tag{8.4}$$

where all derivatives are evaluated at x_0, y_0. When these two forces are added together, the only term surviving is the $\mathcal{T}\xi_{yy}\,\Delta x\,\Delta y$. Similarly, the total force along the right and left edges will be $\mathcal{T}\xi_{xx}\,\Delta x\,\Delta y$. Thus the grand total along all four sides is

$$\mathcal{T}(\xi_{xx} + \xi_{yy})\,\Delta x\,\Delta y = \mathcal{T}\nabla^2\xi\,\Delta x\,\Delta y \tag{8.5}$$

where the extremely important combination of second derivatives ∇^2 is often called the laplacian operator.

If we characterize our membrane by its areal mass density σ (kg/m^2), the mass of the rectangular element will be $\Delta m = \sigma\Delta x\,\Delta y$. When we use this mass in $F = ma$, and (8.5) for F, the area $\Delta x\,\Delta y$ will cancel out. The resulting equation of motion

$$\boxed{\sigma\,\partial^2\xi/\partial t^2 = \mathcal{T}\nabla^2\xi} \tag{8.6}$$

does not depend on our choice of Δx or Δy. Compare this with (7.7) to see how close the parallel is between the string and the membrane. Just as with T/μ in (7.6), we define

$$\boxed{c^2 = \mathcal{T}/\sigma} \tag{8.7}$$

since its units are m^2/s^2. This c can again be interpreted as a wave speed, since expressions like $\xi = f(ct + x)$ or $g(ct - y)$ are solutions of (8.6).

There is no reason why a simple traveling wave could not propagate in any direction, say at a fixed angle θ from the x axis. Such a wave can be mathematically represented by

$$\xi(x, y, t) = g(ct - x\cos\theta - y\sin\theta). \tag{8.8}$$

You may show (in Prob. 8.1) that the crests of this wave are parallel lines all moving together in the direction θ with speed c. The range $0 < \theta < 2\pi$ provides infinitely many such directions, each of which may have some different wave traveling along it. This is a very different situation from the string, where only two directions of wave travel were available. Furthermore, on a string reflections could occur only at a point, while a membrane might have many boundary shapes that would reflect in different ways. Therefore, a traveling-wave description is much less useful in two dimensions than it was in one as a way of describing any general disturbance; we find the relative advantages of standing-wave (or normal-mode) description becoming much greater.

8.2 NORMAL MODES OF A MEMBRANE

A study of the normal modes of any system not only provides a way of describing its free oscillations but will also alert us to the ways in which it may resonate under the action of driving forces and the frequencies at which those resonances will occur. Since a normal mode means a motion in which every part of the system oscillates sinusoidally at the same frequency, it must be represented by a trial solution

$$\xi(x, y, t) = \cos(\omega t - \phi)\, h(x, y). \tag{8.9}$$

This will satisfy (8.6) only if

$$-\omega^2 h = c^2 \nabla^2 h. \tag{8.10}$$

To get a more specific solution we must consider particular boundary shapes. Simplest mathematically is a rectangular boundary, but that is not the case of most practical interest. We relegate details to Prob. 8.3 and only remark here that even in this simple case we have one very striking difference from the string: The list of normal-mode frequencies does not form a harmonic series. The difference between each mode frequency and the next higher one varies irregularly and on average becomes smaller as the mode numbers increase. This is intuitively reasonable, because there is now more than one direction into which an appropriate number of wavelengths can be made to fit between the boundaries.

We have more reason to be interested in circular boundaries, since most drumheads and microphone diaphragms have that shape. In such cases it is much more natural to describe locations on the membrane in polar coordinates r, θ instead of cartesian x, y. It is a standard mathematical exercise (which we will not recount here) to show that the laplacian operator can be written in terms of r and θ as

$$\nabla^2 h = h_{rr} + h_r/r + h_{\theta\theta}/r^2. \tag{8.11}$$

Then (8.10) becomes

$$r^2 h_{rr} + r h_r + h_{\theta\theta} + (\omega/c)^2 r^2 h = 0. \tag{8.12}$$

We want functions $h(r, \theta)$ that not only satisfy this equation but also are well behaved at $r = 0$ and limited to $h(r = a) = 0$ since the membrane cannot move at its boundary at radius a.

Equation 8.12 is very well known in mathematical physics, because it arises in many different problems that involve cylindrical geometry. Its solutions can be found by standard techniques involving power series, and their properties are known in great detail. They are important enough to be given the name Bessel functions; tables of these functions have been published, and they can be considered "known" just as surely as the trigonometric functions. Appendix C contains a brief summary of information about Bessel functions.

Some of the solutions $h(r, \theta)$ depend on θ, while others are circularly symmetric; Fig. 8.2 illustrates a few of the possibilities. Without getting involved in too much detail, let us outline very briefly how these solutions may be written explicitly. Since r occurs in the dimensionless combination $\omega r/c = kr$, we should

Figure 8.2 Photographs of a rubber membrane vibrating in several of its natural modes. Each pair represents the same mode at two times a half cycle apart. (By permission of National Film Board of Canada.)

expect this rather than r alone to serve as argument for the Bessel function, just as kx does in (7.19). We may also hope that in the simplest solutions kr and θ would each affect h independently. If so, a good educated guess would be

$$h(r, \theta) = A \cos m\theta \, J_m(kr), \tag{8.13}$$

where limiting m to integers assures that h is a single-valued function of θ. In this case the term $h_{\theta\theta}$ becomes $-m^2h$, and (8.12) becomes precisely (C.1) with $x = kr$, so that (8.13) is indeed a solution if J_m is taken to be the Bessel function of order m. Note that each different angular dependence $\cos m\theta$ requires use of a different member of the family of Bessel functions.

But remember that not only the differential equation matters; proper boundary conditions must be satisfied as well. For a circular membrane with fixed boundary at radius a, the requirement $h(r = a) = 0$ becomes $J_m(ka) = 0$. That is, only those solutions with ka equal to one of the roots of the Bessel function are allowed. If we write j_{mn} for the nth root of J_m, this requires $k_{mn}a = j_{mn}$, or

$$\omega_{mn} = ck_{mn} = (c/a)\, j_{mn}. \tag{8.14}$$

Note how the boundary conditions allow only a discrete set of values for k or ω, just as with (7.21) for the string. Here, however, it requires a pair of subscripts to identify which mode is involved. As illustrated in Fig. 8.3, these subscripts have the physical interpretation that m tells the number of nodal lines going diametrically across the membrane, and n is the number of concentric nodal circles (including the outer edge). ω_{32}, for instance, refers to a mode in which the membrane is divided like a pie into six wedges, as well as into an inner disk

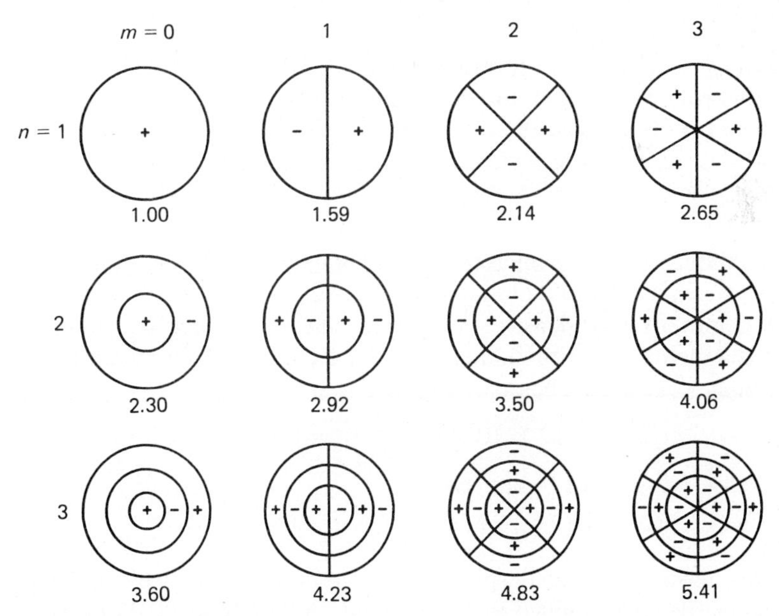

Figure 8.3 Nodal lines and circles for some drumhead modes, when tension in the membrane is the only restoring force. The number under each picture represents the frequency of that mode relative to the first mode. Plus and minus signs are reminders that adjoining sections always move 180° out of phase. Actual shapes along horizontal diameters are given by taking the Bessel function of order m from Fig. C.1 and adjusting the horizontal scale until its nth zero falls at $r = a$.

and outer ring. It has, then, twelve regions each of which moves 180° out of phase with its neighboring sections.

Using the table of values of j_{mn} in Appendix C, we can summarize a few of the important properties of these solutions. First, the list of normal-mode frequencies bears no resemblance to a harmonic series. If all ω_{mn} are expressed as multiples of the fundamental frequency ω_{01}, the second mode vibrates $(3.832/2.405) = 1.59$ times as fast as the first, the third mode 2.14 times as fast, then 2.30, 2.65, ... As with the rectangular membrane, the mode frequencies become crowded closer and closer together for increasing frequency. In the drumhead application, many modes are excited at once, and their combination produces a nonperiodic waveform with little or no sensation of definite pitch. In the microphone application, normal operation will always be at low enough frequency that the $mn = 01$ mode is the only one responding strongly, and from here on we concern ourselves only with that single mode.

As with the string, the fundamental mode is the only one in which every part of the membrane bulges to the same side at the same time; it has no nodes other than at its edge. With $m = 0$, it is completely symmetrical about $r = 0$. The frequency of this most important mode is given by $(c/a)(j_{01}/2\pi)$ with $j_{01} = 2.405$, or

$$\boxed{f_{01} = 0.383c/a.} \tag{8.15}$$

This becomes $f_{01} = 0.766c/d$ if expressed in terms of the diameter $d = 2a$. On comparing this with $f_1 = 0.5c/L$ for the string of length L, we might say the higher frequency indicates that the membrane is effectively a little stiffer than the string because its restoring forces act from every direction instead of only from two ends.

As an example, consider a microphone diaphragm made of very thin metal foil. If it were aluminum (volume density $\rho = 2700$ kg/m^3) with thickness $b = 0.03$ mm, its areal density would be $\sigma = \rho b = 0.081$ kg/m^2. Suppose we want a membrane with diameter $2a = 1.2$ cm to have its lowest resonance at $f_{01} = 12$ kHz. Then c would have to be

$$c = f_{01}a/0.383 = (12 \times 10^3/\text{s})(6 \times 10^{-3}\ \text{m})/0.383 = 188\ \text{m/s},$$

so the tension that must be applied is

$$\mathcal{T} = c^2\sigma = (3.53 \times 10^4\ \text{m}^2/\text{s}^2)(0.081\ \text{kg/m}^2) = 2860\ \text{N/m}.$$

Since this tension is actually spread over a finite thickness, we could say that the tensile stress applied to the material is (2860 N/m)/(0.03 mm) = 0.95×10^8 N/m^2. The limiting tensile strength of aluminum is about 2 or 3×10^8 N/m^2; so this much tension can be applied without rupturing the membrane.

The formulas given above have omitted an effect that is sometimes important, for they assume that the membrane moves in a vacuum. In reality, adjoining air must move along with the membrane. If this air is confined in a small volume, it will be difficult to squeeze it still smaller. The enclosed air will act like a spring, and its additional restoring force will tend to raise the mode frequencies. Mode number 01 is affected most since it is the only one for which downward motion of one part of the membrane is never compensated for by upward motion of some other part. Even the unconfined air on the outside of a drum may constitute a significant additional mass load impeding the membrane motion, and this tends to lower the mode frequencies. The air motion also causes energy to radiate away from the membrane in the form of sound waves, which is important in causing the normal-mode motions to be damped out.

We ought to remark about one thing we will not be able to do with the membrane. For the string we could study its driven motion when forces were applied to specified points. But for a membrane, a finite force applied at an interior point will produce infinite deflection of that point. To be realistic, we must describe any driving forces as being distributed over the area of the membrane. This illustrates one limitation on the usefulness of the concept of input impedance: since the force cannot be concentrated at a single point, we cannot even define $Z = F/v$ in the simple way we did before. When we take up the subject of sound waves in pipes, we will try to find how to reintroduce impedance in a more sophisticated way.

8.3 TRANSVERSE WAVES IN BARS AND PLATES

When we admit that a string is not perfectly flexible, that means its stiffness provides another restoring force to go along with the applied tension. In fact, the system is now capable of vibrating even with the external tension entirely removed. We usually refer to such a system in one dimension as a rod or bar rather than a string, and in two dimensions we call it a plate instead of a membrane.

Writing the equation of motion of a bar or plate is still a matter of satisfying $F = ma$. But now there are shearing stresses and torques to consider, and we will not go through the complete analysis here. The result is that the motion is governed by a fourth-order rather than a second-order differential equation. For a rectangular bar with thickness b made from material of density ρ and stiffness described by the Young's modulus Y,

$$\partial^2\xi(x, t)/\partial t^2 = -K^2\, \partial^4\xi/\partial x^4, \tag{8.16}$$

with K^2 defined as $b^2Y/12\rho$. The width of the rectangle does not matter here; think of increasing width as simply meaning several bars side by side, all moving the same way. For a round rod of diameter d, the factor $b^2/12$ would be replaced by $d^2/16$.

Two very important consequences follow quickly from Eq. 8.16, as outlined in Prob. 8.5. One is that traveling waves of arbitrary shape $g(ct - x)$ are *not* solutions; the only functions that will work this way are sinusoids and real exponentials. Any other waveform must change its shape as it moves along. Since waves tend to spread or disperse as they travel, the bar may be called a **dispersive medium** for this type of wave, in contrast to the string, which was nondispersive.

The other remarkable fact is that even the sine waves do not all travel at the same speed; the higher-frequency waves travel faster, according to

$$v(\omega) = \sqrt{K\omega}. \tag{8.17}$$

Since this equation describes differences in speed that cause spreading of wave packets, it is called a dispersion relation. Equation 8.17 should be believed only in those cases where it predicts v to be less than the speed of longitudinal waves in the bar material. For accurate comparisons at higher frequencies, the more complicated theory of "thick bar" vibrations is required.

For example, a rectangular bar of aluminum ($\rho = 2700\ \text{kg/m}^3$, $Y = 7 \times 10^{10}\ \text{N/m}^2$) with thickness $b = 1$ cm would have

$$K^2 = (10^{-4}\ \text{m}^2)(7 \times 10^{10}\ \text{N/m}^2)/(12)(2700\ \text{kg/m}^3) = 216\ \text{m}^4/\text{s}^2$$

and $K = 14.7\ \text{m}^2/\text{s}$. Then 100-Hz traveling sine waves ($\omega = 628/\text{s}$) would have

$$v = \sqrt{(14.7\ \text{m}^2/\text{s})(628/\text{s})} = 96\ \text{m/s},$$

but 1000-Hz waves would travel at 304 m/s. Note that the corresponding wavelengths $\lambda = v/f$, which are 96 and 30 cm, do not differ from each other by a factor of 10. Since the longitudinal wave speed in aluminum is 6300 m/s, the figures obtained here are reasonable. But for frequencies much above 100 kHz, speeds predicted from (8.17) would be wrong. It is no accident that this occurs when the wavelength is no longer much greater than the bar thickness.

To pursue this subject further, we would try to find the shapes and the frequencies of the normal modes of a bar. Applications could include tuning forks and xylophone bars. We will not present the supporting mathematics here, but in spite of the superficial appearance of Fig. 8.4 the mode shapes are not sinusoidal as they were for a string.

Even without specifying other details, we can deduce a very important "scaling" property from Eq. 8.16. From the simple fact that two powers of time occur on the left but four powers of distance on the right, it follows that two bars identical in every way except for their lengths L_A and L_B must execute similar motions in time periods T_A and T_B such that $(T_A/T_B)^2 = (L_A/L_B)^4$; that is, their frequencies will be related by

$$f_B/f_A = (L_A/L_B)^2. \tag{8.18}$$

This says that a simple doubling of length will lower the pitch of a xylophone bar by two octaves, rather than one octave as for a string.

Where we had two important types of boundary, fixed and free, for a string, here we would have to consider at least three interesting cases: clamped, hinged, and free ends of a bar. You may appreciate how these different degrees of con-

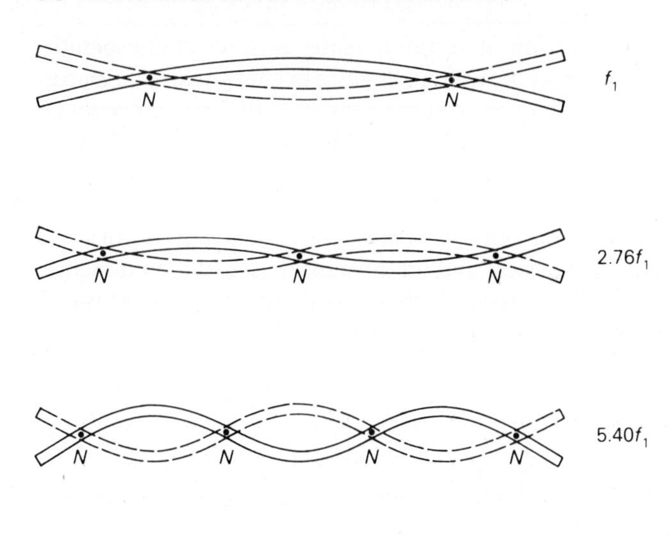

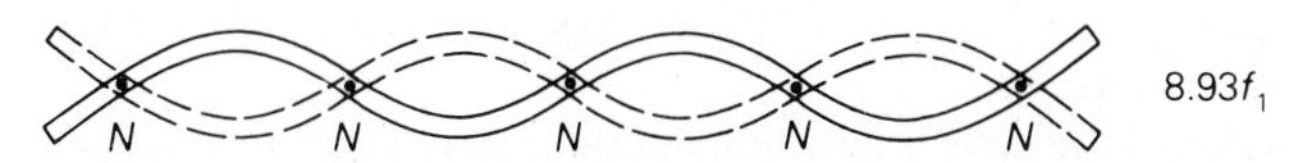

Figure 8.4 First four natural modes of a uniform thin bar free at both ends. (From D. E. Hall, *Musical Acoustics: An Introduction,* Wadsworth, 1980.)

finement affect the difficulty of bending a bar by comparing numerical coefficients α in the formula

$$f_1 = \alpha\, K/L^2 \tag{8.19}$$

for the fundamental frequency. We give here without proof the values $\alpha = 0.56$ for a bar clamped at one end and free at the other (like a tuning-fork prong), $\alpha = 3.56$ for either both ends free (like a xylophone bar) or both ends clamped, and $\alpha = 1.57$ for both ends hinged.

The lists of mode frequencies are generally nonharmonic, but with this interesting difference: The average gap from one mode frequency to the next gets larger and larger as you go to higher mode numbers for a bar, whereas it remains constant for the string and gets smaller and smaller for a membrane.

Vibrating thin plates combine many of the features of rods and membranes. Since stiffness is involved, the governing differential equation is fourth-order in the spatial derivatives. The edges of the plate may be clamped, hinged, or free. The complete solution for circular geometry requires Bessel functions again; but just as the bar required both real exponentials and sinusoids [which are combinations of $\exp(\pm jx)$], so the plate solution involves Bessel functions with both real and imaginary arguments. The fundamental mode for a clamped circular plate with diameter d and thickness b will have frequency

$$f_1 \simeq 6.2K/d^2, \tag{8.20}$$

with K^2 again equal to $b^2Y/12\rho$.

An example of an application of vibrating plates is in telephone mouthpieces, where a thin steel plate may act as the armature of a magnetic transducer. The ruggedness of a plate gives it an advantage in expected lifetime under rough handling compared with a delicate membrane. This can more than offset the disadvantage of requiring greater sound pressure to move the plate, that is, of having lower sensitivity to sound.

This chapter has given you only the barest introduction to a rich and interesting field of study. The physical and mathematical aspects are both challenging and rewarding. It will be well worthwhile if you can find an opportunity to take an entire course on vibrations in solid structures.

PROBLEMS

8.1. Consider the expression for $\xi(x, y, t)$ given in Eq. 8.8, where θ is a fixed angle and g is any well-behaved function.

- **(a)** Verify that this satisfies Eq. 8.6.
- **(b)** If $g(u)$ has a maximum at $u = u_0$, show that the corresponding wave crest lies along a line in the xy plane.
- **(c)** Show that this line keeps the same orientation but shifts its position as time goes by and that it moves with speed c in the direction perpendicular to itself.

8.2. What is the speed of waves on a membrane with density 0.2 kg/m^2 and under tension 2000 N/m? For a circular boundary with radius 0.25 m, what would be the natural frequency of the fundamental mode? How much would you have to increase the tension to raise the frequency by one octave?

8.3. Take the case of a rectangular membrane a little farther. Do this by trying out possibilities like $h(x, y) = h_1(x)h_2(y)$, with $h_1(x) = A \cos k_1x + B \sin k_1x$ and similarly for h_2 with two other undetermined amplitudes C and D. Show first that the equation of motion is satisfied only if k_1 and k_2 are related in a very definite way to ω. Then reject some of these possibilities by requiring that ξ be identically zero whenever $x = 0$ or $y = 0$. Weed out still more by requiring $\xi = 0$ when $x = a$ or $y = b$, which are the other two sides of the rectangle. Show that the surviving solutions provide a list of allowed normal-mode frequencies $f_{mn} = (c/2)\sqrt{(m/a)^2 + (y/b)^2}$ for positive integers m and n, and describe the shape of the mnth mode. As a specific example, list the first 10 frequencies for the case where $a = 40$ cm, $b = 20$ cm, and $c = 36$ m/s.

8.4. **(a)** The drumhead mode with $mn = 11$ is of particular interest. In order to produce a specified frequency $f_{11} = 165$ Hz on a tympani head with diameter $2a = 60$ cm and $\sigma = 0.25$ kg/m^2, what tension would be needed? (Using the information in this chapter we are making only a rough estimate, since we have not shown how to account for the air load that moves along with the membrane.)

- **(b)** What would be the frequencies of the other modes with $n = 1$?
- **(c)** What would be the frequencies of the modes with $m = 0$?

8.5. Insert the trial solution $\xi = g(ct - x)$ into (8.16) and demonstrate that only certain forms of g will work. For the form $\sin(\omega t - kx)$, what relation must hold between ω and k? Using the fact that the waves move with speed ω/k, show that higher-frequency waves move faster in accord with (8.17).

8.6. Suppose a grandfather clock chime is to be made from a circular rod (diameter 3 mm) of material having density 8000 kg/m^3 and Young's modulus 10^{11} N/m^2.

(a) What is K for this rod, and how fast would 200-Hz sine waves travel along an infinite rod?

(b) How long a piece should be used in order that the first normal-mode frequency will be 200 Hz if it is clamped at one end and free at the other?

(c) How long if hinged at both ends?

(d) How long if clamped at both ends?

(e) Does any of these cases match with the guess that L might be $\lambda/2 = v(f)/2f$ for the infinite traveling wave, and can you explain physically why?

8.7. Consider again the microphone diaphragm discussed as an example in the text. If all external tension were removed so that it acted like a clamped plate, what would its fundamental frequency be? (Use $Y = 7 \times 10^{10}$ N/m^2 for aluminum.) By comparing this against 12 kHz, can you justify that the restoring force under the conditions described was indeed due mainly to tension rather than stiffness? Estimate how much tension is sufficient to justfy treating it as a diaphragm rather than a plate. How much does that borderline tension increase if you double the thickness of the material?

LABORATORY EXERCISES

8.A. Choose an appropriate transducer to detect the vibrations of a drumhead or a metal bar. Process the signals with a spectrum analyzer, and note the changes when you strike the object in various places.

8.B. Obtain a strip (something like 3 or 4 cm wide and 50 cm or more long) of thin aluminum sheet. Drive it at one end with short bursts of sine waves with frequencies of several kilohertz. Pick up the resulting traveling transverse waves at various distances down the strip with a magnetic detector. (Even though aluminum is not a magnetic material, it does interact with the detector via induced eddy currents.) Display both signals on a dual-trace oscilloscope. From the time delays find the speed with which the waves travel, and check your results against Eq. 8.17. If you do this carefully, you should find that individual wave crests match that prediction, while the bursts as a whole move twice as fast. These two distinct speeds are called the phase velocity and the group velocity.

Chapter 9

Sound Waves: Basic Properties

It is now time to examine in detail what actually happens in sound waves. This means we want to understand both the basic physical laws dictating what must happen in each part of the fluid and the nature of the motions allowed by those laws. We develop the local equations of motion and begin looking at their simpler solutions in this chapter. The process of examining important general types of sound waves will continue through the following two chapters before we finally begin looking at their practical applications in Chap. 12.

9.1 SOME PRELIMINARY IDEAS

First, how many different aspects of a sound wave may we wish to describe? Since there is motion, we need to specify the displacement of any particular blob of fluid from its undisturbed position. As before, let us use the equilibrium coordinate x as a label identifying each bit of substance, and $\xi(x, t)$ to indicate how far it has moved. Then its present position is $x + \xi$, its velocity is $d\xi/dt$, and its acceleration is $d^2\xi/dt^2$. Waves in two and three dimensions will require that x and ξ both be treated as vectors. The fluid velocity $d\xi/dt$ in ordinary sound waves is always much less than the signal speed c.

Since the longitudinal nature of sound requires regions of compression and rarefaction, it is also important to describe the changes in mass density, pressure, temperature, and internal energy of the fluid. All these quantities have steady nonzero values in the absence of sound waves, but we need to focus on the small deviations from those averages. Let us write the mass density as

$$\rho = \rho_0 + \rho_1(x, t), \tag{9.1}$$

so that ρ_0 is the constant density of the ambient equilibrium conditions and ρ_1 represents the density change due to the wave. Similarly, we use

$$p = p_0 + p_1(x, t), \tag{9.2}$$

$$T = T_0 + T_1(x, t), \tag{9.3}$$

$$e = e_0 + e_1(x, t) \tag{9.4}$$

for pressure, temperature, and energy density, respectively.

Before actually getting involved in the details, let us try to state what we expect will be needed in deriving an equation governing sound waves. (1) Since compression is a direct result of differences in the motion of neighboring bits of fluid, we must construct a relation between ξ and ρ. (2) Since each element of fluid has mass and experiences forces due to the pressure of its neighboring fluid elements, we expect $F = ma$ to provide another relation between the derivatives of p and of ξ. (3) The nature of the substance itself must enter in through its thermodynamic equation of state, which is a relation among p, ρ, and T. (4) The first law of thermodynamics is a statement of the conservation of energy; here it will say that the internal energy of one bit of fluid will change whenever work is done upon it as its neighbors exert pressure during a displacement. This will enable us to obtain e from p and ξ. (5) The thermodynamic nature of the process of compression must provide the final information needed in order that we may have five equations to determine the five variables ξ, ρ, p, T, and e.

Item 5 requires more careful consideration. Two simple extreme cases present themselves as candidates: isothermal and adiabatic processes. In an **isothermal** process heat is conducted as necessary to keep the temperature of the fluid always the same; that would mean here that T_1 is identically zero. An **adiabatic** process is one in which heat conduction does not occur, so that the higher temperature in compressed regions is maintained until they reexpand. If the compression lasts too short a time for heat conduction to carry much energy from the compressed regions to the neighboring cooler rarefactions, the disturbance may be described as adiabatic, whereas a compression that develops slowly enough can remain in temperature equilibrium with its surroundings and thus represent an isothermal process.

It is very easy to draw the incorrect conclusion that sound waves of sufficiently short period (high frequency) would be adiabatic, while low-frequency waves would behave isothermally. Let us demonstrate that exactly the opposite is true, by taking into account not only the time available for heat conduction to operate but also the distance over which the heat must flow. Suppose the amplitude of temperature fluctuation is ΔT in a sound wave with frequency f. The compressions and rarefactions are half a wavelength apart; so the question is whether heat conduction over the distance $\lambda/2$ can appreciably affect ΔT in a time less than about half an oscillation period, or $1/2f$. The maximum temperature gradient dT/dx is roughly $2\Delta T/(\lambda/2)$; so the definition of thermal conductivity K as the ratio of the heat flow to the gradient that causes it enables us to say that the heat flux (power per unit area) will be about $4K\,\Delta T/\lambda$. The region into which it flows has a depth of about $\lambda/4$ (and unit area); so the rate of heat delivery into the rarefaction is roughly $16K\,\Delta T/\lambda^2$ (power per unit volume). This continues at most for half a wave period; so the total heat delivered per unit volume is not more than

$$Q_{\max} = (16K\,\Delta T/\lambda^2)(1/2f) = 8K\,\Delta T/f\lambda^2. \tag{9.5}$$

The definition of heat capacity C (per unit mass) then says that this energy will raise the temperature by about $\delta T = Q/\rho C$, or at most $8K\,\Delta T/\rho C f\lambda^2$. If this change δT is much less than the original ΔT, heat conduction is unable to have much effect and the process should be regarded as nearly adiabatic. That will be the case if $8K \ll \rho C f \lambda^2$.

We can see now the flaw in the naive argument that high frequencies are required to satisfy this condition: high frequencies also mean short wavelengths. That is even more important than the short time scale, for the short wavelength means both a shorter distance for the heat to flow and a stronger temperature gradient to drive that flow. If we replace λ by c/f, with c being the speed of sound, the condition for adiabatic waves becomes

$$f \ll c^2 \rho C/8K. \tag{9.6}$$

For air under ordinary conditions, c = 340 m/s, ρ = 1.2 kg/m^3, C_p = 10^3 J/kg-K, and K = 0.03 W/m-K; so the limiting frequency is about 5×10^8 Hz. (It is no accident that this is roughly how often individual molecules collide with each other, and the corresponding wavelength of about 1 μm is comparable with the average distance they travel between collisions.) For water the corresponding figure is on the order of 10^{12} Hz; so in both cases nearly all imaginable practical problems involve much lower frequencies. Thus we will assume from here on that sound waves involve only adiabatic compression.

9.2 THE WAVE EQUATION FOR SOUND

Now we can proceed to spell out the governing equations for the variables ξ, ρ_1, p_1, T_1, and e_1. Our strategy will be to write three equations involving ξ, ρ_1, and p_1, then eliminate two of these variables to show that the third satisfies a wave equation. It proves convenient to focus on the pressure and to show afterward that once p_1 is determined it is relatively easy to obtain T_1 and e_1 from it.

First, we must write the relation between displacement and density. If every bit of fluid had the same displacement, there would be no change in density. The compression ρ_1 must exist only because ξ differs from place to place, and we can anticipate that ρ_1 may be proportional to a spatial derivative of ξ. To make that more precise, consider the small cube of fluid shown in Fig. 9.1*a*. Suppose for the moment that the motion is only in the x direction. Then the left face of the cube moves under the influence of the sound wave from x to $x + \xi(x, t)$, while the right face moves from $x + \Delta x$ to $x + \Delta x + \xi(x + \Delta x, t)$, as indicated in Fig. 9.1*b*. Its width has increased from Δx to $\Delta x + \xi(x + \Delta x, t) - \xi(x, t)$; so the volume has been multiplied by a factor

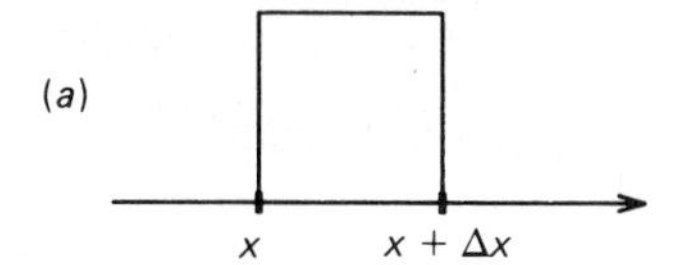

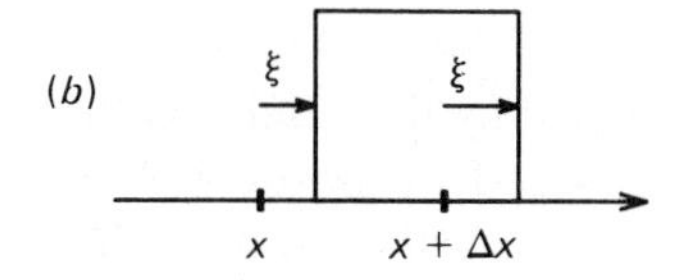

Figure 9.1 A small cube of fluid suffers a change in density if the displacements $\xi(x)$ of its left face and $\xi(x + \Delta x)$ of its right face are not equal.

$$1 + [\xi(x + \Delta x, t) - \xi(x, t)]/\Delta x. \tag{9.7}$$

But in the limit of small Δx this is simply $1 + \partial\xi/\partial x$.

Now let the motion be three-dimensional, so that the label $\mathbf{x}$ becomes a vector (x, y, z), and the displacement $\boldsymbol{\xi}$ is also a vector with components ξ_x, ξ_y, and ξ_z. This means the cube may have expanded by different amounts in all three directions, so that its volume is changed to

$$\begin{aligned} \Delta x\, \Delta y\, \Delta z(1 + \partial\xi_x/\partial x)(1 + \partial\xi_y/\partial y)(1 + \partial\xi_z/\partial z) \\ &\simeq \Delta x\, \Delta y\, \Delta z(1 + \partial\xi_x/\partial x + \partial\xi_y/\partial y + \partial\xi_z/\partial z) \\ &= \Delta x\, \Delta y\, \Delta z(1 + \nabla\cdot\boldsymbol{\xi}). \end{aligned} \tag{9.8}$$

We will be justified in dropping cross terms such as $(\partial\xi_x/\partial x)(\partial\xi_y/\partial y)$ if all the first derivatives of $\boldsymbol{\xi}$ are very small compared with unity; this will be true for waves of sufficiently small amplitude. Though its dimensions have changed, this fluid element must still have the same mass; so its density must have decreased by exactly the same factor by which the volume increased:

$$\rho_0 + \rho_1 = \rho_0/(1 + \nabla\cdot\boldsymbol{\xi}). \tag{9.9}$$

But whenever some quantity σ is very small compared with 1, we may use the binomial expansion to replace $1/(1 + \sigma)$ by the approximation $1 - \sigma$. So, finally,

$$\boxed{\rho_1(x, t) \cong -\rho_0\nabla\cdot\boldsymbol{\xi}.} \tag{9.10}$$

Next is the equation of motion, $\mathbf{F} = m\mathbf{a}$, applied to any small lump of fluid such as the one in Fig. 9.1. The fluid to its left pushes toward the right with pressure $p_0 + p_1[x + \xi_x(x, t), t]$ on a surface with area $\Delta y\, \Delta z$, while the neighboring fluid on the right is pushing in the opposite direction with a slightly different pressure $p_0 + p_1[x + \Delta x + \xi_x(x + \Delta x, t), t]$. In the small-amplitude approximation, the slight difference between x and $x + \xi_x$ will be unimportant in evaluating the function p_1. Then the net force in the x direction is given by

$$F_x = \Delta y\, \Delta z[p_1(x, t) - p_1(x + \Delta x, t)] = \Delta y\, \Delta z(-\partial p_1/\partial x)\, \Delta x. \tag{9.11}$$

Similar arguments about the other two components tell us that the vector force is

$$\mathbf{F} = -\Delta x \, \Delta y \, \Delta z \, \nabla p_1. \tag{9.12}$$

Since the mass of this piece of fluid is $\rho_0 \, \Delta x \, \Delta y \, \Delta z$, the final statement of Newton's law becomes

$$\boxed{-\nabla p_1 = \rho_0 \, \partial^2 \xi / \partial t^2.} \tag{9.13}$$

We need one more relation along with Eqs. 9.10 and 9.13 in order to determine the three functions p_1, ρ_1, and ξ. It is provided by the condition that we have an adiabatic process, and conveniently expressed by saying that the pressure p is uniquely determined as a function of the density ρ (rather than depending separately on both ρ and T). Then the derivative $(dp/d\rho)_{ad}$ is a well-defined and ordinary derivative (rather than a partial derivative), and when it is evaluated at $\rho = \rho_0$ the result is simply a constant number. Then a Taylor-series expansion tells us that small departures from the equilibrium state will obey

$$p \simeq p(\rho_0) + (\rho - \rho_0) \, dp/d\rho. \tag{9.14}$$

But since $p = p_0 + p_1$, this is just

$$\boxed{p_1 \simeq (dp/d\rho)_{ad} \rho_1.} \tag{9.15}$$

Now it is convenient to eliminate ρ_1 by substituting (9.10) into (9.15):

$$p_1 = -\rho_0 (dp/d\rho)_{ad} \nabla \cdot \xi. \tag{9.16}$$

Finally, ξ is most easily eliminated by taking the divergence of (9.13) and the second time derivative of (9.16). This gives two expressions that both equal $-\rho_0 \, \partial^2(\nabla \cdot \xi)/\partial t^2$ and therefore also equal each other:

$$\nabla^2 p_1 = (\partial^2 p_1 / \partial t^2)/(dp/d\rho)_{ad}. \tag{9.17}$$

By comparing with (7.7) and (8.6) we may see that we have once more a simple wave equation,

$$\boxed{\partial^2 p_1 / \partial t^2 = c^2 \nabla^2 p_1,} \tag{9.18}$$

where c can be interpreted as the wave speed and is given by

$$\boxed{c^2 = (dp/d\rho)_{ad}.} \tag{9.19}$$

For specific substances we can say more about $dp/d\rho$. In the case of liquids or solids, data are often available on the compressibility, or fractional density change $\Delta\rho/\rho$ per unit pressure increase Δp. Equivalent information may be presented as the **bulk modulus** $B = \rho(dp/d\rho)$, which is the reciprocal of the compressibility. Then the speed of sound should be related to the bulk modulus by $c^2 = B/\rho$.

For example, water has $B = 2.2 \times 10^9$ N/m^2 and $\rho = 10^3$ kg/m^3. These give

$$c = \sqrt{(2.2 \times 10^9)/(10^3)} = 1.48 \times 10^3 \text{ m/s},$$

in good agreement with measurements. Similarly, aluminum has $B = 7.5 \times 10^{10}$ N/m^2 and $\rho = 2.7 \times 10^3$ kg/m^3, so that $c = 5.2 \times 10^3$ m/s.

In the case of gases at sufficiently low density, their behavior will be well approximated by the ideal gas law. An adiabatic process in an ideal gas is governed by

$$(p/p_0) = (\rho/\rho_0)^\gamma. \tag{9.20}$$

Here $\gamma = C_p/C_v$ is the ratio of specific heat at constant pressure to that at constant volume, and has a definite numerical value for each gas. Air, for instance, being composed almost entirely of diatomic molecules, has $\gamma = 1.40$. From (9.19) and (9.20) it follows that an ideal gas has

$$c^2 = \gamma p_0/\rho_0. \tag{9.21}$$

But the ideal gas law says that $p = \rho RT/M$, where $R = 8314$ J/kg-K is the universal gas constant, T is measured from absolute zero, and M is the average molecular weight of the particular gas involved. So the alternative form

$$\boxed{c^2 = \gamma RT_0/M} \tag{9.22}$$

makes it clearer that the speed of sound is determined uniquely by the temperature of the gas, while neither the pressure nor the density individually really matters.

For instance, in air under ordinary conditions we have

$$\begin{aligned} c^2 &= (1.4)(8314 \text{ J/kg-K})(293 \text{ K})/(29.0) \\ &= 1.174 \times 10^5 \text{ m}^2/\text{s}^2 \end{aligned}$$

and $c = 343$ m/s, again in good agreement with direct measurement. Helium has a molecular weight of only 4.0 and $\gamma = 5/3$; so at the same temperature the speed of sound is $\sqrt{(1.67/1.40)(29/4)} = 2.9$ times as fast in helium as in air. That difference in c causes difficulties in speech intelligibility for deep-sea divers breathing helium-oxygen mixtures.

9.3 RELATING OTHER VARIABLES TO PRESSURE

When we find various solutions of (9.18) to represent sound fields in terms of their pressure distributions, we would sometimes like to extract information about other aspects of the waves. The density increase, for instance, is easily written as

$$\boxed{\rho_1 = p_1/c^2} \tag{9.23}$$

by using (9.15). The displacement ξ would come from integrating (9.13) twice over time, and takes a simple form only for special cases. *If* we consider waves with pure sinusoidal behavior in time, each integration just gives a multiplying factor $1/j\omega$ and we get

$$\xi = \nabla p_1/\omega^2 \rho_0. \tag{9.24}$$

For example a 1-kHz, 80-dB sinusoidal sound wave in air has $\omega = 6283/\text{s}$ and pressure $p_1(\text{rms}) = 0.2$ Pa, since $p_{\text{ref}} = 20\ \mu\text{Pa}$. Then its density fluctuation is

$$\rho_1(\text{rms}) = (0.2\ \text{Pa})/(343\ \text{m/s})^2 = 1.7 \times 10^{-6}\ \text{kg/m}^3$$

or roughly 10^{-6} times ρ_0. The displacement can be estimated by replacing the spatial derivative in (9.24) by $k = \omega/c$ to obtain

$$\begin{aligned}\xi(\text{rms}) &\cong (0.2\ \text{Pa})/(343\ \text{m/s})(6283/\text{s})(1.2\ \text{kg/m}^3)\\ &= 8 \times 10^{-8}\ \text{m}.\end{aligned}$$

The same sound pressure level at a frequency of only 100 Hz would involve 10 times as great a displacement, but both cases would have the same velocity amplitude,

$$v = \omega\xi \cong 5 \times 10^{-4}\ \text{m/s}.$$

The temperature is related to the pressure by $(dT/dp)_{ad}$. Only for the case of an ideal gas will we offer a simpler alternative form. The ideal gas law can be written as $(p/p_0) = (\rho/\rho_0)(T/T_0)$, and this can be combined with (9.20) to eliminate ρ and obtain

$$(p/p_0)^{\gamma-1} = (T/T_0)^{\gamma}. \tag{9.25a}$$

As you may show in Prob. 9.4, this can be further simplified for small disturbances to

$$\boxed{T_1/T_0 = (1 - 1/\gamma)p_1/p_0.} \tag{9.25b}$$

For the example of an 80-dB sound wave in air,

$$1 - 1/\gamma = 1 - 1/1.4 = 1 - 0.714 = 0.286$$

and the temperature fluctuations are only

$$\begin{aligned}T_1(\text{rms}) &= (293\ \text{K})(0.286)(0.2\ \text{Pa})/(10^5\ \text{Pa})\\ &= 1.7 \times 10^{-4}\ \text{K}.\end{aligned}$$

Sound waves involve both internal thermal energy of the fluid (because of its compression) and kinetic energy (because of its motion). The energy of compression is like that of a spring and may be thought of as potential energy, so that the total may be written

$$e = e_p + e_k. \tag{9.26}$$

The kinetic energy density is simply $\rho v^2/2$, and when the wave amplitude is small it will be satisfactory to use ρ_0 in place of ρ. Thus in general we could use (9.13) to write

$$e_k = (\rho_0/2)(\partial\xi/\partial t)^2 = (1/2\rho_0)\left(\int \nabla p_1\, dt\right)^2. \tag{9.27}$$

Only in the special case of a traveling sine wave described by $p_1 = A \cos(\omega t - kx - \phi)$ does this simplify to

$$e_k = (1/2\rho_0 c^2)p_1^2. \tag{9.28}$$

Notice that the wave energy is a "second-order" quantity, proportional to the square of the small amplitude p_1.

> The 80-dB wave discussed above has an average kinetic energy density
>
> $$\langle e_k\rangle = (0.2\text{ Pa})^2/[(2)(1.2\text{ kg/m}^3)(343\text{ m/s})^2]$$
> $$= 1.4 \times 10^{-7}\text{ J/m}^3.$$
>
> This is an average because the 0.2 Pa is an rms average; there are some places where p_1 and e_k are both zero, and others where they attain maximum values 0.28 Pa and 2.8×10^{-7} J/m^3.

The internal energy change can be found by asking how much work is done upon the gas in compressing it. It follows directly from the definition of work as force times distance that a pressure p pushing on an area A through a distance dx does work $dW = pA\,dx$. But $A\,dx$ is the amount by which the volume of an enclosed gas would be reduced by such a piston moving into it; so $dW = -p\,dV$. This remains true even when the volume change dV is the net result of different amounts of motion by various parts of the boundary surface. Since we consider only adiabatic processes, the heat flow $dQ = 0$ and the work dW done upon any parcel of fluid all appears as an increase in its internal energy. Let us avoid a clutter of subscripts by temporarily using the traditional thermodynamic symbols U and u in place of E_p and e_p for total energy and energy density. Then the work dW produces an energy change $dU = dW = -p\,dV$ in this portion of fluid, and the internal energy density change is

$$du = d(U/V) = dU/V - U\,dV/V^2 = -(p + u)\,dV/V. \tag{9.29}$$

Since the product ρV is the mass of this portion of fluid, it is constant. Thus $d(\rho V) = 0$, or $dV/V = -d\rho/\rho$, so that (9.29) can be written as

$$du = (p + u)\,d\rho/\rho. \tag{9.30}$$

As a first approximation, it would appear that we could take $du_1 = (p_0 + u_0)\,d\rho_1/\rho_0$, which integrates to

$$u_1 \simeq [(p_0 + u_0)/\rho_0]\rho_1 \simeq [(p_0 + u_0)/\rho_0 c^2]p_1 \tag{9.31}$$

when we use (9.23). For a perfect gas, which has $u_0 = p_0/(\gamma - 1)$, this reduces simply to $u_1 = p_1/(\gamma - 1)$. It says that the 80-dB wave in air would have an energy density fluctuation of about (0.2 Pa)/(0.4) = 0.5 J/m^3 above and below the equilibrium value u_0 = (10^5 Pa)/(0.4) = 2.5×10^5 J/m^3.

This, however, is not the most interesting aspect of the energy, since the average of u_1 is zero. If we kept terms proportional to wave amplitude squared throughout the derivation, we would discover that u has slightly greater positive values in the compressions than it does negative in the rarefactions. Thus the wave carries a small overall positive energy whose computation is outlined in Prob. 9.6. This second-order contribution turns out to be

$$u_2 = (1/2\rho_0 c^2)p_1^2, \tag{9.32}$$

and it is this that we wish to identify with e_p. Comparing this with (9.28), we see that plane sinusoidal waves exhibit equipartition between the two forms of energy. (Caution: We will find later that this is not true for all wave types.) The portion of the internal energy in (9.32) may be called acoustic potential energy. The total acoustic energy density in both forms for traveling sine waves is

$$\boxed{e = e_p + e_k = p_1^2/\rho_0 c^2.} \tag{9.33}$$

For the 80-dB wave used as an example above, (9.32) gives an average acoustic potential energy density 1.4×10^{-7} J/m^3, identical to the e_k already found. Thus the average total energy density associated with this wave is 2.8×10^{-7} J/m^3.

9.4 PLANE WAVES

The simplest solutions to the wave equation 9.18 are those that depend on only one of the three spatial coordinates, and we may as well call that one x. The laplacian $\nabla^2 p$ reduces to $\partial^2 p/\partial x^2$, and the wave equation becomes mathematically identical to (7.7). Thus we already know the nature of its solutions: any function

$$p_1(x, t) = g(ct \pm x) \tag{9.34}$$

will work. These are waves traveling with speed c parallel to the x axis, and since there is no dependence on y or z the wavefronts are all planes parallel to the yz plane. $\nabla \cdot \xi$ reduces to $\partial \xi/\partial x$ so that ξ can be expressed in terms of p_1 as an integral of g by substituting (9.23) into (9.10):

$$\xi(x, t) = -(1/\rho_0 c^2) \int g(ct \pm x)\, dx. \tag{9.35}$$

It is hard to make any further general statement about this relation without choosing a particular function g. But the acoustic velocity is very simply related to p_1 because $v = \partial \xi/\partial t$ effectively undoes the integral:

$$\begin{aligned} v_1(x, t) &= -(1/\rho_0 c^2) \int c g'(ct \pm x)\, dx \\ &= \mp(1/\rho_0 c) g(ct \pm x). \end{aligned} \tag{9.36}$$

Then the kinetic energy density is

$$e_k = \rho_0 v_1^2/2 = g^2/2\rho_0 c^2. \tag{9.37}$$

But (9.32) and (9.34) give an identical expression for e_p. Thus equipartition of energy, and the simple expression (9.33) for total acoustic energy, are true for all plane traveling waves even if they are not sinusoidal.

At what rate is energy being carried from one place to another by these waves? It is tempting to think simply that the total energy density $g^2/\rho_0 c^2$ moves along with speed c to create an energy flux $g^2/\rho_0 c$ in watts per square meter. This argument succeeds, however, only in the special case of lone plane traveling waves. In general, we must make a more careful argument as follows.

Suppose our attention is focused on a small element of area dA that moves with the fluid, and we ask how much energy is flowing across it. Such a flow exists because the fluid on one side of dA exerts force $p\,dA$ and does work upon that on the other side as it moves. Movement of the fluid through a distance $d\xi$ during time dt will result in work $p\,dA\,d\xi$ if the displacement is perpendicular to the surface. But if the motion is tangent to the surface, there is no energy transfer across it. In general, when we represent $\mathbf{dA}$ as a vector and θ is the angle between it and $\mathbf{d\xi}$ (Fig. 9.2), the work done is

$$dW = p\,\mathbf{dA}\cdot\mathbf{d\xi} = p\,dA\,d\xi\cos\theta. \tag{9.38}$$

Then the intensity of the energy flow across dA is

$$\boxed{I = dW/(dA\,dt) = p(d\xi/dt)\cos\theta.} \tag{9.39}$$

In the plane traveling wave case this does become $I = (g^2/\rho_0 c)\cos\theta = ce\cos\theta$.

But consider the combination of two traveling waves to make a sinusoidal standing wave, as we did with (7.15) and (7.16). If

$$p_1(x, t) = A\cos\omega t\sin kx, \tag{9.40}$$

then $\partial p_1/\partial x = kA\cos\omega t\cos kx$, so that integrating (9.13) over dt twice gives

$$\partial\xi(x, t)/\partial t = -(A/\rho_0 c)\sin\omega t\cos kx \tag{9.41}$$

and

$$\xi(x, t) = (A/\omega\rho_0 c)\cos\omega t\cos kx. \tag{9.42}$$

Notice that the factor $\sin kx$ is zero where $\cos kx$ is maximum and vice versa. Thus it is important that any mention of nodes of an acoustic plane standing

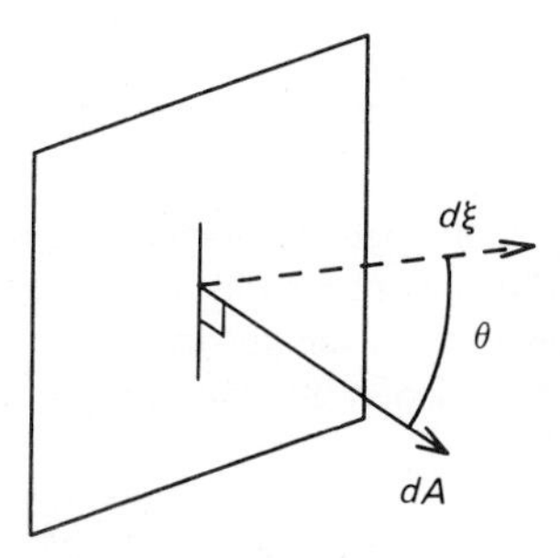

Figure 9.2 Area dA represented by a vector perpendicular to the surface, at an angle θ from direction of fluid motion $d\xi$.

wave specify which variable is zero: pressure nodes are always displacement antinodes, and displacement nodes are pressure antinodes. Without repeating the details (which would be closely analogous to parts of Sec. 7.3), we show in Figs. 9.3 and 9.4 an approximate version of how standing waves may fit in a cylindrical tube. In Sec. 14.4 further consideration will be given to what really happens at the tube ends.

The intensity (9.39) flowing across any plane perpendicular to the x axis, for which $\theta = 0$, becomes

$$I(x, t) = -(A^2/4\rho_0 c) \sin 2\omega t \sin 2kx. \tag{9.43}$$

When we compare this with the energy density

$$e(x, t) = e_p + e_k = (A^2/2\rho_0 c^2)(\cos^2 \omega t \sin^2 kx + \sin^2 \omega t \cos^2 kx), \tag{9.44}$$

we see that I is very different from c times e. In particular, the time-averaged intensity is identically zero for the standing wave even though e has the nonzero average $A^2/4\rho_0 c^2$.

Suppose it were the standing wave (9.40) that produced a maximum reading of SPL = 80 dB at its antinodes. This would mean

$$\langle p_1^2 \rangle = (0.2 \text{ Pa})^2 = (A^2/2) \sin^2 kx = A^2/2$$

wherever kx is an odd multiple of $\pi/2$. The instantaneous energy density fluctuates between zero and

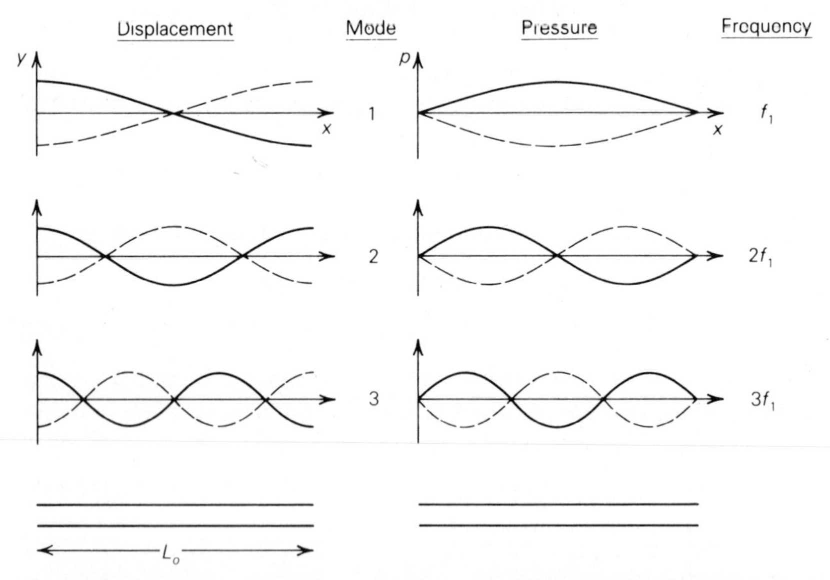

Figure 9.3 Examples of standing waves in a cylindrical tube of length L_o open at both ends, illustrating that pressure nodes and displacement antinodes occur together. The requirement that each open end be a pressure node suffices to limit the possibilities to $n(\lambda_n/2) = L_o$ or $f_n = nc/2L_o$. (From D. E. Hall, *Musical Acoustics: An Introduction*, Wadsworth, 1980.)

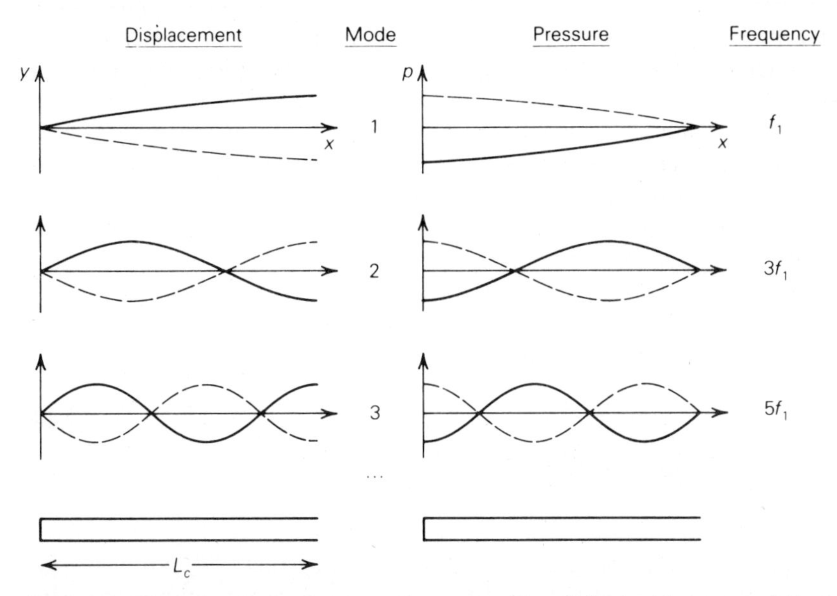

Figure 9.4 Examples of standing waves in a tube of length L_c closed at one end. Requiring a displacement node at the closed end and a pressure node at the open end suffices to determine $(2n - 1)(\lambda_n/4) = L_c$ or $f_n = (2n - 1)c/4L_c$. (From D. E. Hall, *Musical Acoustics: An Introduction,* Wadsworth, 1980.)

$$A^2/2\rho_0 c^2 = (0.04 \text{ Pa}^2)/(1.2 \text{ kg/m}^3)(343 \text{ m/s})^2$$
$$= 2.8 \times 10^{-7} \text{ J/m}^3$$

at both the nodes and antinodes, while halfway in between it remains steadily at $A^2/4\rho_0 c^2 = 1.4 \times 10^{-7}$ J/m^3. The kinetic and potential energy densities fluctuate 180° out of phase with each other in the standing wave, whereas for the traveling wave (9.34) they are in phase. The average intensity of this standing wave is zero, but at some times and places it is as large as

$$A^2/4\rho_0 c = (0.08 \text{ Pa}^2)/(4 \times 415 \text{ kg/m-s}^2) = 48\ \mu\text{W/m}^2.$$

9.5 NONLINEAR EFFECTS

Let us consider briefly how much we are missing by using only "linearized" equations of motion. Several times in Sec. 9.2 we dropped terms of quadratic or higher order in small quantities such as p_1 or ξ. A typical example is our neglect of any term beyond the first in

$$\Delta p = (dp/d\rho)\,\Delta\rho + (d^2p/d\rho^2)(\Delta\rho)^2/2 + \cdots \tag{9.45}$$

If we were careful to retain all second-order terms, we would obtain instead of (9.18) a nonlinear differential equation. Developing solutions for that equation would be much more difficult than anything we want to attempt here. Nevertheless, let us show some very simple arguments about what effects these nonlinear terms may have.

The easiest nonlinear effect to discuss is that represented in (9.45). If the extra term is still only a small correction, it is conveniently viewed as telling us that the speed of sound is not a constant after all:

$$c^2(\rho) \simeq (dp/d\rho)_{ad} + (\rho_1/2)(d^2p/d\rho^2)_{ad}. \tag{9.46}$$

We can use (9.15) and (9.19) to write this as

$$c^2 \simeq c_0^2(1 + \alpha p_1/p_0) \tag{9.47}$$

if we define

$$\alpha = p_0(d^2p/d\rho^2)/2(dp/d\rho)^2. \tag{9.48}$$

In the case of a perfect gas, $(p/p_0) = (\rho/\rho_0)^\gamma$ and

$$\alpha = (\gamma - 1)/2\gamma, \tag{9.49}$$

which is 0.14 for air. As suggested in Fig. 9.5, we interpret (9.47) to mean that the regions of compression (where the temperature is higher) move somewhat faster than do the rarefactions. This picture is analogous to the explanation that water surface waves steepen and break because the crests find themselves on deeper water and so travel a little faster than the troughs and overtake them.

Figure 9.5 shows the result of a calculation based on the approximation of Eq. 9.47 in the case where $\alpha = 0.14$ and the maximum amplitude is $p_1 = 0.4p_0$. After the wave has traveled some distance, the approximation clearly becomes invalid because it predicts three different values of pressure at the same place. More sophisticated calculations show that what actually happens is the development of a **shock wave.** This means a surface where there is a sudden discontinuity in gas properties such as pressure, density, temperature, and velocity. The actual form the waves of Fig. 9.5 would approach is the steady train of shock waves separated by gradual expansions shown in Fig. 9.6.

Let us estimate the circumstances under which we would need to take this

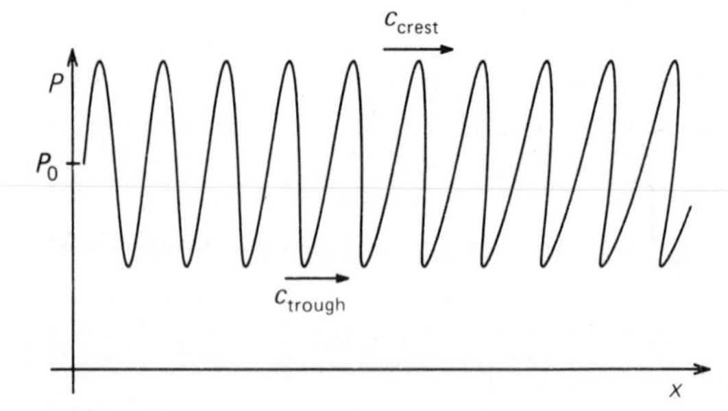

Figure 9.5 For large-amplitude waves, regions of compression have higher temperature and thus higher sound speed than do the rarefactions, and so should travel faster. After the wave travels some distance, the original sinusoidal shape is lost. Where the function becomes multivalued, it no longer gives a correct picture of what is happening physically.

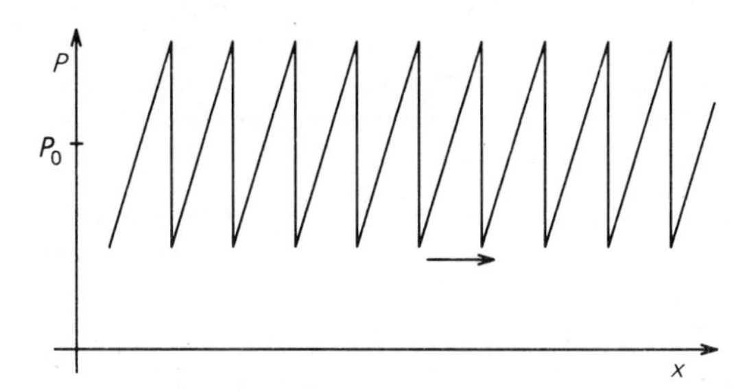

Figure 9.6 Waves traveling as suggested in Fig. 9.5 will in fact form shocks, or sudden jumps in pressure, and this is the waveform that will propagate without further change.

steepening of wave fronts (or other nonlinear effects) into account. A rough guess at what distance D the waves could travel before evolving into shock fronts is obtained by asking how soon a crest could travel the extra distance $\lambda/2$ to catch up with the preceding trough. This requires the common travel time to be

$$t = (D + \lambda/2)/c_{\text{crest}} = D/c_{\text{trough}}, \tag{9.50}$$

so that

$$(c_{\text{crest}}/D)(1 + \lambda/2D)^{-1} = c_{\text{trough}}/D \tag{9.51}$$

or

$$(c_{\text{crest}} - c_{\text{trough}})/D = c_{\text{trough}}\lambda/2D^2. \tag{9.52}$$

For $\lambda \ll D$, c_{trough} is well enough approximated by c_0 on the right-hand side, and the distance required is

$$D \simeq (c_0\lambda/2)/(c_{\text{crest}} - c_{\text{trough}}). \tag{9.53}$$

Using (9.47), the difference in velocities in terms of the maximum wave amplitude p_1 is about $c_0\alpha p_1/p_0$. Then

$$D \simeq (\lambda/2\alpha)[p_0/p_1(\text{max})]. \tag{9.54}$$

For example, a sine wave in air with $f = 1$ kHz and SPL $= 100$ dB has $\lambda = 0.34$ m and rms amplitude 2 Pa. Then the maximum amplitude is 2.8 Pa, so with $p_0 = 10^5$ Pa we get

$$D \simeq [(0.34\text{ m})/(2 \times 0.14)](10^5/2.8) \simeq 40\text{ km}.$$

For $f = 100$ Hz, $\lambda = 3.4$ m, and D would be 400 km. But a 20-dB higher sound level, with $p_{\text{rms}} = 20$ Pa, would reduce both these estimates of D by a factor of 10. Thus nonlinear effects are often unimportant except in cases involving very strong waves or very great propagation distances.

PROBLEMS

9.1. Does the high heat conductivity of metals mean that the limiting frequency for adiabatic behavior is low enough to be of practical importance? Take as an example aluminum, which has $c = 6.3 \times 10^3$ m/s, $\rho = 2.7 \times 10^3$ kg/m^3, $C = 210$ cal/kg-K, and $K = 50$ cal/m-s-K.

9.2. In truth, static measurement of compressibility may be relatively difficult, and the best way to determine that quantity may be to measure sound speed in the substance of interest. If sound travels 2.7×10^3 m/s through a sample of plastic with density 1.2×10^3 kg/m^3, what is its bulk modulus B?

9.3. Use Eq. 9.22 to show that the effect of a small temperature change upon the speed of sound in a gas may be represented by $\Delta c/c = \Delta T/2T$. Use this to find how many m/s c will increase for each degree rise in temperature under ordinary conditions (T around 20°C, which is 293 K).

9.4. Show the details of the derivation of Eqs. 9.25*a* and 9.25*b*. Be sure to maintain a clear distinction between p_0 and p_1, etc.

9.5. What SPL would be necessary in air to cause rms temperature fluctuations of 1 K?

9.6. Prove (9.32) by writing $p = p_0 + p_1$, $\rho = \rho_0 + \rho_1$, and $u = u_0 + u_1 + u_2$ in (9.29). Use the binomial expansion to replace $1/\rho$ by $(1/\rho_0)(1 - \rho_1/\rho_0)$, and use (9.23) wherever convenient. Show first that consistently keeping only first-order terms reproduces (9.31). Once this u_1 has been found, use it to calculate all second-order terms on the right side of (9.29) and obtain u_2.

9.7. For a plane traveling sine wave in water with $f = 5$ kHz and SPL = 160 dB *re* 1 μPa, what are the values of:
(a) displacement amplitude?
(b) velocity amplitude?
(c) density fluctuation amplitude?
(d) average energy density?
(e) intensity?
(f) How would these answers differ for the same SPL with $f = 500$ Hz?

9.8. The atmosphere of Mars is mostly carbon dioxide ($M = 44$, $\gamma = 1.3$), with temperature about −40°C and pressure about 1/100 of that in our atmosphere.
(a) What is the speed of sound in this gas?
(b) For a plane sine wave with $f = 200$ Hz and SPL = 60 dB *re* 20 μPa, what are the fluctuation amplitudes of the pressure, density, temperature, velocity, and displacement?
(c) What is the intensity of the wave?

9.9. Discuss in more detail the relation between (9.43) and (9.44). Show in particular that there are some times when the energy is purely kinetic and others when it is entirely potential. How large a time interval separates those cases, and which of them corresponds to a time of zero intensity everywhere? Show that there are some planes across which energy never flows. How far apart are they?

9.10. Let wave A be a sinusoidal traveling wave and B a standing wave, both with plane wave fronts parallel to the yz plane, $f = 1$ kHz, and SPL = 100 dB in air. What is the average energy density of each? What is the average intensity of energy flow for each, across a surface oriented perpendicular to the x axis? What if the surface is oriented 45° from that direction? Or 90°?

9.11. What SPL would be required for 100-Hz sine waves in air to steepen into shock waves within a distance of only 100 m? What if the frequency is 10 kHz instead?

9.12. Suppose that δT following (9.5) is much less than ΔT, and suppose that $\delta T/\Delta T$ provides a rough estimate of the fractional amplitude loss $\delta p/\Delta p$ in a half cycle of vibration. Show that this would lead to a value of α in (1.21) of about $(140\ K/\rho C c^3) f^2$, and compare that with the "classical" part of α for air shown in Fig. 1.4.

Chapter 10

More About Sound Waves

Thus far we have developed a set of differential equations that describe the local behavior of sound waves. But aside from the simplest possible case of plane waves, we have not yet seen how much variety there may be in the global nature of sound. In this chapter we become acquainted with some other important types of wave behavior associated with spherical and cylindrical geometry. Our emphasis here is on the waves themselves, and in the following chapter we will pursue the interaction of these waves with their sources.

10.1 SPHERICAL WAVES

The evolution of the pressure $p(\mathbf{x}, t)$ in a sound wave is governed by the wave equation derived in the preceding chapter,

$$\partial^2 p/\partial t^2 = c^2\,\nabla^2 p. \tag{10.1}$$

Here we have dropped the subscript from p_1 for the acoustic part of the pressure. Any time we wish to refer to the total pressure, we will explicitly write $p_0 + p_1$.

For waves being radiated outward in all directions from a small source, it should be much more natural to describe p in spherical instead of cartesian coordinates. The customary labeling among physicists and engineers uses θ for the polar angle (colatitude) and ϕ for the azimuthal angle (longitude). Then the components of the radius vector from the origin are

$$\begin{aligned}\mathbf{x} &= (x, y, z)\\ &= (r\sin\theta\cos\phi,\ r\sin\theta\sin\phi,\ r\cos\theta).\end{aligned} \tag{10.2}$$

When written in these coordinates, the laplacian operator becomes

$$\nabla^2 p = (1/r^2)\{\partial[r^2(\partial p/\partial r)]/\partial r + (1/\sin\theta)\,\partial[\sin\theta(\partial p/\partial\theta)]/\partial\theta + (1/\sin^2\theta)\,\partial^2 p/\partial\phi^2\}. \quad (10.3)$$

When (10.3) is used in (10.1), there is a rich variety of possible solutions. Let us begin with the simplest case, where p depends only on r and t, not on θ or ϕ. Then $p(r, t)$ must satisfy

$$\partial^2 p/\partial t^2 = (c^2/r^2)\,\partial[r^2(\partial p/\partial r)]/\partial r. \quad (10.4)$$

From a simple energy-conservation standpoint we have already argued in Chap. 1 that spherical outgoing waves should diminish in amplitude, with p being proportional to r^{-1}. Thus it is reasonable to examine as a possible trial solution

$$p(r, t) = (1/r)f(ct - r). \quad (10.5)$$

Aside from the gradual decrease in amplitude, such a wave would retain its original shape as it moved outward. Any particular feature such as a crest could be followed by setting $\Delta(ct - r) = c\,\Delta t - \Delta r = 0$, which shows that it moves with speed $\Delta r/\Delta t = c$.

Substitution quickly shows that (10.5) does indeed obey (10.4). The objection that this solution has a singularity at the origin can only be met by admitting that in real life all waves come from finite-size sources, so that the wave does not exist all the way in to $r = 0$.

A particularly important case is that of a sinusoidal wave,

$$\boxed{p(r, t) = (A/r)\exp[j(\omega t - kr)],} \quad (10.6)$$

with $\omega = ck$. Then (9.24) gives the associated displacement

$$\xi_r(r, t) = (\partial p/\partial r)/\omega^2\rho_0 = (A/jkrc^2\rho_0)(1 + 1/jkr)\exp[j(\omega t - kr)] \quad (10.7)$$

and velocity

$$\partial\xi/\partial t = (A/rc\rho_0)(1 + 1/jkr)\exp[j(\omega t - kr)]. \quad (10.8)$$

Thus the average potential and kinetic energy densities are

$$e_p = p_{\text{rms}}^2/2\rho_0 c^2 = A^2/4\rho_0 c^2 r^2 \quad (10.9)$$

and

$$e_k = \rho_0 v_{\text{rms}}^2/2 = (A^2/4\rho_0 c^2 r^2)(1 + 1/k^2r^2). \quad (10.10)$$

Notice that these are equal to each other only in the limit where $kr \gg 1$, that is, when $r \gg \lambda/2\pi$.

The intensity of energy flow in the r direction is given, as in (9.39), by

$$I(t) = p\,\partial\xi/\partial t = (A/r)\cos(\omega t - kr)(A/rc\rho_0)[\cos(\omega t - kr) + (1/kr)\sin(\omega t - kr)]. \quad (10.11)$$

We write this out fully to emphasize that p and v are not in phase with each other here, as they were for plane traveling waves. The product of cos and sin of $\omega t - kr$ has a zero time average, and only the first term in the product contributes to

$$\boxed{I_{av} = A^2/2\rho_0 cr^2.} \tag{10.12}$$

This same result would have come directly from applying the phasor prescription $I = \text{Re}(p^* \, \partial\xi/\partial t)/2$.

What is the physical meaning of the other term that averaged to zero? Since this component of the velocity is 90° out of phase with the pressure, energy flows in opposite directions on alternate quarter cycles. This is exactly the same situation we met for standing waves on a string (Eq. 7.41), which also have nonpropagating energy. Note that for $kr < 1$ there is a predominance of oscillating energy flow. Only for $kr \gg 1$ does I_{av} become simply c times the total energy density in (10.9) and (10.10).

A similar situation occurs with electromagnetic waves, and we may borrow some terminology to describe it. The waves at $kr \gg 1$ have practically detached themselves from the source and may legitimately be viewed as a **radiation field.** For $kr < 1$, on the other hand, we have predominantly an **induction field** whose energy remains associated with the source. Equations 10.6 and 10.8 indicate that only for large values of kr will a pressure-sensitive and a velocity-sensitive microphone measure essentially the same signal. The relative strength of signal received will change if r becomes small and will do so in a frequency-dependent way since $kr = \omega r/c$.

Consider, for example, a spherical wave with $A = 0.1$ Pa-m and $f = 1$ kHz in air. At $r = 0.5$ m we would have

$$kr = 2\pi fr/c = (6.28 \times 10^3/\text{s})(0.5 \text{ m})/(340 \text{ m/s}) = 9.2$$

and

$$1 + 1/k^2r^2 = 1 + 1/85 = 1.018.$$

Then at this position the pressure amplitude is

$$p_{max} = A/r = (0.1 \text{ Pa-m})/(0.5 \text{ m}) = 0.2 \text{ Pa},$$

which means $p_{rms} = 0.14$ Pa and SPL = 77 dB. The associated velocity has in-phase and out-of-phase components given by the real and imaginary parts of

$$\begin{aligned} v &= (p/c\rho_0)(1 + 1/jkr) \\ &= [(0.2 \text{ Pa})/(340 \text{ m/s})(1.2 \text{ kg/m}^3)](1 - j/9.2) \\ &= (0.490 - j0.053) \text{ mm/s}, \end{aligned}$$

so

$$v_{max} = 0.493 \quad \text{and} \quad v_{rms} = 0.349 \text{ mm/s}.$$

The associated average energy densities are

$$\begin{aligned} e_p &= p_{rms}^2/2\rho_0 c^2 \\ &= (0.14 \text{ Pa})^2/(2.4 \text{ kg/m}^3)(340 \text{ m/s})^2 = 71 \text{ nJ/m}^3 \end{aligned}$$

and

$$e_k = e_p(1 + 1/k^2r^2) = (71)(1.018) = 72 \text{ nJ/m}^3,$$

and the intensity is

$$I = p_{\rm rms}^2/\rho_0 c = 2ce_p$$

$$= (680 \text{ m/s})(71 \text{ nJ/m}^3) = 48 \ \mu\text{W/m}^2.$$

If we go in to $r = 5$ cm, where $kr = 0.92$, we find $p_{\rm rms}$ simply increased by a factor of 10 to 1.4 Pa (97 dB). But v becomes $4.9 - j5.3$ mm/s, so $v_{\rm rms} = 5.1$ mm/s has increased by a factor $5.1/0.35 = 14.6$. The energy densities $e_p = 7.1\ \mu\text{J/m}^3$ and $e_k = 15.3\ \mu\text{J/m}^3$ are now quite different from each other, and the propagating intensity $I = 4.8$ mW/m^2 is considerably less than $ce_{\rm total} = 7.6$ mW/m^2.

10.2 WAVE IMPEDANCE

The language of impedance may contribute further insight into the effects just described. When considering the motion of massive objects, we defined mechanical impedance as the ratio of force applied to resulting velocity. But here we are dealing with pressure rather than force; so it is appropriate to consider the quantity

$$\boxed{z_a = p/v = p/(\partial\xi/\partial t).} \tag{10.13}$$

This has units Pa/(m/s) = kg/m^{2}s (sometimes called rayleighs) and defines the **specific acoustic impedance.** It must be distinguished both from mechanical impedance and from another type of acoustic impedance still to come in later chapters. By considering p/v we are finding out how hard it is to make the fluid move. This impedance presents a load to the provider of the acoustic pressure p, and that might be either the neighboring fluid or some moving boundary such as a speaker cone.

As an example, consider a plane traveling wave. We found in Sec. 9.4 that if $p(x, t) = g(ct - x)$ then $v(x, t) = g/\rho_0 c$ so that

$$\boxed{z_a = \rho_0 c \equiv z_c.} \tag{10.14}$$

This combination z_c of properties of the wave-carrying medium is called the **characteristic acoustic impedance.**

Air, for instance, has a characteristic impedance

$$z_c = (1.21 \text{ kg/m}^3)(343 \text{ m/s}) = 415 \text{ kg/m}^2\text{s}$$

for $T = 20°$C and $p_0 = 1$ atm, while fresh water at 20°C has

$$z_c = (998 \text{ kg/m}^3)(1481 \text{ m/s}) = 1.48 \times 10^6 \text{ kg/m}^2\text{s}.$$

In both cases, z_c will vary with changes in temperature or pressure.

It is important to avoid the misconception that all waves in a given medium will simply conform to (10.14); so let us show that z_a is a property *not* of the medium alone but also of the particular wave that is present. Already if we merely consider $p(x, t) = g(ct + x)$, for which $v = -g/\rho_0 c$, we find that a plane wave traveling to the left has $z_a = -z_c$. In general, z_c may serve conveniently to provide the right units, but z_a will involve a further dimensionless factor depending on the geometry of the waves involved. We can emphasize this in the plane-wave case just by considering standing instead of traveling waves. The ratio of the pressure and velocity given in (9.40) and (9.41) for a sinusoidal standing wave is best considered with phasor notation. Then

$$p = A \sin kx e^{j\omega t},$$

$$v = (kA/\rho_0) \cos kx(1/j\omega)e^{j\omega t},$$

and

$$z_a = p/v = j\rho_0 c \tan kx. \tag{10.15}$$

Thus the magnitude of the specific impedance depends on which part of the wave you are examining. It may be either much smaller or larger than z_c, depending on whether x is near a node or an antinode of the pressure. The fact that this z_a is purely imaginary is a reminder that p and v are 90° out of phase and that there is no net long-term power flow through the medium.

What is the specific impedance of an outgoing spherical sine wave? Equations 10.6 and 10.8 above give the answer:

$$z_a = z_c/(1 + 1/jkr) = jkrz_c/(jkr + 1). \tag{10.16}$$

Rationalizing the complex fraction by multiplying both numerator and denominator by $(-jkr + 1)$ gives

$$z_a = z_c(k^2r^2 + jkr)/(k^2r^2 + 1). \tag{10.17}$$

We might choose to describe this either as a specific resistance and reactance $z_a = r_a + jx_a$ or in terms of its magnitude and phase angle $z_a = |z_a| \exp(j\delta)$. Then (10.17) gives

$$r_a = z_c k^2r^2/(k^2r^2 + 1) \qquad x_a = z_c kr/(k^2r^2 + 1) \tag{10.18}$$

and

$$|z_a| = z_c kr/\sqrt{k^2r^2 + 1} \qquad \delta = \arctan (1/kr). \tag{10.19}$$

These functions are shown in Fig. 10.1.

In the spherical wave example above, we first considered $kr = 9.2$ and $k^2r^2 = 85$. If $z_c = 415$ kg/m^{2}s, we then have

$$z_a = (415)(85 + j9.2)/(86) = 410 + j44 \text{ kg/m}^2\text{s}$$

or

$$|z_a| = (415)(9.2)/(9.25) = 413 \text{ kg/m}^2\text{s}$$

and

$$\delta = \arctan (1/9.2) = 0.113 \text{ rad} = 6.5°.$$

This differs only slightly from the plane-wave impedance. But closer to the

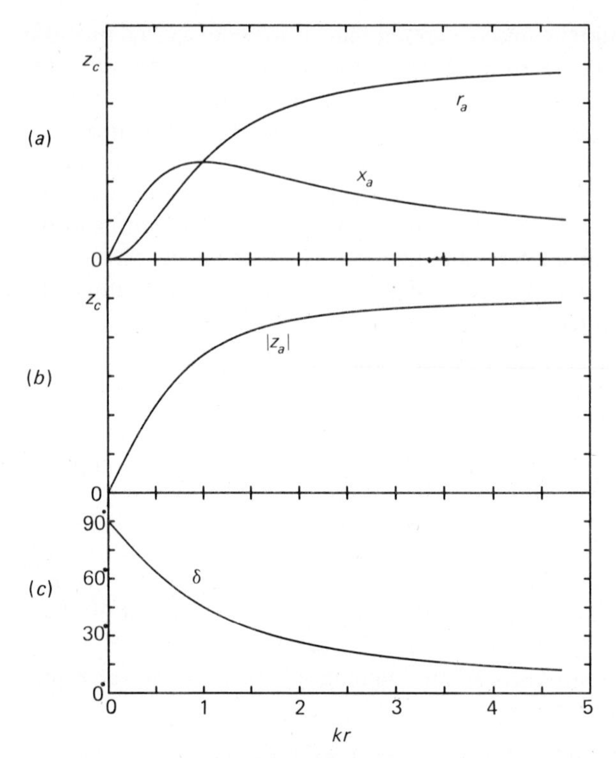

Figure 10.1 Specific acoustic impedance of a spherical wave. (*a*) Real and imaginary parts, $z_a = r_a + jx_a$. (*b, c*) Magnitude and phase, $z_a = |z_a|\exp(j\delta)$.

source, with $kr = 0.92$, we have $z_a = 190 + j206$ kg/m²s, $|z_a| = 280$ kg/m²s, and $\delta = 47°$.

We emphasize again that the specific acoustic impedance depends on the frequency, as well as on the medium, the wave type, and the geometric position. Neither the frequency nor the distance alone is so important as is the dimensionless combination $kr = (\omega/c)r = 2\pi fr/c = 2\pi r/\lambda$. For $kr \gg 1$, z is predominantly resistive, meaning that energy fed into such waves propagates away from the scene, never to return. The larger the value of kr, the less curved are the wave fronts and the more nearly do these waves act just like plane waves. But for $kr < 1$ the reactive component is larger, and it is positive (i.e., masslike). This means that pushing on such a wave is mainly a matter of storing kinetic energy in motion of the gas, where (like energy in an electrical inductor) it can be recovered later by the source. The geometry of the diverging outward flow effectively reduces the confinement of the gas by its surroundings and makes it easy to produce large velocities (and associated kinetic energy) with relatively little pressure.

There is an alternative view that suggests a simpler insight into the spherical wave. If we had considered the ratio v/p instead of p/v, the resulting admittance would be a much simpler function. Using (10.16),

$$y_a = 1/z_a = (1 + 1/jkr)/z_c$$
$$= 1/\rho_0 c + 1/jkr\rho_0 c. \qquad (10.20)$$

This suggests that (10.18) and (10.19) are complicated because we tried to insist that the resistance and reactance be viewed as in series in order to add them together. If we were willing to treat them as in a parallel circuit instead, for which admittances are additive, the parallel resistance would merely be the constant $\rho_0 c$. And the parallel reactance, being positive and proportional to $\omega = ck$, would be that of an inductance $\rho_0 r$ which is also independent of frequency. Thus it is possible (Fig. 10.2) to view the radiation resistance seen by a spherical source as a constant load, with the amount of motion produced in that load depending on whether the parallel inductance offers an easier path or not.

In the last example, we would say that the effective impedances found there are the same as if a resistance $z_c = 415$ kg/m²s is in parallel with an inductance of

$$\rho_0 r = (1.2 \text{ kg/m}^3)(0.5 \text{ m}) = 0.6 \text{ kg/m}^2$$

at the first position, but 0.06 kg/m² at $r = 0.05$ m. For $\omega = 6280$/s the reactance of this parallel inductance would be (6280/s)(0.6 kg/m²) = 3760 kg/m²s at $r = 0.5$ m but only 376 kg/m²s at $r = 0.05$ m.

10.3 CYLINDRICAL WAVES

Sound diffracting through a long, narrow slit provides an example of a situation where waves radiate from something acting more or less as a line source rather than a point source. Then it is natural to describe such waves in cylindrical coordinates r, ϕ, z, such that

$$\mathbf{x} = (x, y, z) = (r \cos \phi, r \sin \phi, z). \qquad (10.21)$$

Note that this $r = \sqrt{x^2 + y^2}$ is different from the spherical r used above. Now the laplacian operator becomes

$$\nabla^2 p = (1/r^2)\{r\, \partial[r(\partial p/\partial r)]/\partial r + \partial^2 p/\partial \phi^2\} + \partial^2 p/\partial z^2. \qquad (10.22)$$

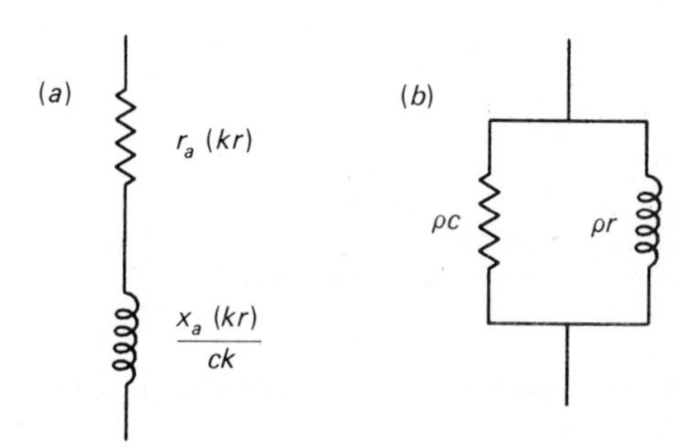

Figure 10.2 Two ways of representing the specific acoustic impedance of a spherical wave. (*a*) If $z_a = r_a + jx_a$, the equivalent series circuit elements must be functions of frequency according to (10.18). (*b*) If a parallel circuit is used instead, the equivalent resistance and inductance in (10.20) are independent of frequency.

The simplest solution using these coordinates will be when p does not depend on ϕ or z, so that

$$\partial^2 p/\partial t^2 = (c^2/r)\, \partial[r(\partial p/\partial r)]/\partial r. \tag{10.23}$$

It seems logical again to guess on simple grounds of energy conservation that we could solve this equation with a traveling wave whose amplitude decreases appropriately with distance:

$$p(r, t) = (A/\sqrt{r})g(ct - r). \tag{10.24}$$

It is rather shocking to discover upon substitution that this is *not* a solution of (10.23)! In fact, there is not even any specific form of g that will work in this way.

This has a rather remarkable consequence for the problem of transient response in loudspeakers. When working with plane or spherical waves, it is reasonable to undertake the technical problem of trying to design a speaker so that it can accurately produce pressure waveforms corresponding to specified voltage input waveforms. Once properly produced, such a wave would retain its shape as it propagated. But in cylindrical geometry, no matter what waveform you generate, it is guaranteed to change in shape as well as in amplitude as it moves out to larger r. A short pressure pulse, for example, unavoidably develops a wake following along behind it, as illustrated in Fig. 10.3.

If we ask specifically about steady traveling waves with sinusoidal dependence in time, (10.23) requires Bessel functions to describe the spatial dependence.

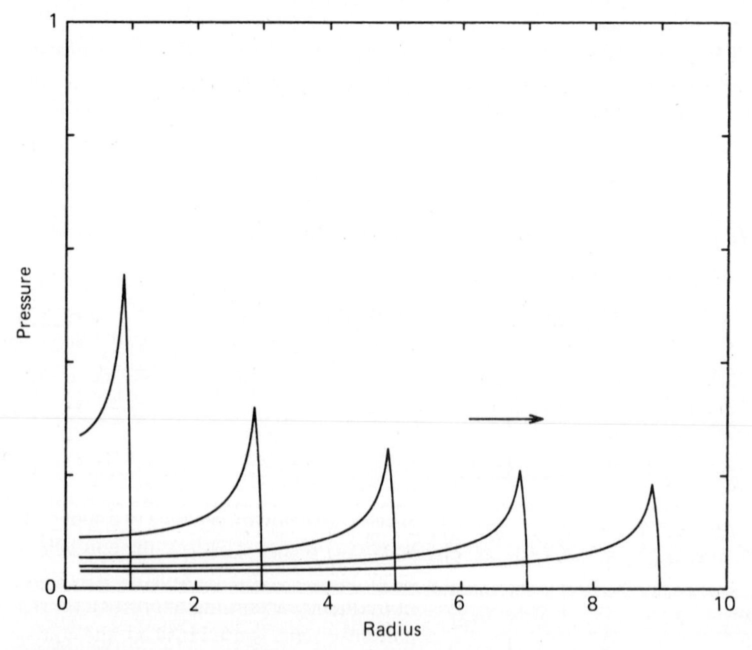

Figure 10.3 An example of the unavoidable shape change of an outward-traveling cylindrical wave.

We will not examine those solutions in detail. But we may remark that, like the spherical waves, they take on a much simpler form when $kr = \omega r/c$ becomes large. In fact, they asymptotically approach $(1/\sqrt{r})\exp[j(\omega t - kr)]$, in accordance with (10.24). In this limit the specific impendance z_a again approaches z_c, as it must since very large cylindrical wave fronts will be practically indistinguishable from plane waves.

10.4 ANISOTROPIC RADIATION

In looking for the simplest possible waves, we have limited the discussion so far to the isotropic case. That means the waves travel outward equally in all directions from $r = 0$, which is reasonable if their source is spherically symmetric. But certainly in many practical cases sound is radiated with unequal strength in different directions from an unsymmetrical source. Let us attempt to write some equations describing such waves.

First consider the possibility that sound could move outward from the origin in any given direction in accord with Eq. 10.6, yet with a different amplitude in different directions. That is, take as a trial solution

$$p(r, \theta, \phi, t) = [A(\theta, \phi)/r] \exp[j(\omega t - kr)]. \qquad (10.25)$$

When this is substituted into (10.1), it leads to

$$(-\omega^2/c^2)(A/r) = (jk - jk - k^2 r)(A/r^2)$$
$$+ [\cos\theta\, \partial A/\partial\theta + \sin\theta\, \partial^2 A/\partial\theta^2 + \partial^2 A/\partial\phi^2]/r^3 \sin\theta. \qquad (10.26)$$

This will be true only if some special form of $A(\theta, \phi)$ can be found that will make the last bracketed expression zero. But rather than search for such an exact solution, let us notice the possibility of a very useful approximate solution. If $\omega = ck$, all terms on the first line of (10.26) cancel each other out regardless of the form of $A(\theta, \phi)$. And those terms are of order r^{-1} and r^{-2}, whereas each term on the second line is proportional to r^{-3}. Thus in the limit where r is sufficiently large, (10.26) is nearly satisfied in the sense that the terms that do balance each other are much larger than the ones that are left over.

Any reasonably smooth $A(\theta, \phi)$, then, provides an expression that asymptotically matches an actual solution of the wave equation as kr goes to infinity. Physically, this says that once the waves are very far out from the source they become nearly plane, and what is happening at one angle can no longer be appreciably affected by what is happening at some other angle because that is many wavelengths away. Mathematically, it is equivalent to saying that the Fraunhofer limit exists in a meaningful way again here, as it does in the diffraction and interference problems described in Chap. 4. This argument will justify our using an $A(\theta, \phi)$ in the following chapter as an informative way of summarizing the radiation patterns created by different sources of sound. The rate of energy flow for these waves is represented, to the same accuracy that (10.25) is true, simply by

$$I_r(\theta, \phi) = A^2(\theta, \phi)/2\rho_0 c r^2. \qquad (10.27)$$

We still have not shown an exact solution $p(r, \theta, \phi, t)$ that is valid everywhere. There does exist a family of such solutions, called **multipole** radiation fields. The first and simplest is the spherically symmetric solution (10.6) already studied above. It is called **monopole** radiation, because it is given off by a single point source of oscillating pressure at $r = 0$.

The next simplest case (and the only other one we consider here) is that of **dipole** radiation. The name suggests a two-sided nature, perhaps like that of a bare loudspeaker cone that is always pulling away from air on one side just as much as it pushes on the other. We examine more closely in the following chapter how to describe the kind of source that might produce it. Suppose for now that previous acquaintance with static electric dipoles suggests that it could be of interest to try a pressure that is independent of ϕ (azimuthally symmetric about the z axis) but proportional to $\cos\theta$. Our first inclination would be simply to insert that factor into (10.6), but the resulting expression fails to satisfy the wave equation. Suppose then we try

$$p(r, \theta, t) = h(r)(A\cos\theta/r)\exp[j(\omega t - kr)], \qquad (10.28)$$

where $h(r)$ is still to be determined but should approach unity as r becomes very large.

Substitution shows (see Prob. 10.6) that (10.28) is a solution of (10.1) only if

$$r^2\, d^2h/dr^2 - 2jkr^2\, dh/dr - 2h = 0, \qquad (10.29)$$

which is satisfied by

$$h(r) = 1 + 1/jkr. \qquad (10.30)$$

The associated velocity comes again from (9.24):

$$\mathbf{v} = j\omega\xi = j\,\nabla p/\omega\rho_0. \qquad (10.31)$$

This is a vector, with components

$$v_r = (A\cos\theta/c\rho_0 r)(1 + 2/jkr - 2/k^2r^2)\exp[j(\omega t - kr)] \qquad (10.32)$$

and

$$v_\theta = (A\sin\theta/c\rho_0 r)(1/jkr - 1/k^2r^2)\exp[j(\omega t - kr)]. \qquad (10.33)$$

The out-of-phase parts of p and v represent a complicated inductive field, involving nonradial flow and energy densities oscillating as in a standing wave. Only the in-phase parts represent traveling waves; in spite of the lengthy expressions for p and v, the average intensity is again simply

$$I_r(\theta) = \mathrm{Re}(p^*v)/2 = A^2\cos^2\theta/2\rho_0 cr^2 \qquad (10.34)$$

for all values of r. (Compare Eq. 10.12.) This time-averaged energy flow is only in the radial direction.

The nature of the dipole field is perhaps best represented by a polar plot of the intensity, measured in units of $I(0)$, its value on the axis of symmetry:

$$I(\theta)/I(0) = \cos^2\theta. \qquad (10.35)$$

This function is plotted on a linear scale in Fig. 10.4*a*, and again on a logarithmic scale in Fig. 10.4*b*. The points at which $I(\theta)/I(0) = 1/2$ are marked on both. The

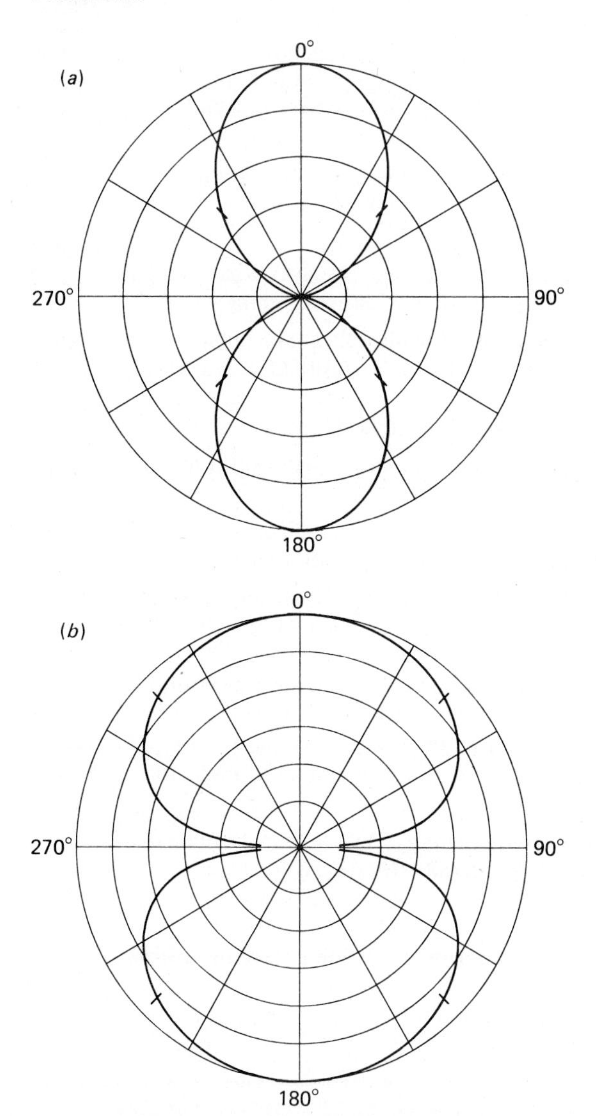

Figure 10.4 Intensity $I(\theta) = I(0)\cos^2\theta$ for a dipole wave, plotted on (*a*) a linear scale and (*b*) a logarithmic scale, with 5-dB intervals.

dipole radiation field may be described as concentrated in two conical beams, each with a half-power (or 3 dB down) beam diameter of 90°.

PROBLEMS

10.1. **(a)** Consider an outgoing spherical sine wave with frequency 200 Hz and pressure amplitude 5 Pa at $r = 2$ m. What is its displacement amplitude, its velocity amplitude, and its average energy density and average intensity at this point?

(b) What value will all these quantities have at $r = 0.2$ m, assuming the source is small enough for the wave to exist there?

10.2. What is the specific acoustic impedance for the spherical wave in the preceding problem at both positions? Express your answers both in terms of resistance and reactance, and in terms of magnitude and phase. What is the limiting behavior of z_a as r becomes very large? Very small? Large or small compared with what?

10.3. Show that the specific impedance z_a for a spherical wave traces out a half circle in the complex plane as kr goes from 0 to ∞, with diameter z_c.

10.4. Show how to calculate the characteristic impedance of air at 1 atm pressure and 0°C temperature. (*Hint:* Remember the perfect gas law.) What about air at 2 km altitude where the pressure is about 0.78 atm, again assuming $T = 0°\text{C}$?

10.5. How far away must you be from a small loudspeaker to be in the "radiation field" if the frequency involved is **(a)** 100 Hz, **(b)** 1 kHz, **(c)** 10 kHz? Is a velocity-sensitive microphone placed much closer than 1 m to a typical musical instrument picking up the actual radiated sound or the induction field?

10.6. Substitute the trial function (10.28) into the wave equation 10.1 to obtain (10.29). Then show that (10.30) provides a valid solution.

10.7. If you preferred to characterize the beam width of the dipole radiation by the angle at which it is 10 dB below maximum, instead of 3 dB, what angle would that be?

10.8. **(a)** Consider a dipole sine wave with frequency 200 Hz and pressure amplitude 5 Pa at $r = 2$ m and $\theta = 0°$. What is its velocity amplitude and its average intensity at this point?

(b) For the same wave (same A), what value will all these quantities have at $r =$ 0.2 m, assuming the source is small enough for the wave to exist there? In each case, compare the product of averages $p_{\text{rms}}v_{\text{rms}}$ with the average product $\langle pv \rangle$ and comment on the difference. If you have also done Prob. 10.1, comment on the difference in answers here.

10.9. **(a)** What is the ratio of total powers emitted by a monopole and a dipole source if they both provide the same intensity $I(0)$ on the dipole axis?

(b) What fraction of the total power radiated by a dipole source is concentrated within 45° of its axis, that is, at angles where I is within 3 dB of maximum? What fraction of 4π steradians is contained within these two cones?

Chapter 11

Sources of Radiation

In the preceding chapter we examined mathematical expressions representing several different kinds of sound waves. It is important to relate the properties of these waves more closely to those of the sources that radiate them. Specifically, we would like to connect the intensity of outgoing radiation in all directions with the total power output of the source and with some measure of the source's own motion.

If we were concerned only with plane waves, this would be a relatively trivial matter. Consider a plane boundary located at $x = 0$ and radiating into the half space $x > 0$. When the boundary moves with speed $v(t)$ it must carry the adjoining air along at the same speed. According to Sec. 9.4 the air motion must be accompanied by a pressure $p(0, t) = \rho_0 c v(t)$ at the boundary and $p(x, t) = p(0, t - x/c)$ elsewhere. The wave retains exactly the same amplitude and intensity, however far it travels. No matter what the waveform or frequency, the driving boundary sees the characteristic impedance $z_c = \rho_0 c$. The power it provides per unit area is simply the intensity being carried away by the wave, $I = p(t)v(t) = z_c v^2(t) = p^2(0, t)/z_c$.

Let us see now to what extent we can say similar things about the motion of other types of sound-wave sources, the effective impedance into which they radiate, and the power carried away by the waves.

11.1 MONOPOLE SOURCES

The next simplest case is the spherically symmetric outgoing sine wave described in Sec. 10.1. We found there that a pressure

$$p(r, t) = (A/r) \exp[j(\omega t - kr)] \tag{11.1}$$

is accompanied by a fluid velocity

$$v(r, t) = (A/z_c r)(1 + 1/jkr) \exp[j(\omega t - kr)]. \tag{11.2}$$

If at any specific radius $r = b$ we had a radially moving boundary that was responsible for generating the wave, we could say that (11.1) and (11.2) apply only for $r > b$ and avoid any embarrassment over the singularity at $r = 0$. This must make no difference to the behavior of the wave farther out, as long as the boundary is capable of exerting the appropriate pressure to go along with its velocity. That is, the ratio $p(b, t)/v(b, t)$ must be the same for both the source and the wave where they join.

You might think of the source as a balloon that is rhythmically expanding and contracting. Any given spherical wave might be generated either by a source with large radius b and small velocity amplitude or by a smaller source moving more vigorously (Fig. 11.1). It is interesting to describe the source in terms of the total amount of air it displaces. The expansion of a sphere by an amount db displaces a volume $dV = 4\pi b^2\, db$; so the total **flow rate** U (volume of fluid per unit time, in cubic meters per second) is

$$U(b, t) = dV/dt = 4\pi b^2\, db/dt = 4\pi b^2 v(b, t)$$

$$= (4\pi bA/z_c)(1 + 1/jkb) \exp[j(\omega t - kb)]. \tag{11.3}$$

Since pressure multiplied by volume change gives work done by the source on the surrounding fluid, the average product of $p(b, t)$ and $U(b, t)$ will be the rate at which total power is given off by the source:

$$P = \langle pU \rangle = (1/2)\,\mathrm{Re}(p^* U) = 2\pi A^2/z_c. \tag{11.4}$$

This agrees with what we would get by integrating the intensity

$$I = \langle pv \rangle = A^2/2\rho_0 c r^2 \tag{11.5}$$

over a distant spherical surface with area $4\pi r^2$.

We would also like to consider the impedance presented to the source by the surrounding fluid. If we examine the specific acoustic impedance $z_a = p/v$, Eq. 10.17 shows that it simply approaches zero along with r, so that z_a has no finite value in the limit of a point source. The total volume flow U is a better measure than v of the motion produced by the pressure. So we define the **acoustic impedance** seen by a source of radius b as

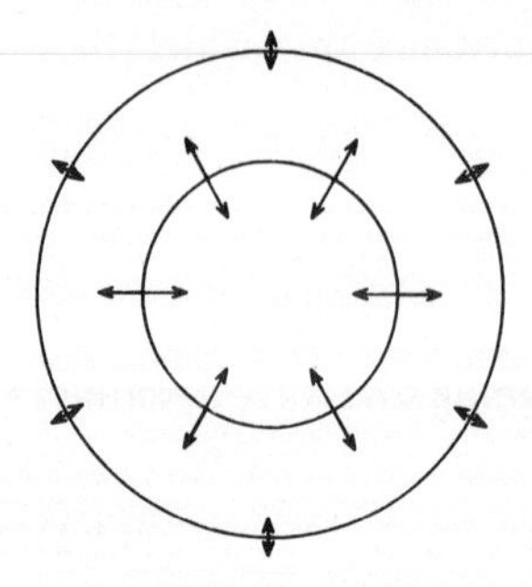

Figure 11.1 Boundary motions of two spherical sources, either of which could have launched the same outgoing wave.

$$\boxed{Z_a = p(b)/U(b)}$$

$$= p(b)/4\pi b^2 v(b) = z_a(b)/4\pi b^2. \tag{11.6}$$

Using (10.17), we can write this explicitly as

$$Z_a = (z_c k^2/4\pi)(1 + j/kb)/(1 + k^2 b^2). \tag{11.7}$$

In the limit as b approaches zero, Z_a has a very large positive imaginary component. This means that the load on the source is primarily a masslike reactance that absorbs no net power. The relatively small real part of the impedance represents the loss of energy into radiation, and it approaches the finite limit $z_c k^2/4\pi$. This is important because we can say unbounded space presents a series **radiation resistance**

$$\boxed{R_{a,\mathrm{rad}} = z_c k^2/4\pi = \pi f^2 \rho_0/c} \tag{11.8}$$

to any sufficiently small source, regardless of its exact size. For this to be true, the largest dimension of the source should satisfy $kb \ll 1$ or $b \ll 2\pi/\lambda$.

Another quantity that approaches a finite limit as b goes to zero (for fixed wave amplitude A) is the volume flow rate. According to (11.3),

$$Q(t) = U(b = 0, t) = (4\pi A/jkz_c)\exp(j\omega t). \tag{11.9}$$

The symbol Q is commonly used because the German word for source is *Quelle,* but you may also think of it as standing for "quantity of flow." One reason this is a good way to characterize a source is that even a source lacking spherical symmetry may still have a well-defined value of Q. Without presenting a formal proof, we may argue that any such source will produce radiation exactly the same as would an ideal monopole source with the same value of Q, as long as all source dimensions are much smaller than $\lambda/2\pi$ so that $kb \ll 1$. The physical reason for this is that over a short distance in space, half a cycle of oscillation allows plenty of time for mutual adjustments to be communicated among different bits of fluid. Thus the wave can smooth itself out in the induction zone and be practically spherically symmetrical by the time it is radiated away, regardless of the original source shape.

In terms of the source amplitude Q_0, the constant A above may be replaced for a sinusoidal source by

$$A = jkz_c Q_0/4\pi = jkz_c\sqrt{2}Q_{\mathrm{rms}}/4\pi. \tag{11.10}$$

Then the intensity (11.5) and total power (11.4) radiated by any monopole source into all space may be written as

$$\boxed{\begin{aligned} I(r) &= k^2 z_c Q_{\mathrm{rms}}^2/16\pi^2 r^2 \\ &= f^2 \rho_0 Q_{\mathrm{rms}}^2/4cr^2 \end{aligned}} \tag{11.11}$$

and

$$P = k^2 z_c Q_{\text{rms}}^2 / 4\pi$$
$$= \pi f^2 \rho_0 Q_{\text{rms}}^2 / c. \quad (11.12)$$

Note how for a given flow amplitude Q, a small source becomes less efficient for low frequencies.

As an example consider a small source in air with $Q_{\text{rms}} = 0.01\ \text{m}^3/\text{s}$ and $f = 500$ Hz. Then the total power is

$$P = (3.14)(500/\text{s})^2(1.21\ \text{kg/m}^3)(0.01\ \text{m}^3/\text{s})^2/(340\ \text{m/s})$$
$$= 0.28\ \text{W},$$

and the intensity 5 m away is

$$I = (0.28\ \text{W})/(12.57)(25\ \text{m}^2) = 0.9\ \text{mW/m}^2,$$

for which SPL = 89.5 dB. To produce as much power at 100 Hz would require five times as great a value of Q.

11.2 BAFFLES AND REFLECTORS

Suppose a small source is set into a wall so that it can only radiate into the half space on one side, as shown in Fig. 11.2*a*. A hard flat surface separating the space in front of a source from that behind it is commonly called a **baffle.** Here we are considering the case of an infinite baffle.

The spherically symmetric wave $p = (A/r) \exp[j(\omega t - kr)]$ is still a valid solution in the region $0 < \theta < \pi/2$ (with $\theta = 0$ chosen along the perpendicular to the baffle). It should be the relevant solution in any case where the source dimensions are much smaller than $\lambda/2\pi$. Nearly all the preceding discussion of monopole sources applies again, except that when the wave reaches radius r it covers only a hemisphere with area $2\pi r^2$ instead of $4\pi r^2$. Then

$$A = jkz_c Q_0 / 2\pi \quad (11.13)$$

instead of $jkz_c Q_0/4\pi$, and a monopole source radiating into a half space will have

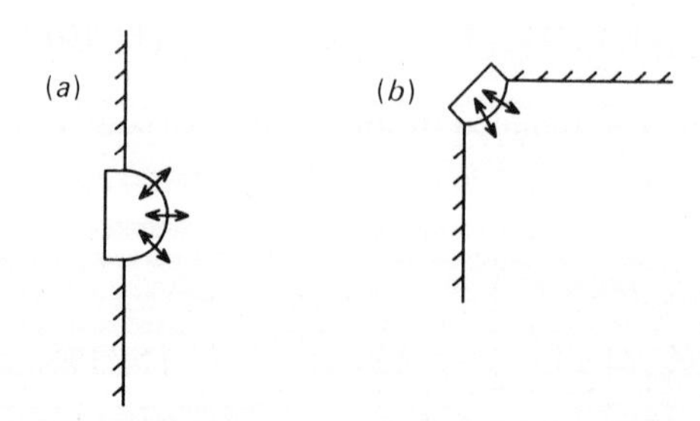

Figure 11.2 A simple source mounted (*a*) in a wall and radiating on one side only, and (*b*) in a corner.

$$I(r) = k^2 z_c Q_{\text{rms}}^2 / 4\pi^2 r^2$$
$$= f^2 \rho_0 Q_{\text{rms}}^2 / c r^2 \tag{11.14}$$

and

$$P = k^2 z_c Q_{\text{rms}}^2 / 2\pi$$
$$= 2\pi f^2 \rho_0 Q_{\text{rms}}^2 / c. \tag{11.15}$$

Since the same pressure is associated with only half as much volume flow here as in Sec. 11.1, the radiation resistance

$$R_{a,\text{rad}} = \text{Re}(Z_a) = k^2 z_c / 2\pi = 2\pi f^2 \rho_0 / c \tag{11.16}$$

of the half space is twice that of totally unbounded space. This provides a way of understanding why a source with constant Q will radiate twice as much power $P = Q_{\text{rms}}^2 R_{\text{rad}}$ when baffled. Then since that power need only fill half as much space, the intensity for a given distance is quadrupled. Thus addition of the baffle is expected to raise the sound level everywhere by 6 dB. Remember that none of these simple statements will be justified if the wavelength is not much larger than the source size.

It is also interesting to consider a monopole source located in a corner where two walls (or three) meet, as in Fig. 11.2*b*. You may provide details (by doing Prob. 11.3) to back up claims that a source of given Q will radiate 4 (or 8) times as much power and that the resulting sound levels will be 12 (or 18) dB higher. This is one of the reasons why home hi-fi consumers are routinely advised to place their speakers in the room corners, or at the very least close against a wall. If we are thinking of a speaker assembly with dimensions like 0.3 m, this advice is mainly relevant to sounds with wavelength larger than $2\pi(0.3\text{ m}) = 1.9$ m, or frequency below 180 Hz.

What if you place a monopole source some distance B out in front of an infinite wall, as in Fig. 11.3? The spherical wave previously considered cannot by itself be the proper solution to this problem, for the wall will not allow the air to move with the velocity described in Eq. 11.2. From our experience with mirrors, however, we might guess that a sufficiently hard and smooth wall will act as a perfect reflector. Then two identical spherical waves should be present,

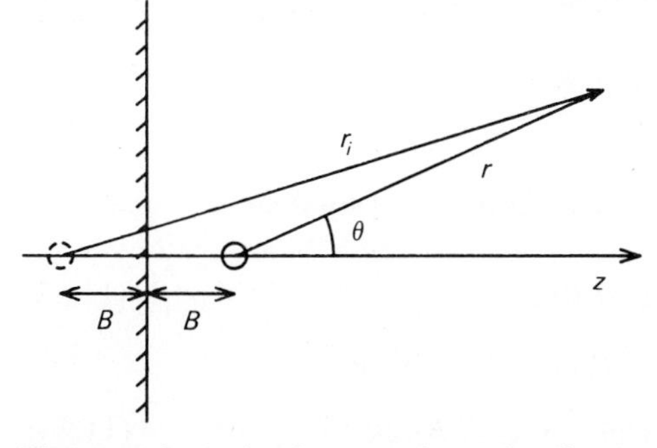

Figure 11.3 A simple source located a distance B in front of an infinite baffle, its image source an equal distance behind, and definitions of r, θ, and r_i.

one coming directly from the actual source and one from a fictitious "image source" located an equal distance behind the wall. At the boundary, any velocity component of the direct wave pointed into the wall will be exactly matched by an outward velocity component of the reflected wave. So the two waves together require no motion across the wall and provide a solution compatible with its presence.

There are two ways of computing the total power radiated by the source, and both give interesting insights. First and most straightforward is to find the intensity of radiation far from the source and add this over all directions to get the total power. This is essentially a matter of studying the interference pattern created by a pair of identical sources. Let r be measured from the actual source to the point where the sound is received, and let the polar axis of a spherical coordinate system pass through both sources, as shown in Fig. 11.3. Then the distance from the image source is

$$r_i = \sqrt{r^2 + (2B)^2 + 2(2B)r \cos \theta}. \tag{11.17}$$

The calculation can be kept simple by considering only the Fraunhofer limit, in which $r \gg 2B$. Then the square root is accurately approximated by

$$r_i \simeq r + 2B \cos \theta. \tag{11.18}$$

The signal received is the total of the two waves,

$$p = (A/r)e^{j(\omega t - kr)} + (A/r_i)e^{j(\omega t - kr_i)}. \tag{11.19}$$

When $r \gg 2B$, it is adequate as far as the amplitudes are concerned to approximate r_i by r. But in the phase factor, we must retain the difference because it is not necessarily small compared with λ. Thus

$$p \simeq (A/r)e^{j(\omega t - kr)}(1 + e^{-jk2B\cos\theta}), \tag{11.20}$$

for which

$$p_{\text{rms}}(r, \theta) = \sqrt{p^*p/2} = (A/r)\sqrt{1 + \cos (2kB \cos \theta)}. \tag{11.21}$$

Then the intensity is p_{rms}^2/z_c, or

$$I(r, \theta) = (A^2/z_c r^2)[1 + \cos (2kB \cos \theta)]. \tag{11.22}$$

The total power radiated into the entire half space is

$$\begin{aligned} P &= \int_0^{\pi/2} I(r, \theta) 2\pi r^2 \sin \theta \, d\theta \\ &= (2\pi A^2/z_c) \int_0^{\pi/2} [1 + \cos (2kB \cos \theta)] \sin \theta \, d\theta \\ &= (2\pi A^2/z_c)[1 + (\sin 2kB)/2kB]. \end{aligned} \tag{11.23}$$

In terms of the source strength $Q_{\text{rms}} = 2\sqrt{2}\pi A/jkz_c$, this is

$$P = (k^2 z_c Q_{\text{rms}}^2/4\pi)[1 + (\sin 2kB)/2kB]. \tag{11.24}$$

On comparing this with (11.12), we see that the presence of the wall multiplies the power that otherwise would have been radiated by the last factor in brackets. This is illustrated in Fig. 11.4. Notice that if B goes to zero we recover the previous result that the power output is doubled when a source is put flush with an infinite baffle, and the radiation is isotropic. For very large B the wall has no effect on the total power, as we would expect. In the latter case the intensity fluctuates on a very small angular scale between zero and $2A^2/z_c r^2$ but is just $A^2/z_c r^2$ when averaged over any substantial range of θ. This is twice the intensity without the wall, just as if the real and image sources were independent, incoherent sources. It is the doubled intensity over only half space that gives the same total power as without the wall.

The other way of understanding what the wall does is to consider the impedance seen by the source. In order to inject the volume flow Q, the source must work against the total pressure in the immediately adjoining fluid. This pressure now includes not only that associated with the original radial flow but the pressure of the reflected wave as well. Letting the source have for the moment a finite radius b, the total pressure is

$$p(b, t) = (A/b)\exp[j(\omega t - kb)] + (A/2B)\exp[j(\omega t - k2B)]. \tag{11.25}$$

Here we are assuming $b \ll B$ and $b \ll \lambda$, so that the pressure from the image source is practically constant on all sides of the real source. The acoustic impedance is this total pressure divided by the volume flow given in (11.3):

$$Z_a = p/U = (z_c/4\pi b^2)[1 + (b/2B)e^{jk(b-2B)}]/(1 + 1/jkb). \tag{11.26}$$

As you may show by doing Prob. 11.5, the real part of this impedance has a finite limit as b approaches zero:

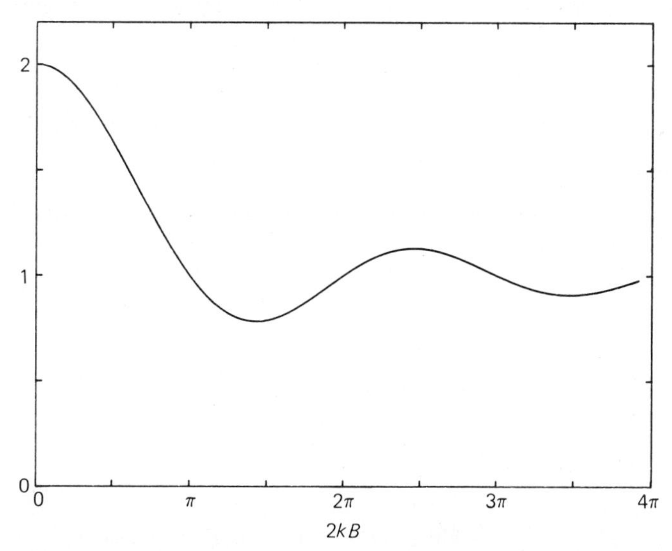

Figure 11.4 Graph of the function $1 + (\sin 2kB)/2kB$. This is the ratio of power output in the situation of Fig. 11.3 to that with the baffle removed. The same graph also gives the ratio of radiation impedances.

$$R_{a,\mathrm{rad}} = (k^2 z_c/4\pi)[1 + \sin(2kB)/2kB]. \tag{11.27}$$

Thus the characteristic radiation resistance of free space for the point source is multiplied by precisely the same factor as the power in (11.24) above. Physically this says that for large kB the image source is so far away that it has no effect on the operation of the real source. For very small kB the image source contributes an in-phase component that doubles the pressure against which the flow must work. For intermediate values of kB, the reflected wave pressure may make it either harder or easier to move the air, according to whether it is in or out of phase with the volume flow.

*11.3 DIPOLE SOURCES

The simplest sources are those which create a nonzero total outward flow of fluid. But in other cases air may simply be moved back and forth, with no net volume change for the source. An example is a bare speaker cone, where the volume of air pushed out by one side of the cone is exactly compensated by new volume into which air may be pulled on the other side (Fig. 11.5). This suggests that the airflow here may be just about the same as if we had two simple sources of equal strength $+Q$ and $-Q$ placed a short distance apart, operating at the same frequency but always 180° out of phase with each other. Both this description and the flow lines pictured in Fig. 11.5 should remind you of how electric field lines are arranged in the vicinity of a pair of opposite charges, and suggest that the mathematical description of this **acoustic dipole** will be very much like that of an electric dipole. Since the dipole gives opportunity for the air to simply swirl around the sides instead of pushing always outward, we might anticipate that dipole sources are relatively inefficient. Let us provide some quantitative backing for that suspicion.

We have already written some expressions for dipole fields in the preceding chapter. In repeating them, let us suppress the factor $\exp[j(\omega t - kr)]$ and just write the phasor amplitudes for the pressure

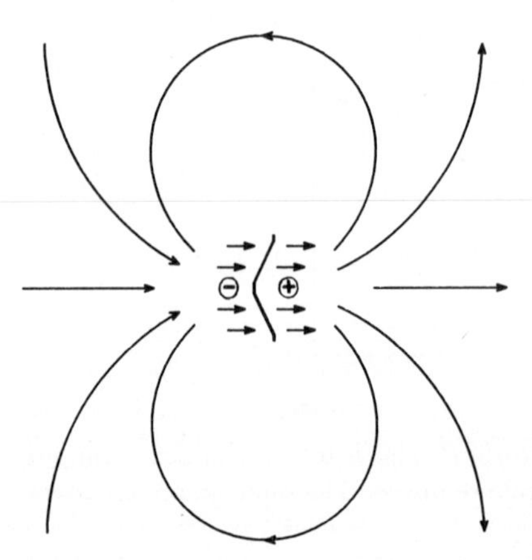

Figure 11.5 A bare speaker cone acting as a dipole source. The distant flow is the same as if one positive and one negative monopole are placed close together.

$$p(r, \theta) = (1 + 1/jkr)(A \cos \theta/r) \tag{11.28}$$

and the radial velocity

$$v_r(r, \theta) = (1 + 2/jkr - 2/k^2r^2)(A \cos \theta/\rho_0 cr). \tag{11.29}$$

The intensity of outward energy flow

$$I(\theta) = A^2 \cos^2 \theta/2\rho_0 cr^2 \tag{11.30}$$

follows directly from p and v and has already been given in (10.34). The total radiated power was to be calculated in Prob. 10.9 simply by integrating the intensity over the surface of a sphere:

$$P = \int_0^{\pi} I(\theta) 2\pi r^2 \sin \theta \, d\theta = 2\pi A^2/3\rho_0 c. \tag{11.31}$$

All these expressions are in terms of the constant A, and we must devise a way of relating that to some appropriate measure of dipole source strength. We cannot take the total outward flow because that is identically zero, but we could try calculating the volume flow only over the hemisphere where it is positive. For hemisphere radius b, this is

$$\begin{aligned} Q_+ &= \int_0^{\pi/2} v_r(\theta) 2\pi b^2 \sin \theta \, d\theta \\ &= (\pi b A/\rho_0 c)(1 + 2/jkb - 2/k^2b^2). \end{aligned} \tag{11.32}$$

But in the limit as b approaches zero this diverges, and in fact suggests that Q multiplied by b might be the finite quantity we are looking for. This reminds us of the electrical case, where the monopole moment of a distribution of charges is $\int dq$ but the proper measure of dipole moment along the z axis is $\int z \, dq$. The analogy here would be to multiply each bit of outward flow dQ by $z = b \cos \theta$ before integrating. We may now also extend the integration to $\theta = \pi$ since the extra factor results in positive contributions from both halves of the dipole. Finally, a factor 3/2 is arbitrarily included in order to agree with a differently motivated definition that is widely used. The result is the **dipole moment** of the flow distribution:

$$\begin{aligned} D &= (3/2) \int_0^{\pi} v_r(\theta) 2\pi b^3 \cos \theta \sin \theta \, d\theta \\ &= (2\pi b^2 A/\rho_0 c)(1 + 2/jkb - 2/k^2b^2) = 2bQ_+. \end{aligned} \tag{11.33}$$

This now is a quantity that has the finite limit

$$D_0 = -4\pi A/\rho_0 ck^2 \quad \text{or} \quad D_{\text{rms}} = -2\sqrt{2}\pi A/\rho_0 ck^2 \tag{11.34}$$

as b goes to zero, so is a measure of dipole source strength not dependent on exact details of the dipole structure, as long as it is small compared with $\lambda/2\pi$.

Now we can eliminate A in favor of D in the expressions above to obtain

$$p(r, \theta) = -(\rho_0 ck^2 D_0/4\pi)(1 + 1/jkr)(\cos \theta/r), \tag{11.35}$$

$$I(\theta) = \rho_0 c k^4 D_{\mathrm{rms}}^2 \cos^2\theta / 16\pi^2 r^2 \tag{11.36}$$

and

$$\boxed{\begin{aligned} P &= \rho_0 c k^4 D_{\mathrm{rms}}^2 / 12\pi \\ &= 4\pi^3 f^4 \rho_0 D_{\mathrm{rms}}^2 / 3c^3. \end{aligned}} \tag{11.37}$$

Comparison with (11.12) suggests an important conclusion about dipole sources: For given source dimensions and amplitude of displacement, that is, given monopole strength Q_{rms} or dipole strength D_{rms}, the ability of the dipole source to radiate power depends more strongly on frequency than does that of the monopole. Specifically, the two additional powers of f say that the dipole output will fall off toward low frequencies 6 dB/octave faster than if one half of the dipole could radiate alone as a monopole. This is why bare speaker cones are very inefficient radiators for $kb < 1$, and mounting them in cabinets or baffles to convert them to monopole sources is crucial to the success of any bass speaker, as we will study in Chap. 16.

Suppose, for instance, that the speaker cone of Fig. 11.5 has radius b = 0.1 m and radiates a signal with f = 100 Hz. Then

$$kb = 2\pi fb/c = (6.28)(100/\mathrm{s})(0.1\ \mathrm{m})/(340\ \mathrm{m/s}) = 0.185$$

and this source may be considered small compared with the wavelength radiated. Suppose the cone motion has maximum amplitude ξ_0 = 1 mm, so that the maximum rate at which it moves air to one side is

$$\begin{aligned} Q_+ &= \pi b^2 \omega \xi_0 \\ &= (3.14)(0.01\ \mathrm{m}^2)(628/\mathrm{s})(0.001\ \mathrm{m}) = 0.02\ \mathrm{m}^3/\mathrm{s}. \end{aligned}$$

According to (11.33),

$$D_0 = 2bQ_+ = (0.2\ \mathrm{m})(0.02\ \mathrm{m}^3/\mathrm{s}) = 0.004\ \mathrm{m}^4/\mathrm{s},$$

so the power radiated is given by (11.37) as

$$\begin{aligned} P &\simeq \frac{(4)(3.14)^3(100/\mathrm{s})^4(415\ \mathrm{kg/m\text{-}s}^2)(0.004\ \mathrm{m}^4/\mathrm{s})^2}{(3)(340\ \mathrm{m/s})^4(2)} \\ &= 1.03\ \mathrm{mW}. \end{aligned}$$

If the cone were mounted so that Q_+ could radiate alone as a monopole source with no cancellation from Q_-, its power output according to (11.12) would be

$$\begin{aligned} P &\simeq \frac{(3.14)(100/\mathrm{s})^2(1.21\ \mathrm{kg/m}^3)(0.0002\ \mathrm{m}^6/\mathrm{s}^2)}{(340\ \mathrm{m/s})} \\ &= 22\ \mathrm{mW}. \end{aligned}$$

This is

$$10 \log (22/1.03) = 13\ \mathrm{dB}$$

greater than the dipole. You may show by doing Prob. 11.17 that the comparison when $kb \ll 1$ is always given by

$$P(\text{dipole})/P(\text{monopole}) = 4(kb)^2/3.$$

If we attempt to write an acoustic impedance for the dipole to compare with (11.8) for the monopole, there are two difficulties. One is that the imaginary part of p/U diverges as b goes to zero, indicating a large inductive load on the source; but we might excuse that as we did for the monopole and go on. The other difficulty is that the real part goes to zero along with b, indicating an inability to account for any radiated power. That forces us to realize that the proper analog of impedance for the point dipole is not the ratio of pressure to volume flow. Rather, the proper measure of motion is the dipole moment defined above, and that is associated not merely with a pressure but with a pressure difference. If we take $[p(r = b, \theta = 0) - p(r = b, \theta = \pi)]/2b$ as characteristic of the pressure gradient created by the dipole, then (as you may show in Prob. 11.8) there is a finite limit as b goes to zero for the ratio

$$\text{Re}[(\Delta p/2b)/D_0] = \rho_0 ck^4/8\pi = 2\pi^3\rho_0 f^4/c^3. \tag{11.38}$$

Comparison with (11.16) shows again the extra factor of f^2 working against the dipole source at low frequencies.

11.4 RADIATION FROM A BAFFLED PISTON

We have learned thus far that quite simple and general expressions can be written for the radiation given off by sources whose characteristic dimension b satisfies $kb \ll 1$. But when kb is greater than 1, the monopole or dipole descriptions can apply only if the source possesses a high degree of symmetry. Unsymmetrical sources will be able to produce a great variety of radiation patterns depending on their exact shape for frequencies well above $c/2\pi b$.

The most important example is a piston-type source, in which a flat circular solid disk moves along its axis of symmetry. The disk by itself would act as a dipole source at low frequencies and thus would be a very inefficient radiator. So in most practical applications the piston is set in a baffle. In the case of an infinite plane baffle, analytic formulas can be obtained to describe the distant radiation field at all frequencies; so this is used as a model for other, more practical baffles. This model is also idealized in that most real air loudspeakers have either a shallow cone or a domed shape rather than being flat disks. The differences in radiation for these other shapes have been studied in the research literature by both analytic and numerical techniques (see, for instance, H. Suzuki and J. Tichy, *JASA*, **69**, 41, 1981).

Before we attempt a complete solution for the radiation from the baffled piston, it is best to state as much as we can about what properties that solution must have in various limiting cases. Suppose the piston radius is b and that it moves sinusoidally with frequency $\omega = ck$, phasor displacement ξ, and velocity v. Then the volume flow rate is

$$Q = \pi b^2 v = \pi b^2 j\omega\xi. \tag{11.39}$$

In the limit where $kb \ll 1$ the piston must act as a simple monopole source (Fig. 11.6); so the discussion in Sec. 11.2 shows that the radiation is isotropic with

$$I = k^2 z_c Q_{\text{rms}}^2 / 4\pi^2 r^2$$

$$= 4\pi^4 b^4 f^4 \rho_0 \xi_{\text{rms}}^2 / cr^2, \tag{11.40}$$

$$P = k^2 z_c Q_{\text{rms}}^2 / 2\pi$$

$$= 8\pi^5 b^4 f^4 \rho_0 \xi_{\text{rms}}^2 / c, \tag{11.41}$$

and

$$\boxed{R_{a,\text{rad}} = k^2 z_c / 2\pi = 2\pi f^2 \rho_0 / c.} \tag{11.42}$$

In view of (11.7), we may also use (j/kb) times R_{rad} as a crude guess at the radiation reactance; that is,

$$\boxed{X_{a,\text{rad}} \sim kz_c / 2\pi b = f\rho_0 / b.} \tag{11.43}$$

But it requires a more careful calculation to make this truly correct, since b does not have the same meaning here as it did in (11.7). We indicate in the following section how it is shown that (11.43) should be multiplied by the factor $16/3\pi = 1.70$.

In the opposite limit where the piston is very large compared with a wavelength (Fig. 11.7), the waves must be nearly plane waves as they are launched; only in a ring within about $\lambda/4$ of the edge will the fluid motion be affected much by that edge. Although the beam of radiation will later spread out somewhat, its total power must already be determined as simply the piston area πb^2

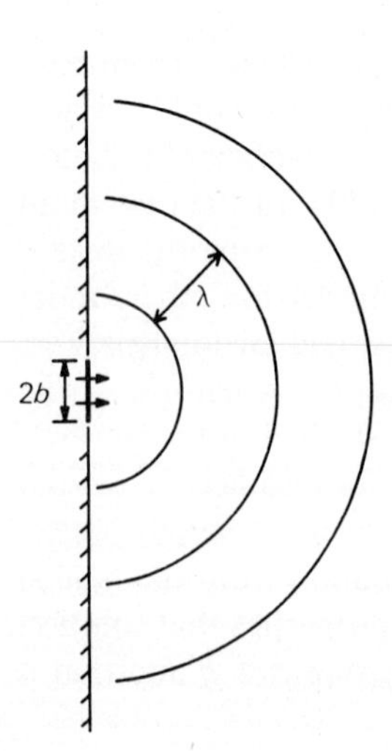

Figure 11.6 A baffled piston source small compared with a wavelength acts as a simple monopole source.

times the intensity $z_c v_{\mathrm{rms}}^2 = z_c \omega^2 \xi_{\mathrm{rms}}^2$ for plane-wave radiation. That is, for $kb \gg 1$,

$$P = 4\pi^3 b^2 f^2 z_c \xi_{\mathrm{rms}}^2. \tag{11.44}$$

Since this total power equals Q_{rms}^2 times the radiation resistance, (11.39) gives

$$\boxed{R_{a,\mathrm{rad}} = z_c/\pi b^2.} \tag{11.45}$$

Diffraction must later cause the narrow beam to spread out, and we can surmise that its angular width will be roughly the same as that of waves diffracted through a slit of width $W = 2b$. Equation 4.22, as discussed in Prob. 4.11, gives this estimate as

$$\boxed{\theta_0 \sim \lambda/2b = c/2bf = \pi/kb} \tag{11.46}$$

for the edge of the beam. Some practical considerations follow immediately from (11.46) when we specify what kind of radiation pattern we need in different applications. For location of underwater objects with sonar, a rather narrow beam is needed, and we must have $2bf > c/\theta_0$. That is, we must either have a large-diameter projector or use a sufficiently high frequency. For instance, to achieve a beam radius of 6° = 0.1 rad with c = 1480 m/s requires $2bf > 15$ km/s; so a 10-kHz source would need diameter 1.5 m. Alternatively, 0.3 m diameter would suffice if the frequency could be raised to 50 kHz, though that has the disadvantage that higher-frequency waves suffer greater absorption if they must travel long distances. Loudspeakers in air are usually intended to radiate into a large angle, say $\theta_0 \sim 1$ rad. But a disk or cone of fixed size can only do this for sufficiently low frequencies; for instance, with diameter $2b$ = 13 cm and c = 340 m/s the upper limit would be $f \sim c/2b \sim 2.6$ kHz. A common

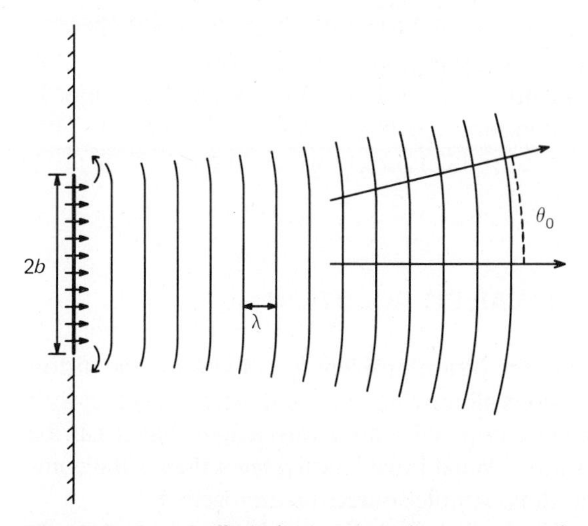

Figure 11.7 A baffled piston source large compared with a wavelength can confine its radiation largely within a beam of half width θ_0.

solution to this problem is to use an electrical filter network called a crossover to send only signals with lower frequencies to this speaker, and those of higher frequency to another smaller speaker. Wide-range single speakers have also been made by segmenting the cone so that it would move more or less rigidly at low frequencies, but only the inner part would respond much at high frequencies.

At distance r, a beam with angular radius θ_0 covers an area of roughly $\pi(r\theta_0)^2$. The intensity is then approximately the total power divided by this area, or

$$I \sim 16\pi^2 b^4 f^4 \rho_0 \xi_{\rm rms}^2 / cr^2, \qquad (11.47)$$

for $\theta < \theta_0$ and zero outside this beam. Again, the exact calculation to follow will give a more accurate result to replace this estimate.

Finally, what about the radiation reactance when $kb \gg 1$? Insofar as the piston effectively radiates pure plane waves, their impedance is entirely real. The only fluid with much chance of providing an out-of-phase load is that in a ring of circumference $2\pi b$ and cross section very roughly $(\lambda/2)(\lambda/4)$, as indicated by the little swirls at the edge of Fig. 11.7. The total mass in this ring is about $\rho_0 \pi b \lambda^2/4$, and its mechanical impedance would be $j\omega$ times this mass. But the acoustic impedance we are concerned with here is defined in terms of pressure rather than force, and volume flow rather than ordinary velocity. Each of those divides the mechanical impedance by a factor of area, so we get the estimate

$$\boxed{X_{a,\rm rad} \sim \omega \rho_0 \pi b \lambda^2 / 4(\pi b^2)^2 = \rho_0 c^2 / 2b^3 f.} \qquad (11.48)$$

We will see below that this is a considerable overestimate of the numerical coefficient, which should be 0.032 instead of 1/2. But it does correctly indicate how X depends on b and f.

The impedance provides a good way of summarizing both how much radiation the piston produces (namely, $R_{a,\rm rad} Q_{\rm rms}^2$) and what kind of load it represents for whatever mechanism drives the piston. We have been able here to give arguments that determine all the asymptotic behavior shown in Fig. 11.8. In particular, we can say confidently that the impedance is small and mostly reactive for $kb \ll 1$, and nearly constant and mostly resistive for $kb \gg 1$. That constant value $z_c/\pi b^2$ provides a convenient unit in which to measure all other impedances, as has been done in Fig. 11.8.

*11.5 BAFFLED PISTON: DETAILED SOLUTION

The key to a complete solution of the piston problem lies in viewing the piston as made up of many very tiny pistons all moving in unison. Any one piece, with area dS, has dimensions so small compared with a wavelength that it can be treated as a simple monopole source. What must be calculated then is the combined sound field created by all those simple sources taken together.

To describe this situation, let the disk lie in the xy plane with its center at the origin, as shown in Fig. 11.9. Since the radiation must have the same axial

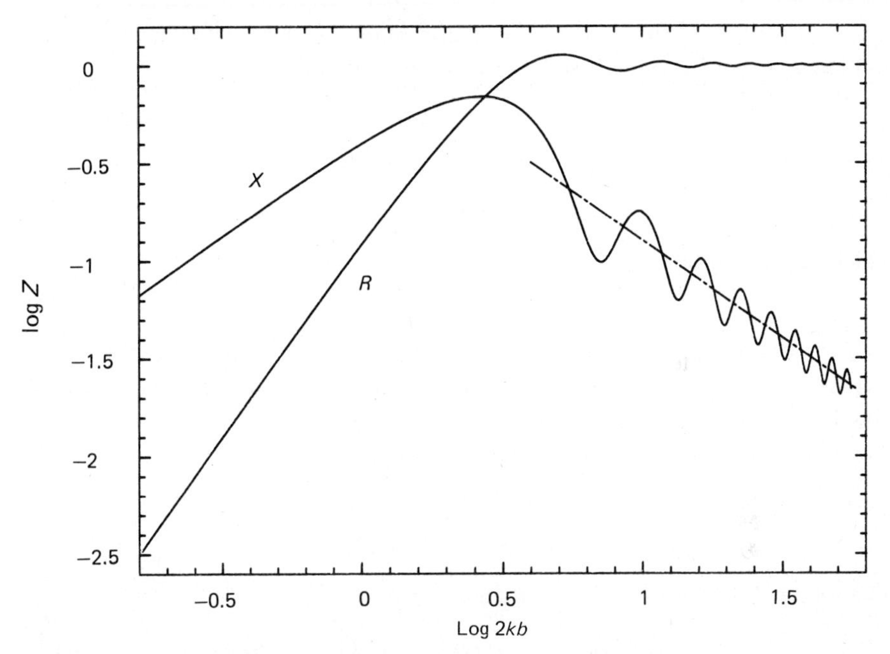

Figure 11.8 Dependence of radiation impedance from a piston of radius b upon the dimensionless quantity $2kb = 4\pi fb/c$. The dimensionless function $Z_1 = R_1 + jX_1$ equally well represents the acoustic impedance Z_a in units of $z_c/\pi b^2$, or its equivalent mechanical impedance Z_m in units of $z_c\pi b^2$. The logarithmic plot clearly shows the asymptotic behavior discussed in Sec. 11.4 for both $kb \ll 1$ and $kb \gg 1$. For kb between 0.5 and 2.5, Table C.1 provides accurate values.

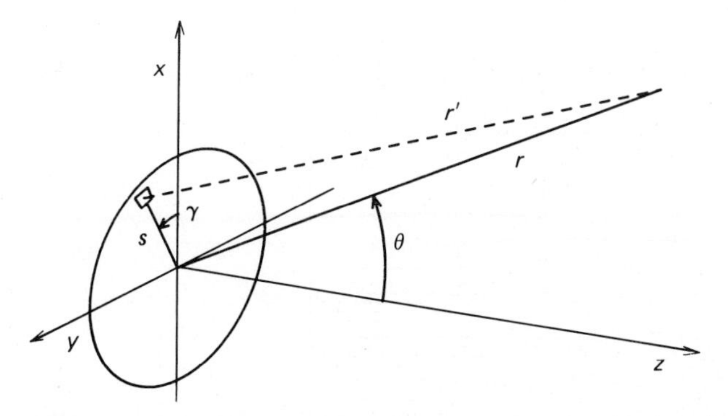

Figure 11.9 Definition of coordinates used to describe location of elementary source (s, γ) and observer (r, θ) of piston radiation.

symmetry as the source itself, we are free to orient our coordinate system so that the point of observation lies in the xz plane. It is described in spherical coordinates by r and θ (both variable) and $\phi = 0$, and so its cartesian components are

$$\mathbf{r} = (r \sin \theta, 0, r \cos \theta). \tag{11.49}$$

A typical portion of the source is located in plane polar coordinates at radius s and angle γ (both variable), and its cartesian components are

$$\mathbf{s} = (s \cos \gamma,\ s \sin \gamma,\ 0). \tag{11.50}$$

What is physically important is the vector from the source to the observer,

$$\mathbf{r} - \mathbf{s} = (r \sin \theta - s \cos \gamma,\ -s \sin \gamma,\ r \cos \theta), \tag{11.51}$$

which has magnitude

$$r' = \sqrt{r^2 + s^2 - 2rs \sin \theta \cos \gamma}. \tag{11.52}$$

That portion of the source lying between s and $s + ds$ and between γ and $\gamma + d\gamma$ has area $dS = s\, d\gamma\, ds$. So for piston velocity v, the associated volume flow is

$$dQ = v\, dS = vs\, ds\, d\gamma. \tag{11.53}$$

According to (11.1) and (11.13), the pressure produced by this flow is

$$dp = (jkz_c/2\pi r')(v\, dS)e^{j(\omega t - kr')}. \tag{11.54}$$

The total pressure is obtained by integrating over all parts of the source:

$$p(r, \theta) = (jkz_c v/2\pi)e^{j\omega t} \int_0^b s\, ds \int_0^{2\pi} d\gamma e^{-jkr'}/r'. \tag{11.55}$$

For general values of r and θ this integral is very difficult to evaluate, for it must describe the rather complicated sound field near the piston (Fig. 11.10). There are three limiting cases where we can make further progress: (1) For $\theta = 0$ the integral can be done in terms of elementary functions (see Prob. 11.12), and the result makes it clear that the picture of simple outgoing radiation with intensity

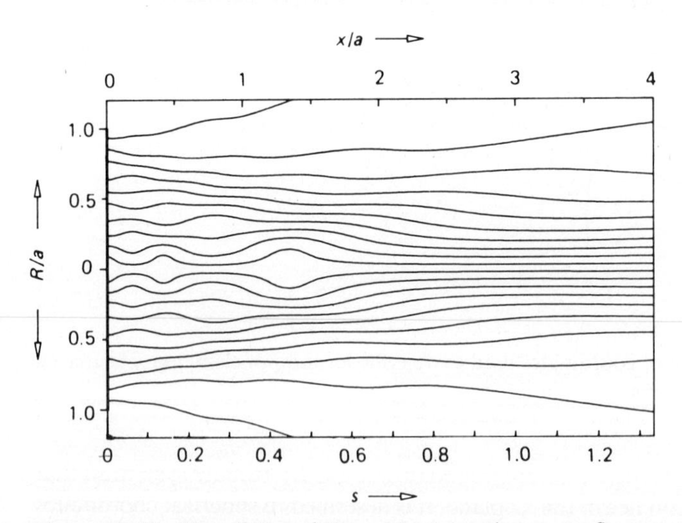

Figure 11.10 Plot of lines followed by acoustic energy flow near a large piston (kb = 18.85 or $b = 3\lambda$), illustrating that only for $r \ll b^2/\lambda$ (= $3b$ in this case) does the radiation field become smooth and simple. (From K. Beissner, *JASA,* **71,** 1406, 1982.)

$A^2(\theta)/2z_cr^2$ from Sec. 10.4 cannot fully apply unless $r \gg b^2/\lambda$. (2) For $\theta = 90°$ and $r < b$, the pressure developed on the surface of the piston itself can be studied, and this is what must be done to determine the source impedance. (3) For sufficiently large r we can use the Fraunhofer approximation to find the radiated intensity $I(\theta)$, and from it the total power P; this is what we will do next.

When $r \gg b$, the binomial approximation enables us to replace (11.52) by

$$r' \simeq r - s \sin\theta \cos\gamma. \tag{11.56}$$

As before, the whole idea of the Fraunhofer approximation is that $r' \simeq r$ is adequate for the amplitude factor, even though (11.56) must be used in the phase factor. Then we may write

$$p(r, \theta) = (jkz_cv/2\pi r)e^{j(\omega t - kr)}\pi b^2 H(\theta), \tag{11.57}$$

with

$$H(\theta) = (1/\pi b^2)\int_0^b s\, ds \int_0^{2\pi} d\gamma e^{jks \sin\theta \cos\gamma}. \tag{11.58}$$

We have chosen to insert the factor πb^2 this way because in the limit $\theta = 0$ this will make $H(0) = 1$. For other values of θ, the integral over γ can be done by replacing the exponential by its power-series representation and integrating term by term. This is the special case with $m = 0$ of Eq. C.12, which gives the Bessel function $2\pi J_0(ks \sin\theta)$. With the aid of Eq. C.10, the second integral can then also be done to give

$$\boxed{H(\theta) = 2J_1(kb \sin\theta)/kb \sin\theta.} \tag{11.59}$$

For sufficiently large r the outgoing waves will be nearly plane, and we will be able to use $I(\theta) = p_{\text{rms}}^2/z_c$. Then, since $H(0) = 1$, (11.57) gives a very simple expression for the on-axis intensity,

$$\boxed{\begin{aligned} I_0 = I(0) &= k^2b^4v_{\text{rms}}^2z_c/4r^2 \\ &= 4\pi^4f^4b^4\xi_{\text{rms}}^2\rho_0/cr^2, \end{aligned}} \tag{11.60}$$

and the intensity at all other angles is conveniently expressed in terms of it as

$$\boxed{I(\theta) = I_0H^2(\theta).} \tag{11.61}$$

The total power is obtained by integrating the intensity over a hemisphere:

$$\begin{aligned} P &= \int_0^{\pi/2} d\theta \int_0^{2\pi} d\phi\, r^2 \sin\theta\, I(r, \theta) \\ &= 2\pi r^2 I_0 \int_0^{\pi/2} H^2(\theta) \sin\theta\, d\theta. \end{aligned} \tag{11.62}$$

It is again possible with an identity from Appendix C to perform this integration and get

$$P = (4\pi r^2 I_0/k^2b^2)[1 - J_1(2kb)/kb]$$
$$= 4\pi^3 b^2 f^2 z_c \xi_{\text{rms}}^2 [1 - J_1(2kb)/kb]. \tag{11.63}$$

In the limit where $kb \gg 1$, the bracketed factor approaches unity, and this reproduces our earlier estimate (11.44). In the opposite limit $kb \ll 1$, $J_1(2kb) \simeq kb - k^3b^3/2$, and this gives $P = 2\pi r^2 I_0$ in agreement with the previous discussion of the low-frequency limit at (11.41).

The angular distribution of radiation is described by $H^2(\theta)$, which is plotted for several cases in Fig. 11.11. It has the expected property of being nearly isotropic for low frequencies but concentrated near the axis for high frequencies. We may conveniently characterize the degree of spreading by asking at what angle $I(\theta)$ first falls to zero. According to (11.59), this will simply be where $kb \sin\theta$ equals the first root of J_1, which is $j_{11} = 3.83$; thus

$$\theta_0 = \arcsin(3.83/kb) = \arcsin(0.61\lambda/b). \tag{11.64}$$

For wavelengths greater than $b/0.61$ (meaning $f < 0.61c/b$) this equation has no solution, indicating that there is no direction with zero intensity in the range $0 < \theta < \pi/2$. When $f \gg 0.61c/b$, (11.64) gives $\theta_0 \simeq 0.61c/bf$, with the factor 0.61 improving on our earlier estimate of 0.50 in Eq. 11.46.

Finally, we want to calculate the impedance seen by the piston. This requires evaluating (11.55) for $\theta = \pi/2$ to obtain the pressure felt by the disk. That pressure is then integrated over the disk to get the total radiation reaction force, which can be divided by the assumed velocity v to get the mechanical impedance. Or the total force can be divided by πb^2 to get the average pressure, and the ratio of that pressure to the assumed volume velocity Q is the acoustic impedance of the outgoing wave. Unfortunately, that calculation is quite lengthy; to see the details, you may consult Sec. 5-4 in Pierce's advanced text. The results have already been shown graphically in Fig. 11.8, to which we now add some final comments.

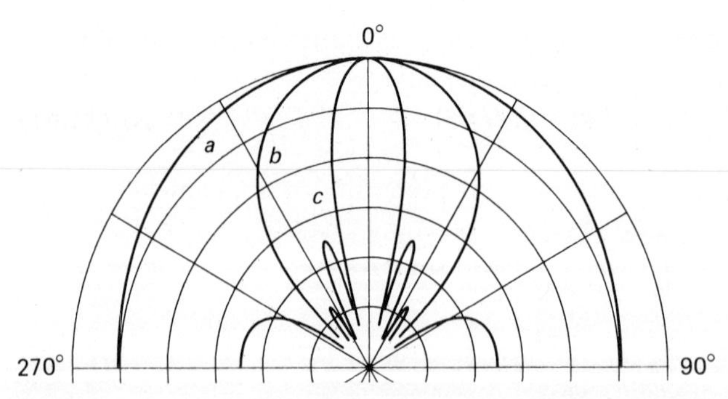

Figure 11.11 Polar plot of the directional radiation pattern $H^2(\theta)$ for a plane piston of radius b, on a logarithmic scale with 5-dB intervals. Cases shown have (*a*) $kb = 2$, (*b*) $kb = 5$, and (*c*) $kb = 15$. The patterns have rotational symmetry about the $\theta = 0$ axis.

The radiation impedance is customarily written in terms of a dimensionless function $Z_1 = R_1 + jX_1$ that depends only on the relative size of the piston and the sound wavelength. The actual impedance is

$$\boxed{Z_{m,\text{rad}} = \pi b^2 z_c Z_1(2kb)} \tag{11.65}$$

or equivalently

$$\boxed{Z_{a,\text{rad}} = (z_c/\pi b^2) Z_1(2kb).} \tag{11.66}$$

The resistive part of Z_1 comes out in terms of the Bessel function of order 1 as

$$R_1(2kb) = 1 - J_1(2kb)/kb, \tag{11.67}$$

so the direct calculation agrees exactly with the argument that the total power given by (11.63) has to equal $R_{\text{rad}}Q_{\text{rms}}^2$. The reactive part can be written in terms of a Struve function (a cousin to the Bessel functions) as

$$X_1(2kb) = H_1(2kb)/kb. \tag{11.68}$$

These are the expressions that are represented in Fig. 11.8. For our purposes it is best simply to accept Fig. 11.8 or Table C.1 as the way to determine these quantities for any value of kb in the intermediate range around unity. For small values of kb, power-series expansions provide the approximate formulas

$$R_1 = (kb)^2/2 - (kb)^4/12 + \cdots \tag{11.69}$$

and

$$X_1 = (8/3\pi)kb - (32/45\pi)(kb)^3 + \cdots, \tag{11.70}$$

which justify (11.42) and the corrected version of (11.43). For large kb, R_1 simply approaches unity, as we have already discussed following (11.63); how much it differs can be found from the next term in an asymptotic expansion,

$$R_1 = 1 - \cos(2kb - 3\pi/4)/\sqrt{\pi k^3 b^3} + \cdots. \tag{11.71}$$

Similarly,

$$X_1 = 2/\pi kb + \sin(2kb - 3\pi/4)/\sqrt{\pi k^3 b^3} + \cdots, \tag{11.72}$$

in which the leading term gives

$$X_{a,\text{rad}} = 2z_c/\pi^2 kb^3 = \rho_0 c^2/\pi^3 b^3 f. \tag{11.73}$$

This shows that (11.48) needed to be multiplied by an additional factor $2/\pi^3 = 0.065$.

As before for the monopole source, an important reason why R_1 and X_1 depend so strongly on frequency is that we assumed Z_1 to consist of the series combination. If we considered a parallel combination of some other R_2 and X_2 instead, we would have to add admittances:

$$1/Z_1 = 1/R_2 + 1/jX_2. \tag{11.74}$$

It turns out that R_2 is nearly constant over the entire range of frequencies. Similarly, if jX_2 is written as $j\omega M_2$ as one might expect for a simple mass, then M_2

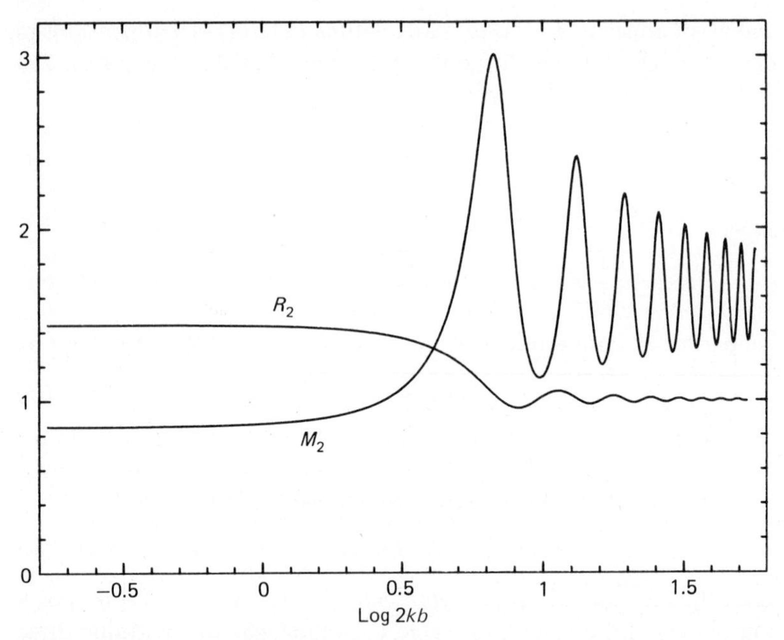

Figure 11.12 Values of $R_2(kb)$ and $M_2(kb)$ when piston radiation impedance function Z_1 is represented as parallel combination of R_2 and $j\omega M_2$. M_2 is given as a multiple of b/c, so that ωM_2 will be dimensionless like R_2.

also does not vary more than about a factor of 2 at any frequency. R_2 and M_2 can more readily be obtained accurately from a graph (Fig. 11.12) than could R_1 and X_1, though the advantage may be lost in having to calculate (11.74). This does suggest, however, that there may be more physical insight in visualizing the piston as having the choice of causing *either* radiative or inductive motion (or both), rather than being constrained to put the same amount of current into each in turn.

PROBLEMS

11.1. The expression (11.2) for velocity in a spherical wave can be trusted only if the wave has small amplitude so that nonlinear effects are not important. This will be true only if $v \ll c$; suppose you set a limit $v_{max} = 0.01c$. For waves of frequency **(a)** 100 Hz and **(b)** 10 kHz in air and source size $b = 0.05$ m, what is the limiting SPL at b? What is the corresponding total radiated power?

11.2. Consider a simple monopole source radiating into free space (perhaps a loudspeaker high above a football stadium rather than in a living room). In order to produce sound of frequency $f = 500$ Hz and level SPL $= 90$ dB at a distance $r = 10$ m, what must be the source strength expressed as Q_{rms}? What total acoustic power is being radiated by the speaker? If the effective source size is $2b = 12$ cm, what acoustic impedance does the speaker see? Is the amplitude $\xi(b)$ much smaller than b itself, so that we are justified in ignoring nonlinear effects? How would these answers differ if $f = 50$ Hz?

11.3. **(a)** Consider a small source of strength Q located on an edge where two infinite plane walls meet. Write expressions in terms of Q for $p(r)$, $v(r)$, $I(r)$, P, and $\mathrm{Re}(Z)$, and verify the statements made about this case in the text.
(b) Do the same for a simple source in a corner where three walls meet.

11.4. What is the radiation resistance seen by a point source in free space for frequency $f = 200$ Hz? For $f = 2$ kHz? What if the source is mounted on an infinite baffle? For a volume flow $Q_{rms} = 10^{-2}$ m^3/s, how much acoustic power is radiated in each case?

11.5. Take the limit of Eq. 11.26 as kb goes to zero, and explicitly show its real and imaginary parts. Show that the imaginary part still has the same infinite value as does (11.7), plus a finite but relatively unimportant contribution from the image source, while the real part gives (11.27).

11.6. Consider a male singer whose lowest notes may have fundamental frequencies as small as 80 Hz. Remember that the voice also involves many harmonic components, and there may be important amounts of these as high as 4 or 5 kHz.
(a) In view of the size of a typical head, estimate for what portion of the total frequency range it will be legitimate to treat the voice as a point source.
(b) In view of the dependence of R_{rad} upon k for a point source, explain why the bass is at a disadvantage compared with a soprano in radiating strong sounds, especially at the low end of his range.
(c) For what range of frequencies would it aid the bass appreciably to be standing within 0.3 m of a hard wall?

11.7. In order to produce a sound level of 70 dB at a distance of 3 m on-axis, what dipole strength D_0 would be required for frequency **(a)** 5 kHz, **(b)** 500 Hz. What total power is being radiated?

11.8. Take the ratio of $\Delta p/2b$ and D_0 and examine its limit as b approaches zero to verify Eq. 11.38. You should find that the imaginary part of this ratio diverges as b^{-3}.

11.9. **(a)** In order to produce radiation with SPL = 70 dB and $f = 500$ Hz at a distance $r = 10$ m and angle $\theta = 0$ from a baffled piston of diameter $2b = 20$ cm, what source strength Q is required?
(b) What velocity and what displacement amplitude for the piston?
(c) What impedance is seen by the piston?
(d) How wide is the beam of radiation produced?

11.10. Repeat Prob. 11.9 for **(a)** $f = 50$ Hz, **(b)** $f = 5$ kHz.

11.11. If a baffled piston of radius 10 cm operates in air with $2kb = 4$, what is the mechanical impedance of the air load? If the piston's own mass is 4 g, how much additional impedance does that contribute? In order to produce a velocity amplitude of 0.2 m/s, how much force must be applied?

11.12. Carry out the integral (11.55) in the case where $\theta = 0$ and r is not necessarily very large. Show that if $b > \lambda$ there are some values of r for which the pressure amplitude is zero. Explain physically how such nulls can occur. Find the largest such value of r in terms of b and λ, and in the case where $b \gg \lambda$ show that it provides the criterion $r \gg b^2/\lambda$ for being in the simple "far field" of the piston.

11.13. In order to produce a signal with intensity 1 mW/m^2 at a distance of 1 km from a baffled hydrophone in sea water (neglecting any loss of strength due to absorption), what total acoustic power is required if kb is **(a)** 0.2, **(b)** 2.0, and **(c)** 20? **(d)** If

$b = 4.7$ cm, what are the three frequencies, and how much do your answers change when you take absorption into account using Fig. 1.5?

11.14. Using the graph of the Bessel functions in Appendix C, estimate $H(\theta)$ at its first maximum beyond θ_0. Then how many decibels below I_0 is the intensity in the strongest side lobe of Fig. 11.11?

11.15. Notice in Fig. 11.8 how much more prominent the oscillations are in the value of X than in R for $kb \gg 1$. Use Eqs. 11.71 and 11.72 to estimate how large kb must be in each case for the second term to be no more than a 10 percent correction to the leading term.

11.16. For a piston with radius $b = 10$ cm in air, consider the parallel-circuit representation of its radiation impedance in Fig. 11.12. What are the largest and smallest possible values of the equivalent mechanical resistance and mass, R_m and m, and at what frequencies do they occur? Explain how these ultimately give all the same results as does Fig. 11.8.

11.17. Consider as a model of an acoustic dipole two monopoles of strength $+Q_0$ and $-Q_0$ located on the z axis at $z = +d/2$ and $-d/2$, respectively. Write the combined pressure $p(r, \theta)$ due to both in a similar manner to Eqs. 11.17 through 11.24, but making necessary sign changes and noting that d now plays the same role $2B$ did there. By taking the limit as kd approaches zero, show that you obtain for large kr the dipole field described by (11.35) to (11.37) if you define D_0 to be $Q_0 d$.

11.18. Use (11.33) to relate the dipole strength D_0 of a symmetrical flow like that in Fig. 11.5 to the monopole strength Q_+ of one side. Show then from (11.37) and (11.12) that the ratio of power radiated as a dipole to that radiated by the monopole (radiating into unbounded space) is $4(kb)^2/3$. What is the ratio if the monopole radiates only into a half space? Which estimate is more relevant if the isolation of Q_- from Q_+ is accomplished by enclosing the back side of the speaker with a box that is small compared with a wavelength?

11.19. What is the value of the dimensionless piston impedance function Z_1 for **(a)** $kb = 0.2$, **(b)** $kb = 2$, and **(c)** $kb = 20$? In each case, obtain Z_1 in three ways: from (11.69) and (11.70), from (11.71) and (11.72), and then from Fig. 11.8 or Table C.1. Indicate which answers are incorrect and why.

Chapter 12

Room Acoustics

Most of our daily activities take place in enclosed rooms, where sound waves may be reflected back and forth among the walls, floor, and ceiling, as well as from obstacles within. Our study of simple sound sources provides only a starting point, and much more must be added to gain any real understanding of sound behavior in rooms.

We begin by considering how best to describe the complex sound fields that occur indoors. Then we must try to specify the properties we most often wish to achieve in order that a room may be acoustically pleasant or useful, and to recognize the relations between those acoustical properties and the way the room is designed and constructed. A simple quantitative theory of reverberant sound fields will be presented. In the final section we consider a few aspects of the use of electronic amplification systems to supplement natural room acoustics.

12.1 DESCRIBING SOUND IN ENCLOSED SPACES

A great variety of spoken and musical sounds may be of interest in studying classroom or auditorium acoustics. We might also be concerned with noises created by machinery in an office or factory, or by milling crowds in an exhibition hall. It is useful to put all such sounds under two main headings. **Steady sounds** may be either periodic (like a test signal from a sine-wave generator), quasi-periodic (like a chord played by musical instruments), or constant only in a statistical sense (like crowd noise). **Impulsive sounds** may be isolated events (such as a single impact of a stick upon a drumhead) or may represent the transient beginning or ending generated when a steady sound source is turned on or off.

Our description of these sounds also involves two possibilities. Just as was the case for waves on a string in Chap. 7, so for waves in three dimensions it is again true that any sound may in principle be described either as a superposition of many **traveling waves** going in all possible directions or as the sum of many **standing waves.** Both pictures are equally valid, and each contributes in a different way to our physical understanding. Thus for an impulsive sound we may envision a brief wave pulse bouncing back and forth among the walls of a room, losing some energy each time, and presenting a listener with a rapid train of repetitions of the original sound event. But that same sound may also be described as a sudden excitation of all the normal modes of motion of the air in the enclosed space, with each mode then proceeding to oscillate at its own natural frequency and decay with its own damping rate. Similarly, the traveling-wave picture for a steady sound shows the listener being bombarded simultaneously by many waves coming from different directions after originating at various times in the past and undergoing any number of wall reflections, whereas the standing-wave picture says each normal mode has a steady motion with amplitude and phase determined by the relation between its natural frequency and the frequency of the driving source.

Let us describe each of these frameworks in a little more detail, beginning with the traveling impulse sequence indicated schematically in Fig. 12.1. For any given sound event and any specific location of the listener in the room, the first-arriving signal is the **direct sound.** This comes along a straight-line path from source to receiver. Its strength is determined entirely by the source output and the inverse-square law, since the walls play no role. The direct sound is soon followed by several **early reflections,** each of which has traveled a somewhat longer path involving reflection from one of the walls. Soon enough time has passed that components arrive only after multiple reflections from several surfaces. These later reflected sounds become more and more numerous, while each

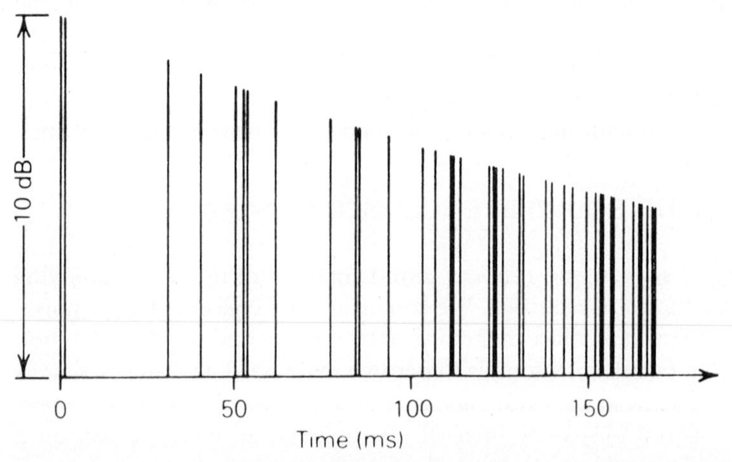

Figure 12.1 After a short sound pulse is created at one point in a room, its various reflections will arrive spread over time at another point. At later times, the strength of each individual impulse becomes less but the number arriving per unit time continually increases. (From H. Kuttruff, *Room Acoustics,* Applied Science Publishers, 2d ed., 1979, p. 278.)

one individually becomes weaker and weaker, both because of inverse-square-law spreading and because some energy is lost in the reflection process. So the later components blend into a seemingly smooth and continuously decaying **reverberant sound.**

Figure 12.2 points out the need to consider two different types of reflection. A flat hard wall produces **specular reflection,** where the word "specular" simply means "mirrorlike." The reflected wave is the same in every respect as if it were a duplicate of the incident wave, except for originating at a fictitious image source located behind the wall. Each part of the incident wave has effectively bounced off the wall in a specific single direction, namely, so that the reflected wave leaves the wall at the same angle as the incident wave approached (Fig. 12.3*a*). But any sharp point sticking out of a wall produces a new disturbance that spreads outward from that point, and a wall with many such points must create **diffuse reflection** (Fig. 12.3*b*). It is helpful to be more specific: a wall will be effectively smooth or rough, thus creating either specular or diffuse reflection, according to whether the irregularities in its surface are small compared with a wavelength or not. Note that a single wall might well be smooth as far as bass sounds are concerned, but rough for treble.

The standing-wave picture has components that are each spread throughout the whole room. In a hard-walled rectangular room defined by $0 < x < X$, $0 < y < Y$, and $0 < z < Z$, the proper boundary condition is that the velocity must have a node (and thus that the pressure has an antinode) at every wall. This requirement is easily met by constructing the normal-mode solutions

$$p_{lmn}(x, y, z, t) = \cos(l\pi x/X) \cos(m\pi y/Y) \cos(n\pi z/Z) e^{j\Omega_{lmn}t}, \qquad (12.1)$$

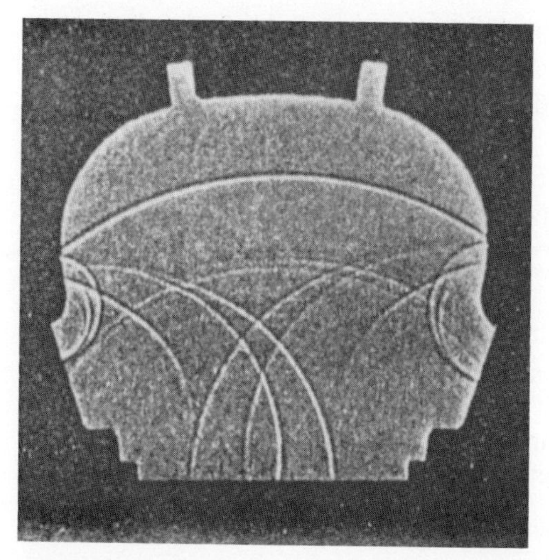

Figure 12.2 Photograph of waves from a spark source in a scale model of a New York theater, from an early paper of Sabine (*J. Franklin Inst.*, **179,** 1, 1915). Note how several of the early reflections spread out in all directions from projecting corners, while another represents reflection from the flat part of the side wall and appears centered on an image source behind the wall.

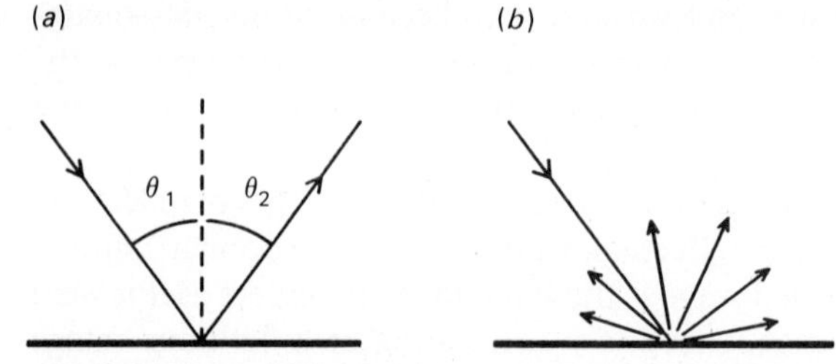

Figure 12.3 (*a*) In specular reflection from a smooth surface, all energy is concentrated at an angle of reflection θ_2 equal to the angle of incidence θ_1. (*b*) In diffuse reflection from a rough surface, energy is scattered more or less in all directions.

where *l, m,* and *n* are nonnegative integers. These simply require that an integral number of half wavelengths fit between each pair of walls. You can visualize the room as divided into many small rectangular chunks, much as a shipping carton may be divided to provide an individual compartment for each cup in a set of china. The dividers represent nodal surfaces where the air does not move, separating adjoining regions where the motion is in opposite directions.

Only if the pressure given in (12.1) actually satisfies the wave equation

$$\partial^2 p/\partial t^2 = c^2 \nabla^2 p \tag{9.18}$$

can such a mode exist. You may verify by substitution that this requirement is fulfilled by allowed modes with frequencies

$$\Omega_{lmn} = ck_{lmn} = \pi c\sqrt{(l/X)^2 + (m/Y)^2 + (n/Z)^2}. \tag{12.2}$$

(Note that this is a fairly simple generalization of Eq. 7.21.) In order to interpret how the room responds to a driving source, we want to know how many such modes exist with frequency $\Omega_{lmn}/2\pi$ less than some specified value f. By squaring both sides of (12.2), we can pose the question this way: how many sets of allowed values of *lmn* satisfy

$$(l/X)^2 + (m/Y)^2 + (n/Z)^2 < (2f/c)^2? \tag{12.3}$$

The form of this equation suggests that we think of *l, m,* and *n* as cartesian coordinates, and interpret it as specifying the interior of an ellipsoid whose principal axes extend to $l = 2fX/c$, $m = 2fY/c$, and $n = 2fZ/c$ from the origin. Since *l, m,* and *n* can have only integer values, the allowed combinations occupy the corners in a lattice of unit cubes. There is one such point per unit volume in *lmn* space; so (remembering the limitation to nonnegative integers) the total number of modes is estimated by the volume of one-eighth of the ellipsoid:

$$\begin{aligned} N(<f) &\simeq (1/8)(4\pi/3)(2fX/c)(2fY/c)(2fZ/c) \\ &= (4\pi/3c^3)Vf^3. \end{aligned} \tag{12.4}$$

Here $V = XYZ$ is the total room volume, and it can be shown that this is still a good estimate for most nonrectangular shapes. That is, for some purposes only the total room volume is important, and the shape does not matter.

It is particularly interesting to consider the average number of modes per unit frequency range:

$$dN/df \simeq (4\pi/c^3)Vf^2. \tag{12.5}$$

The reciprocal of this would be the average frequency range per mode, that is, the typical frequency difference between each mode and its nearest neighbor in frequency. Let us call this mode spacing Df; then

$$\boxed{Df \simeq (c^3/4\pi V)f^{-2}.} \tag{12.6}$$

It appears that a combination of low frequency and small room volume might make Df large enough to indicate that the room response would be quite different according to whether the source frequency matches one of the mode frequencies or falls between them. In Sec. 12.3 we will find a way to make this criterion more specific.

Take as an example a room with $X = 5$ m, $Y = 8$ m, and $Z = 3$ m, and thus with $V = 120$ m^3. The lowest natural mode frequency corresponds to air motion only in the y direction, with a single half wavelength in that 8-m distance:

$$f_{010} = (c/2)(1/Y) = (170 \text{ m/s})/(8 \text{ m}) = 21 \text{ Hz}.$$

An example of a higher mode, which fits five half wavelengths into the width, three into the length, and six into the height of the room, is

$$f_{536} = (170)\sqrt{(5/5)^2 + (3/8)^2 + (6/3)^2} = 386 \text{ Hz}.$$

The total number of modes below $f = 1$ kHz is approximately

$$N \simeq [(4.19)/(340 \text{ m/s})^3](120 \text{ m}^3)(10^3/\text{s})^3 \simeq 12{,}800$$

and the mode density near 1 kHz is about

$$dN/df \simeq [(12.57)/(340 \text{ m/s})^3](120 \text{ m}^3)(10^3/\text{s})^2$$

$$\simeq 38 \text{ modes/Hz}.$$

Thus there is on average another mode for each increment of 0.026 Hz near 1 kHz. But near 100 Hz the spacing is about 2.6 Hz.

12.2 CRITERIA FOR DESIRABLE ACOUSTICS

There is no single way to provide an ideal acoustical environment, because there are several distinct purposes that a room might serve. At one extreme we could consider a factory or a gymnasium, with multiple sources of sound and the goal of simply keeping the total noise down to a bearable level. On the other hand, an auditorium for theater or musical performance may call for carefully hoarding the delicate sound of a whisper or a soft flute note, and directing it so that it may be effectively heard throughout the room. Let us devote most of our attention now to the needs of speech and music, and defer some of the problems of office or factory noise to the following chapter. Here then are several goals for auditorium acoustics, and comments about how they are likely to be achieved.

Clarity We should like to hear each word or note clearly; most obviously with speech we do not want one word to be mistaken for another. This property of clarity (or good "definition" of sound) depends somewhat on having an adequate total strength of sound, but even more on having relatively strong direct sound. The less a new direct sound has to compete with continuing reverberation of an old sound, the easier it will be to understand.

One way to achieve clarity is to minimize the reverberant sound component by having nonreflective walls. Since the resulting "dead" sound is rather unpleasant, especially for music, it is important to pursue clarity from the other side as well by having the direct sound as clear and strong as possible. This will occur if (1) the listener is not too far from the source and (2) the line of sight along which this direct sound travels is not blocked by any obstacle. Here we have a principal reason for balconies (in which a greater number of listeners can be accommodated within a given maximum distance from the source) and for sloped seating areas (so that each listener can see and hear clearly over the heads of those in the row ahead).

An objective measurement that can be correlated with clarity is the ratio of early to late sound strength. This measurement recognizes that the early reflections may also contribute to clarity along with the direct signal. With modern analyzers it is possible to measure separately how much sound energy arrives before and after some time $t = \tau$, where $t = 0$ is taken as the arrival time of the first direct sound from an impulsive source (such as a shot from a starter's pistol). The ratio

$$R = E(0 < t < \tau)/E(t > \tau) \qquad (12.7)$$

measures the relative prominence of the early sound. With a time window $\tau =$ 80 ms (a choice that has often been made in such work), values of $R > 2$ will ensure maximum clarity while $R < 1$ will correspond to significant lack of clarity.

Envelopment Early sound arriving mainly from in front of the listener or from the ceiling overhead may leave a feeling of remoteness from the source. A more warm or intimate environment will be felt if the first few early reflections include good **lateral reflections** from side walls. Traditional narrow, rectangular concert halls have an advantage in this regard over the fan-shaped halls that have become so common in the twentieth century under the economic pressure to maximize total seating capacity. Note how Fig. 12.2 suggests that a wall broken into several staggered sections can create several reflections in quick succession in place of a single one. The importance of lateral reflections has been emphasized within the last decade, and several experimental ways of enhancing such reflections are still being developed.

An objective measurement associated with lateral reflections is the interaural disparity, or nonidentical nature of signals arriving at a listener's right and left ears. A pair of microphones mounted on a dummy head can provide signals to an autocorrelation analyzer. Too high a degree of correlation then suggests a lack of enough lateral reflections.

Reverberation Relatively greater amounts of reverberant sound are usually associated with perceived qualities described by adjectives such as warm, alive, or majestic. Where acoustical qualities seem cold, dead, or brittle, that may indicate that more reverberation could help. Both the strength and the time duration of the reverberant sound need to have appropriate values, but of course these are closely related since a longer reverberation time will allow a greater strength of reverberant sound to accumulate. Reverberation will be relatively strong in rooms with larger dimensions and with hard walls that absorb very little of the sound incident on them.

One measure of reverberant sound is its strength relative to a steadily continuing direct sound. It may be possible to separate them by taking advantage of their different spatial distributions. The direct sound falls off with distance from the source, whereas the reverberant sound is supposed to have undergone multiple reflections and so be spread rather uniformly throughout the room (Fig. 12.4). If you find fairly similar sound-level readings throughout all parts of a room far enough from the sound source, you can take those as representative of the reverberant sound pressure. By taking another reading sufficiently close

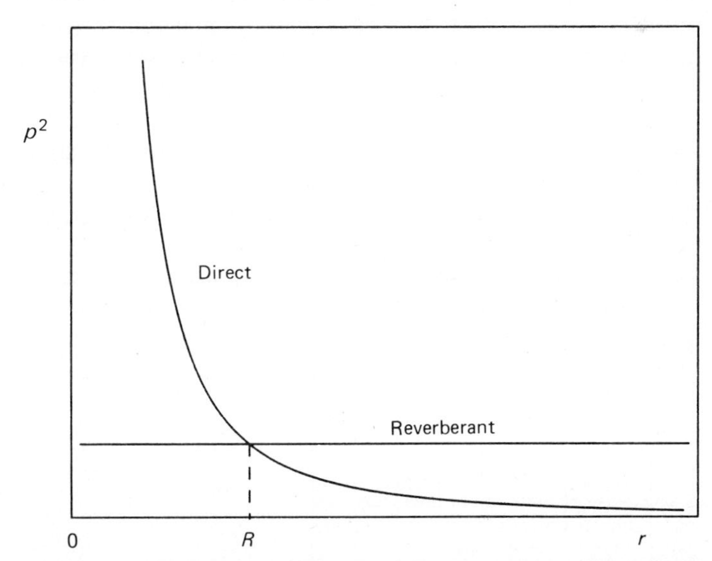

Figure 12.4 The intensity of the direct sound component falls off with the inverse square of distance r from the source, while the reverberant component is ideally uniform throughout the room. As shown in Prob. 12.13, the distance R at which the two are equal is about $0.1\sqrt{V/\pi T_r}$ for a room of volume V and reverberation time T_r.

to the source for the direct sound to predominate, you may calculate from the inverse-square law how strong that direct sound would be at any other distance and compare it with the reverberant sound. Figure 12.5 shows estimates from one study of what relative amounts of direct and reflected sound are considered most pleasing for different types of music and speech.

> Suppose, for instance, you measured SPL = 85 dB at r = 1 m from the source and SPL = 72 dB everywhere beyond about r = 5 m. Since $10 \log (1/8)^2 = -18$ dB, you could infer that a listener at r = 8 m receives 67 dB of direct sound and that this listening position is characterized by a 5-dB excess of reverberant over direct sound. This would be good for symphonic music, but rather muddy for speech.

We may also describe reverberation in terms of its time duration. Since continuing multiple reflections get weaker and weaker but never totally disappear, we cannot define reverberation time in terms of physical vanishing of sound. But perceptually we find that a decaying sound seems to human ears to disappear after it reaches a level around 40 to 60 dB lower than its starting point. It has become a widely accepted standard that **reverberation time** T_r is defined as the time required for any sound to decay 60 dB, that is, to one-millionth of its original intensity.

How long should the reverberation last to be most pleasing to the ear? That depends on what kind of sounds are involved. Accumulated experience is often summarized in a form somewhat like Fig. 12.6, suggesting that not only different types of music but also different room sizes affect our preferences. (Note that some sources advocate that the room volume should have little or no bearing on the matter.) Such figures generally refer to midrange frequencies, with measurements most commonly in the 500- or 1000-Hz octave band. Most rooms will have a somewhat shorter high-frequency T_r, but if the 4-kHz reverberation is less than 80 percent of that in the midrange the room response may seem to be lacking in brilliance. Similarly, a longer reverberation in the bass (perhaps

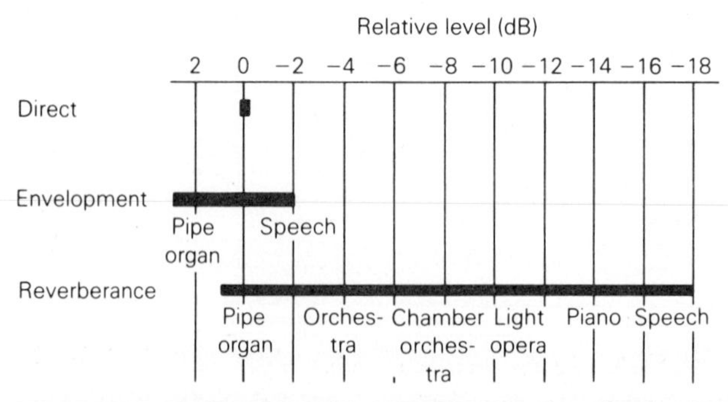

Figure 12.5 Preferred relative strengths of early reflections and long-term reverberation, compared with direct sound, for several types of music and speech. Relative strength is greater toward the left. (From an article by Veneklasen in *Auditorium Acoustics,* ed. R. Mackenzie, Wiley-Halsted, 1975.)

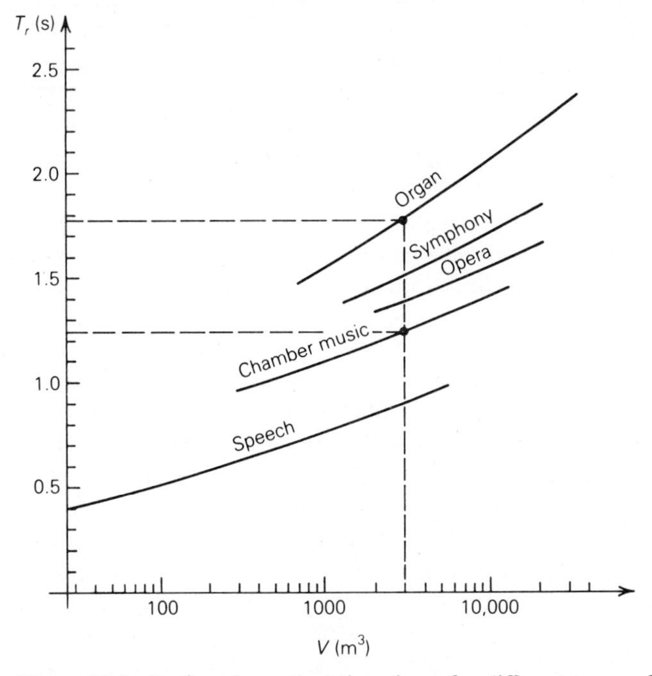

Figure 12.6 Preferred reverberation times for different types of music and speech as a function of room volume. These numbers are only intended as typical, with a 10 to 20 percent spread to be expected in individual cases. Example marked with dashed lines suggests that $T_r = 1.2$ s would work well for chamber music in a 3000-m^3 recital hall, but an organist would prefer well over 1.5 s. (From D. E. Hall, *Musical Acoustics: An Introduction,* Wadsworth, 1980.)

130 percent or so of the midrange value in the 125-Hz band) is desirable; if it is not present the room may seem to lack warmth and fullness.

Smoothness There are three distinct ways in which the sound might fail to be pleasingly smooth in its arrival. One is irregularity in the way the late reverberant sound dies away; we will see some possible reasons for that in the following section. Another is the presence of a distinct **echo,** perceived as a separate event from the original sound. This may occur if a large hard wall (in the back of an auditorium, for instance) returns a reflection significantly stronger than the others that preceded it, especially if this arrives 100 ms or more after the direct sound. The most spectacular cases occur when concave surfaces focus the sound, concentrating more energy back toward a particular part of the auditorium instead of allowing it to spread further. Echoes may usually be avoided by properly slanting such surfaces or changing their curvature or breaking them up into multiple panels facing in various directions, and also of course by simply covering them with sound-absorbing material.

A third way in which a sound may be rather rough and unpleasant, even though without evoking a consciously noticed echo, is to have excessive time gaps among the early reflections. As long as the first reflection arrives sooner than 30 to 40 ms after the direct sound, and every reflection is followed by another within a similar time, all the early reflections up to 100 ms or so will have the psychological effect of seeming to simply be part of the direct sound, thus strengthening it. Any gap of more than 40 ms between one component and the next will tend to ruin this desirable effect. Such a problem generally calls for additional reflecting surfaces to be installed. It has become fairly common for acoustical consultants to have computer programs that will calculate the early-reflection sequence for various locations in a room of specified shape, to ensure that design flaws that would leave reflection-time gaps can be spotted in advance and corrected.

Uniformity We have already meant implicitly that all the properties described above should apply for all listener locations throughout a room. Absolutely perfect uniformity is unattainable, of course, but with care we can avoid "hot spot" seats where focusing effects make sound quality quite obviously different from other locations, or differences much over 10 dB in overall sound level in different parts of a room. There is one sense in which this optimism in unjustified: Considering the traveling-wave picture for a steady source of sine waves, it appears that we have a combination of many component waves with various phases determined by how far each one has traveled. As either the source or the listener moves about the room, the phase relations change, and we would expect to observe varying degrees of constructive or destructive interference in the way the components combine. It is tempting to argue that if there is a sufficiently large number of component waves there will always be some coming in phase as others go out so that statistical chance will ensure nearly the same total signal everywhere. While this may hold pretty well for room averages, it is not really true for every position. In fact, more sophisticated analyses show that instead of a precisely uniform sound field we should expect for any given frequency to find variations from place to place whose standard deviation is 5.5 dB or even more. Only by averaging over a range of frequencies can we hope ever to measure a really uniform spatial distribution of SPL; fortunately, such a frequency average is what matters for many practical purposes.

One of the more common problems with uniformity is a dead area underneath a balcony. A common rule of thumb is that the vertical opening into such an area must be at least half of the horizontal depth. The situation may also be helped if the underside of the balcony can be used to help reflect more sound to the rear seats (Fig. 12.7).

Performer Satisfaction Not only must the quality of sound for the audience be considered; the speaker or musician should also receive feedback from the room that gives an accurate and pleasing feeling for what he or she is producing. For example, a strong echo from a concave rear wall or a feeling of isolation on the front of a large stage can both be serious problems. Musicians in an orchestra need to be positioned so that they can hear each other well, not only directly but also with early reflections. For this reason, reflecting panels are often provided closer to the group than the permanent walls.

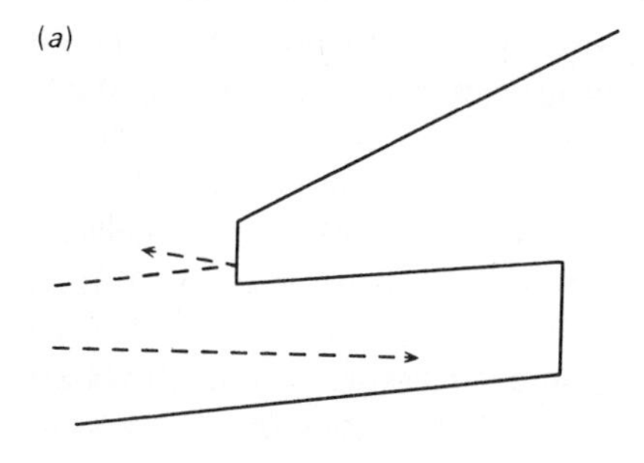

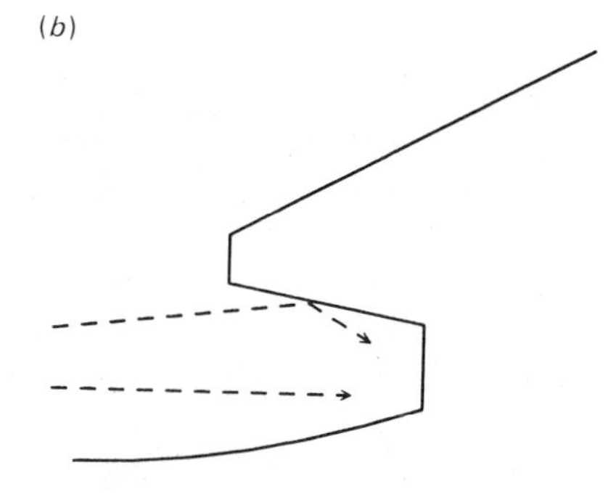

Figure 12.7 (*a*) A seating area under a balcony that would be expected to have poor sound because the depth is too great for the height. (*b*) Possible improvements include less depth, greater height, and slanted ceiling to provide helpful reflections.

Freedom from Noise Listeners should not be disturbed by extraneous sounds, either from street traffic or subway trains outside the building, or from noisy lighting or ventilation systems inside. Some material in the following chapter will be relevant to such problems.

All these criteria depend somewhat on the activity involved. For speech the clarity may take preference over other properties, whereas for some kinds of music it may give way to warmth and reverberation. On the other hand, many concerts of popular music today rely on amplification systems rather than natural room acoustics, and they may do just as well in a room that would be quite unsatisfactory for a symphony orchestra. Perhaps the greatest challenge is in the design of multipurpose rooms for schools and community centers, where it can be quite difficult to reach a compromise that provides adequate acoustics for assembly speeches, band concerts, and basketball games.

12.3 REVERBERANT SOUND FIELDS

There are several distinct questions to be considered about reverberant sound. In the preceding section we have already given two ways of quantitatively defining reverberance, namely, the SPL of the steady sound field and the reverberation time T_r required for 60-dB decay of an impulsive sound. We have also given some information about what values of these quantities are most pleasing. In this section we first describe briefly how T_r may be measured and then consider at some length how the reverberation characteristics can be calculated directly from information about the design and furnishings of the room. You should keep in mind all the while that other room properties may be just as important as reverberation.

Measurement of T_r is done either with a very short sound impulse such as that from a starter's pistol or with a narrow-band noise source that is run long enough to establish a steady sound field in the room and then is suddenly turned off. The decaying sound can be measured by a microphone connected through a logarithmic converter (in order to measure on a decibel scale) to a strip-chart recorder. Ideally you would simply measure the distance along the time axis to the 60-dB-down point. But more often, as in Fig. 12.8*a,* that point lies below a

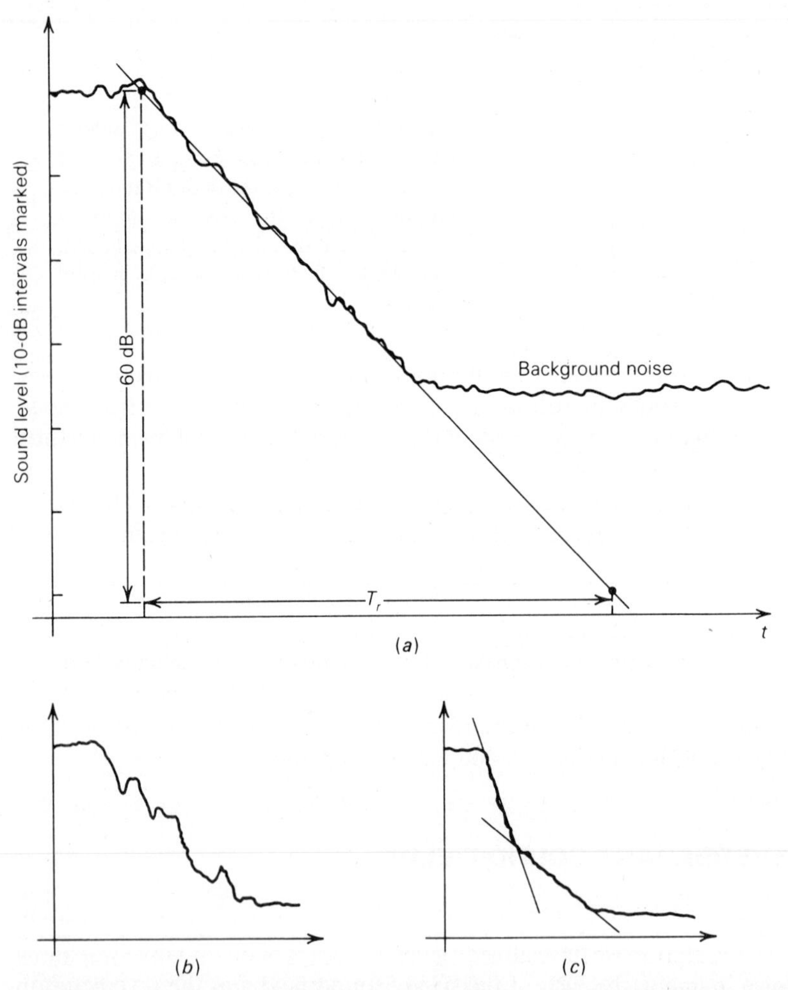

Figure 12.8 (*a*) Chart record of measurement of T_r at 1 kHz in CSUS recital hall, showing need to extrapolate 60-dB decrease. (*b*) Ragged decay associated with nonuniform sound distribution. (*c*) Double-slope decay resulting from some modes being damped more rapidly than others. (From D. E. Hall, *Musical Acoustics: An Introduction,* Wadsworth, 1980.)

minimum measurable level that may be due either to background noise in the room or to electronic noise limitations in your microphone amplifier. So commonly the initial downward slope is extrapolated, and T_r is taken in practice to be the time after which the sound level would have decreased 60 dB if it had continued to decay at the same rate (expressed in decibels per second). It is possible to purchase equipment that performs this extrapolation automatically and provides a digital display of the resulting T_r. Some difficulty in recognizing a well-defined T_r may occur (Fig. 12.8*b, c*) if the reverberant sound is poorly distributed either in space or in frequency, or if concentration of absorbing material in some parts of the room causes sound components traveling in different directions to decay at substantially different rates.

To relate reverberation time and strength to room design, let us try to draw information from both the standing-wave and traveling-wave pictures. The standing-wave picture emphasizes that the air contained inside a room has certain normal modes of motion, each with its own natural frequency. If the walls were perfectly rigid, these modes would be the three-dimensional analogues of the standing waves on a string stretched between rigid supports. But the acoustic impedance of real walls is not infinite; so energy continually leaks away from the air motion. The simplest picture is that each normal mode acts independently as a damped simple harmonic oscillator, so that its energy will decay exponentially after the driving source is turned off (see Sec. 6.1). That is, if E_n is the amount of energy in mode number n at time $t = 0$, we expect the total sound energy in the room to vary as

$$E(t) = \sum_n E_n e^{-2\beta_n t}. \tag{12.8}$$

Without attempting to relate each β_n exactly to the wall properties, we can see that the total energy decays in a simple exponential way if and only if all the damping constants are the same:

$$E(t) = E_0 e^{-2\beta t} \quad \text{if every} \quad \beta_n = \beta. \tag{12.9}$$

Since sound level is proportional to the logarithm of sound energy, this is the idealized situation that will have SPL decreasing linearly with time to give a well-defined T_r. It ought to occur if all mode patterns, even though pushing mainly on different parts of the walls, encounter about the same impedance. But if some modes make the air move against highly absorbing material while others push mostly against hard walls, the former will have higher values of β_n than the latter. Then the total sound will decay rapidly at first, but later the modes with small β_n will persist longer to create a multiple-slope decay as in Fig. 12.8*c*. Such prominent changes in decay rate are generally perceived as a strange and undesirable acoustical property; so we are led to conclude that it is a good thing to distribute absorbing materials fairly evenly over all surfaces in a room.

The standing-wave picture is also useful in thinking about whether the room responds uniformly at all frequencies. We showed above that the typical separation between one mode frequency and the next is

$$Df \simeq (c^3/4\pi V)f^{-2}. \qquad (12.6)$$

This tells us that the modes are crowded closer and closer together as we go toward higher frequencies. At what point are they close enough to effectively overlap in their response, so that any arbitrarily chosen frequency would be pretty sure to have one or more modes vibrating strongly in response to its driving force? Recall from Chap. 6 that the half-power bandwidth of a simple oscillator is $\Delta\omega = 2\beta$, or

$$\boxed{\Delta f = \beta/\pi = 2.2/T_r,} \qquad (12.10)$$

where the last equality is proved in Prob. 12.3. The work of Manfred Schroeder (*JASA* **34,** 1819, 1962) and David Lubman (*JASA* **56,** 523, 1974) suggests that $Df \leqslant \Delta f/3$ is a good criterion for ensuring fairly smooth room response. Using (12.6) and (12.10) we find this condition is satisfied only for $f > f_s$, where the Schroeder frequency

$$\boxed{f_s \simeq \sqrt{3c^3T_r/8.8\pi V} \simeq 2000\sqrt{T_r/V}} \qquad (12.11)$$

marks the transition point at which we anticipate a change in behavior. For instance, a room with $V = 10^3$ m^3 and $T_r = 1.6$ s would have $f_s \simeq 2000\sqrt{0.0016} = 80$ Hz. For $f \gg f_s$ many modes effectively overlap and the room response should be quite smooth. But for $f < f_s$ the individual resonances may make the room "ring" noticeably more at some frequencies than at others nearby. Since V appears in the denominator, we see that very small rooms are the ones where this may be troublesome (as in Prob. 12.5). Many readers may be reminded of the extreme case of a tiled bathroom, in which it is common for one or a few notes to resonate spectacularly during singing.

Now let us use instead a traveling-wave picture, as did Wallace Sabine when he laid the foundations of reverberation theory around 1900. Suppose we assume either that the room has a complex shape with parts of its surface oriented in many different directions, or equivalently that the surface materials are rough enough to ensure diffuse scattering. Then after a few reflections it might be reasonable to suppose that the reverberant sound field everywhere in the room would be isotropic (i.e., would consist of equal fluxes of sound energy in all directions) and also that the sound energy density would be uniform throughout the room. These claims are not entirely obvious, and it is possible to construct special room shapes for which they are not true. (For a more complete discussion see H. Kuttruff, *Room Acoustics,* Chap. V; note also that the validity of the theory to follow has been examined critically by W. B. Joyce in *JASA,* **58,** 643, 1975; **64,** 1429, 1978; and **67,** 564, 1980.) But for most practical purposes this is a useful approximation, and we write explicitly

$$E = eV \quad \text{and} \quad dI/d\Omega = ec/4\pi. \qquad (12.12)$$

Here E is the total sound energy in the whole room and e the local energy density, and $dI/d\Omega$ means the flux or intensity of traveling waves whose direction is within the small range of solid angle $d\Omega$ close to any specified direction. Since a complete sphere occupies a solid angle of 4π steradians, the last equation just says the total energy is distributed evenly over all possible directions.

Now consider how the total energy E may change with time. The room is like a reservoir with any source of sound power within the room acting as an input and any process removing sound energy acting as an outlet. Sabine's hypothesis is that no sound is lost in traveling around the room interior; it will only be lost insofar as it is absorbed at the walls. (See, however, Prob. 12.17.) If we knew enough about the acoustic impedance of the wall material, we might deduce something about the absorption from that, but let us take a more practical view. Let us claim that we will later be able to describe a measuring process in which the absorbing characteristics of any material can be determined in an entirely empirical way. We would like to summarize what the material does with an **absorption coefficient,** or **absorptivity,** which is defined as the fraction of energy lost by the traveling waves in making a single reflection from that material:

$$\boxed{\alpha = \text{(energy lost)/(energy incident).}} \tag{12.13}$$

(Please do not confuse this dimensionless α with that in Chap. 1, which had units of decibels per meter.) We expect α always to have values somewhere between 0 and 1 for real materials, with $\alpha = 0$ corresponding to perfect reflectors and $\alpha = 1$ to perfect absorbers. Strictly speaking, α depends on the angle at which the waves strike the wall, and what we must actually use here is an absorptivity that has been averaged over all possible angles of incidence.

The rate at which sound energy is lost at each wall should then be given by the fraction α of the rate at which sound is reaching that wall. Now any small wall area dA has sound coming toward it from all directions in a hemisphere, that is, from all angles $\theta < 90°$ in Fig. 12.9. A beam of sound approaching from angle θ and just filling this surface dA has an intrinsic cross section (perpendicular to its own direction of travel) of $dA \cos\theta$. Hence its incident energy flux is $(dI/d\Omega)\, dA \cos\theta$. Then the total energy flux onto dA from all directions is

$$\int_0^{2\pi} d\phi \int_0^{\pi/2} \sin\theta\, d\theta (dI/d\Omega)\, dA \cos\theta = \pi(ec/4\pi)\, dA = (Ec/4V)\, dA. \tag{12.14}$$

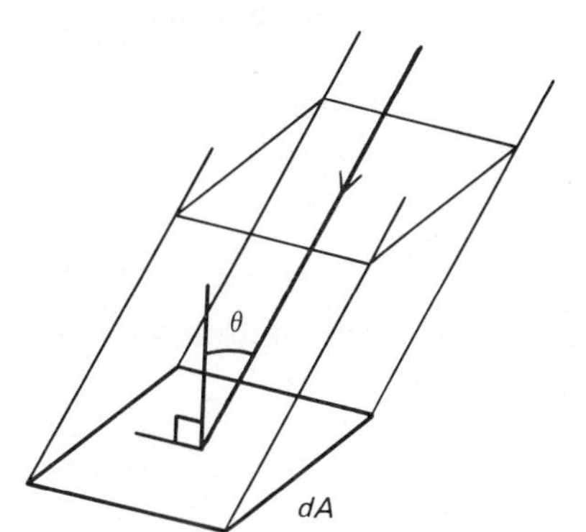

Figure 12.9 Sound striking a small area dA. Its direction of arrival θ is measured from the perpendicular to the surface.

Setting the fraction α of this equal to the rate of energy loss, and integrating it over all surfaces in the room, we have energy being absorbed at the total rate

$$(Ec/4V) \int \alpha \, dA = (Ec/4V)A_e \tag{12.15}$$

where

$$\boxed{A_e = \alpha_1 A_1 + \alpha_2 A_2 + \alpha_3 A_3 + \cdots = \bar{\alpha} A_{\text{tot}}.} \tag{12.16}$$

Here we use $\bar{\alpha}$ to represent an average value of α for all surfaces in the room, and A_e is called the **effective total absorption area.** Finally, representing the sound source strength by its power P, we obtain the desired equation for E:

$$\boxed{dE/dt = P - (A_e c/4V)E.} \tag{12.17}$$

From this equation we can read off almost directly the two pieces of information we most want. First, in the case where a steady source has been operating long enough for its input to reach equilibrium with the wall losses, $dE/dt = 0$ and the reverberant energy must have the constant value

$$E_r = 4VP/A_e c. \tag{12.18}$$

Using $E/V = e$ and (from Sec. 9.3) $e = p^2/\rho c^2$, we can write the equivalent pressure

$$p_r^2 = (4\rho c/A_e)P, \tag{12.19}$$

so that the reverberant sound level is

$$\boxed{\begin{aligned} \text{SPL}_r &= 10 \log (4\rho c P/A_e p_{\text{ref}}^2) \\ &= 10 \log (P/A_e) + 126 \text{ dB}. \end{aligned}} \tag{12.20}$$

The last step makes use of $p_{\text{ref}} = 2 \times 10^{-5}$ Pa and requires that the units of P and A_e be watts and m^2, respectively. Strictly speaking, the relation $e = p^2/\rho c^2$ applies only to plane sine waves. Our justification for using it here is that the total sound field at any point consists of a great many plane waves traveling in all directions and that they are statistically independent. This theory does not attempt to deal with the spatial fluctuations of SPL that are caused by constructive or destructive interference.

Our second main deduction from (12.17) concerns the case where the source has been turned off. With $P = 0$ we clearly have an equation for exponential decay: $dE/dt = -\gamma E$, so that

$$E(t) = E_0 e^{-\gamma t} \quad \text{with} \quad \gamma = A_e c/4V. \tag{12.21}$$

Since the reverberation time $t = T_r$ is defined in terms of a 60-dB drop, we must have

$$10^{-6} = e^{-\gamma T_r} \quad \text{or} \quad -6 \ln 10 = -\gamma T_r. \tag{12.22}$$

Then, since $\ln 10 = 2.303$,

$$T_r = 13.82/\gamma = 55.3V/A_e c. \tag{12.23}$$

Finally, using $c = 344$ m/s, this is usually written as

$$\boxed{T_r = \eta V/A_e \quad \text{with} \quad \eta = 55.3/c = (0.161 \text{ s/m}).} \tag{12.24}$$

For example, consider a room with volume $V = 4000$ m^3 and surface area $A = 2000$ m^2. If half this surface has $\alpha = 0.1$ and the other half 0.3, the total effective absorption area is

$$A_e = (0.1)(1000) + (0.3)(1000) = 400 \text{ m}^2,$$

and the average absorptivity is $\bar{\alpha} = 0.2$. Any transient sound in this room would decay with a characteristic time

$$T_r = (0.161 \text{ s/m})(4000 \text{ m}^3)/(400 \text{ m}^2) = 1.6 \text{ s}.$$

A steady source with power $P = 20$ mW in this room could maintain a reverberant sound level

$$\text{SPL}_r = 10 \log [(0.02 \text{ W})/(400 \text{ m}^2)] + 126 = 83 \text{ dB}.$$

In order to make full use of (12.24), we need to have available the values of the absorptivity α of various materials. Those shown in Table 12.1 should be taken only as typical possibilities; considerable variation may occur according to the particular material manufacturer or product line (of acoustic tile, for instance) and according to the exact manner of mounting or rigidity of construction. The porous materials, such as fabric or fiberglass, convert sound energy directly to heat by presenting resistance to the airflow through them; they are usually more absorptive for higher frequencies. In the case of solid materials the sound energy is first converted primarily into vibration of a wall structure, which then may be internally damped or even transmitted into an adjoining room; this mechanism often is more effective at low frequencies for which the massive wall can yield more easily. It is also possible to design panels with layers, slots, or holes whose dimensions are "tuned" to provide preferential absorption for chosen midrange frequencies.

The most straightforward use of Eq. 12.24 is in predicting a value of T_r when a room is being designed. If the complete geometry is given, together with full specification of what material will cover each surface, an expected reverberation time can be calculated. Since the material absorptivities α_i differ from one frequency to another, one would in fact calculate several different values of T_r for different frequency bands. If a preliminary plan yields poor expected reverberation times, revisions in geometry or in materials may be considered before construction proceeds.

You may of course make similar calculations for an existing room and compare them with actual measurements of T_r, in hopes of seeing which parts of the room furnishings are most responsible for its acoustical behavior. Even more often, your practical interest might be in doing such a calculation to see what effect possible changes in the room might have upon T_r. It is not always necessary in such a case to make a complete inventory of surfaces, if some use is made of measurements of the present room properties, as we can illustrate

TABLE 12.1 APPROXIMATE TYPICAL ABSORPTION COEFFICIENTS α FOR VARIOUS MATERIALS

Surface material	Frequency band 125	250	500	1000	2000	4000
Acoustic tile, rigidly mounted	0.2	0.4	0.7	0.8	0.6	0.4
Acoustic tile, suspended ceiling	0.5	0.7	0.6	0.7	0.7	0.5
Acoustic plaster, sprayed	0.1	0.2	0.5	0.6	0.7	0.7
Glass wool, 5 cm mounted on concrete	0.1	0.3	0.7	0.8	0.8	0.8
Glass wool covered with perforated panel	0.1	0.4	0.8	0.8	0.5	0.4
Plaster, on lath	0.2	0.15	0.1	0.05	0.04	0.05
Gypsum wallboard, ½ in, on studs	0.3	0.1	0.05	0.04	0.07	0.1
Plywood sheet, ¼ in, on studs	0.6	0.3	0.1	0.1	0.1	0.1
Concrete block, unpainted	0.4	0.4	0.3	0.3	0.4	0.3
Concrete block, painted	0.1	0.05	0.06	0.07	0.1	0.1
Concrete, poured	0.01	0.01	0.02	0.02	0.02	0.03
Brick	0.03	0.03	0.03	0.04	0.05	0.07
Vinyl tile, on concrete	0.02	0.03	0.03	0.03	0.03	0.02
Heavy carpet, on concrete	0.02	0.06	0.15	0.4	0.6	0.6
Heavy carpet, on felt backing	0.1	0.3	0.4	0.5	0.6	0.7
Platform floor, hard wood	0.4	0.3	0.2	0.2	0.15	0.1
Heavy plate glass	0.2	0.06	0.04	0.03	0.02	0.02
Ordinary window glass	0.3	0.2	0.2	0.1	0.07	0.04
Draperies, medium velour	0.07	0.3	0.5	0.7	0.7	0.6
Upholstered seating, unoccupied	0.2	0.4	0.6	0.7	0.6	0.6
Upholstered seating, occupied	0.4	0.6	0.8	0.9	0.9	0.9
Wood or metal seating, unoccupied	0.02	0.03	0.03	0.06	0.06	0.05
Wooden pews, occupied	0.4	0.4	0.7	0.7	0.8	0.7

Source: J. Backus, *The Acoustical Foundations of Music,* 2d ed. (W. W. Norton, New York, 1977), p. 172; L. Doelle, *Environmental Acoustics* (McGraw-Hill, New York, 1972), p. 227; H. Kuttruff, *Room Acoustics,* 2d ed. (Applied Science Pub., London, 1979), p. 154.

with the next example. A similar calculation will also serve to convert a measured T_r of an occupied room to an expected T_r when people fill the seating area.

Suppose a small church has $V = 4000$ m^3 and a measured midrange T_r of 1.9 s, and it is proposed that carpet be added to cover 200 m^2 of aisle and rostrum area that are presently concrete or equivalent. This is enough information to say that the present total absorption is

$$A_{e_1} = 0.16V/T_r = 640/1.9 = 337 \text{ m}^2,$$

without needing to know just how that total absorption is distributed. Suppose the carpet is expected to increase α from 0.02 to 0.4 for those 200 m^2. Then that surface alone increases its contribution to A_e from 4 m^2 to 80 m^2, while $337 - 4 = 333$ m^2 from the remaining surfaces is unchanged. The new total

$$A_{e_2} = 333 + 80 = 413 \text{ m}^2$$

is predicted then to give

$$T_r(\text{modified}) = 640/413 = 1.55 \text{ s},$$

enough less to constitute a radical change in acoustical properties.

The same approach underlies a standard method for determining absorptivities of various materials. A special reverberation chamber is built out of hard-surfaced materials, so that even with small volume it may have a reverberation time of several seconds. That T_r is measured first with the chamber empty, then again with a sample of absorbing material inside. From the reduction in T_r, the increase in A_e can be calculated, and for known sample area A_s this readily gives

$$\alpha_s = \Delta A_e/A_s = (\eta V/A_s)\,\Delta(1/T_r). \tag{12.25}$$

A somewhat more careful derivation is sometimes used to suggest that Sabine's formula,

$$T_r = \eta V/A\bar{\alpha}, \tag{12.26}$$

should be replaced by the Eyring-Norris formula,

$$T_r = \eta V/A[-\ln(1 - \bar{\alpha})]. \tag{12.27}$$

The latter equation will always give a somewhat shorter value of T_r than the former, so may be viewed as a more conservative estimate. But both are approximations to reality, and it is possible to construct bizarre cases where the Eyring-Norris form is worse than Sabine instead of better. It may be useful just to let the difference suggest how much they should both be distrusted. Since the first terms in a Taylor-series expansion of $-\ln(1 - x)$ for small x are $x + x^2/2$, we see that for sufficiently small $\bar{\alpha}$ the two formulas agree. Thus for rooms with small $\bar{\alpha}$ and relatively long T_r for their size, this theory is most successful. But for "dead" rooms (and even for very large "live" ones where $\bar{\alpha}$ must be fairly large to prevent excessive reverberation) the formulas disagree more and must not be trusted too readily because the reflected waves weaken so quickly that the assumed isotropic and uniform distribution of sound is never really achieved.

12.4 SOUND REINFORCEMENT

For any given sound source, we know from experience that it is more difficult to produce enough sound to fill a larger room. We can verify this by rewriting (12.24) as $A_e = \eta V/T_r$ and substituting that in (12.20) to obtain

$$\boxed{\begin{aligned} \text{SPL}_r &= 10 \log (4\rho c P T_r/\eta V p_{\text{ref}}^2) \\ &= 10 \log (P T_r/V) + 134 \text{ dB.} \end{aligned}} \tag{12.28}$$

Again, the last form is true only if V is in m^3. This can be used to study how much sound amplification might be needed in any given room. It tells us that more powerful natural sources in smaller rooms with longer reverberation times can best deliver an adequate SPL. But longer reverberation times cannot always accomplish this alone without causing other difficulties. The following example may be considered to show that sound-amplification systems will often be indispensable in rooms with much more than a thousand cubic meters of volume.

A moderately vigorous speaking voice can produce acoustic power on the order of $P = 100\ \mu$W. Consider a series of rooms with volumes $V = 10^2$,

10^3, 10^4, and 10^5 m^3, each with a similar value of $T_r \simeq 1.2$ s that would not be excessive for speech purposes. The reverberant sound level in the first room would be

$$SPL_r = 10 \log [(10^{-4})(1.2)/(10^2)] + 134 = 75 \text{ dB},$$

which is quite adequate. But volumes 10^3, 10^4, and 10^5 m^3 would correspond to SPL = 65, 55, and 45 dB, respectively, and the last two levels are certainly not enough. A good opera singing voice might do another 10 dB better in each case.

Put the other way around, how much acoustic power must a public-address system be able to put out in order to create SPL = 100 dB for a musical show in the 10^5 m^3 auditorium? Now we require

$$100 = 10 \log (1.2P/10^5) + 134,$$

$$1.2 \times 10^{-5}\, P = 10^{-34/10} = 4.0 \times 10^{-4},$$

$$P = (4.0 \times 10^{-4})/(1.2 \times 10^{-5}) = 33 \text{ W}.$$

We will see in Chap. 15 that most loudspeakers are not very efficient converters, so that electrical power of at least hundreds of watts is required. In order to allow "headroom" for unusual events, such an installation in practice would probably have amplifiers rated for total output of several kilowatts.

Let us describe briefly a couple of interesting points about sound reinforcement, without pretending to fully cover the subject of sound-system design. First, let us mention the importance of avoiding strong feedback loops. Some sound from the system loudspeakers unavoidably reaches the microphone, and the amplifier immediately sends the same signal through the speakers again. If the microphone is right in the mouth of a loudspeaker, even a small amplifier gain is enough that a self-sustaining squeal can be set up. But if the loudspeaker sends most of its energy elsewhere in the room and very little to the microphone, much higher amplifier gain (and correspondingly higher total sound output) can be used before the critical threshold for runaway ringing is encountered. This implies that sound systems will have a wider practical operating range if (1) microphones are kept as far as possible from loudspeakers, (2) loudspeakers are chosen and placed with directional patterns that point away from microphone locations, and (3) directional microphones are used and are placed so as to be relatively insensitive in the direction toward the nearest loudspeaker.

Directional speakers, judiciously placed, can also relieve a little of the conflict between the conditions for clarity and for warmth or ambience. Since a seating area filled with people is quite absorbent, much of the sound originally directed toward that surface will never contribute to the later reverberation. Thus loudspeakers carefully chosen and aimed to cover the audience but little else can strengthen the direct sound and the clarity of speech even in a fairly live room. Music performed in the same room using only the natural acoustics can have a higher proportion of reverberation, since a large fraction of the direct sound falls upon hard walls or ceiling instead of on the audience. Another minor adjustment that may help the clarity of amplified speech in a large reverberant room is to remove the low-frequency end with a high-pass filter. The low frequencies, which generally reverberate longest, are not missed much, since the range between 300 and 5000 Hz includes nearly all the information content of speech sounds.

Another interesting issue involves the listeners' perception of where the sound originates. Without making any conscious effort, people ordinarily can judge the direction from which a sound arrives quite well. This ability stems mainly from our possession of two ears apiece; differences in intensity and in arrival time at the right and left ears both provide a physical basis for such judgments. However, while this makes left-center-right judgments relatively easy, it remains much more difficult to distinguish in the front-above-behind plane. Thus a loudspeaker located in the same vertical plane with the original source and listener makes it fairly easy to accept the illusion that all the sound came from the source, which is psychologically pleasing. But a strong loudspeaker signal from well off to one side of the listener-source direction can be irritating and distracting; the eyes and ears are in conflict about where attention should be focused. In small to medium-sized installations this is a strong reason to choose a single large speaker or cluster located more or less above the microphone, in preference to speakers in the front corners or (even worse) distributed along side walls.

But in some situations (for instance, very large auditoriums, especially underneath balconies) it may be unavoidable to have some of the audience receive sound from loudspeakers that are closer to them than the original source, and perhaps off to one side as well. The associated problems can sometimes be considerably reduced by taking advantage of the **precedence effect** (also called the Haas effect). This refers to the fact that directional judgment depends most strongly on first-arriving sound: even though early reflections may come from several other directions, it may still seem to you that the source direction was well defined and corresponded to the direct sound arrival. Even when some of the early reflections are as much as 5 to 10 dB stronger than the direct sound, your directional perception may still be controlled by the first-arriving sound.

But if a loudspeaker is closer to the listener than is the original source, the precedence effect could strengthen the impression that all the sound originated at the loudspeaker. To overcome that, we need somehow to delay the electrical signal to the loudspeaker long enough that the direct sound traveling through the air can be first to arrive and then be quickly followed by the amplified component (Fig. 12.10). One way of delaying the signal is to record it on a small loop of magnetic tape and recover it with a playback head a short distance farther along the tape. But such systems can no longer compete in cost, flexibility, or reliability with digital-delay units. These convert the signal from analog to digital form, and store it in a circulating memory from which it can be recovered after the desired time lag. Several signals with differing amounts of delay may be used to feed loudspeakers at different distances from the source. When the delay system is properly adjusted, the reinforced sound will arrive more than 5 but

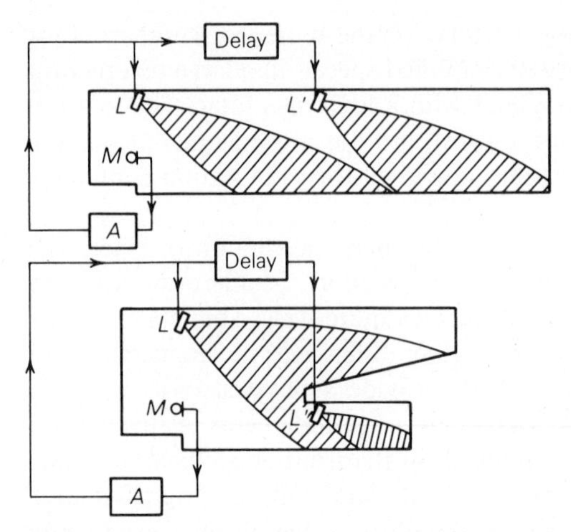

Figure 12.10 Two situations where speakers should be fed through delay lines. (From H. Kuttruff, *Room Acoustics,* by permission of Applied Science Publishers.)

less than 30 ms after the direct sound, with a 10- to 20-ms gap being optimum. The loudspeaker signal then acts like an early reflection in contributing to the strength and clarity of the direct sound without ruining the direct sound's control over directional perception. In a very large room, the role of direct sound might actually be played by the signal from a central speaker cluster.

PROBLEMS

12.1. Why is there practically no reverberant sound in an outdoor amphitheater? What value will the ratio R have in such a situation? For what reasons might the clarity of sound still be less than perfect?

12.2. A rectangular room occupies the space -10 m $< x < +10$ m, $0 < y < 30$ m, $0 < z < 12$ m. Suppose all four walls and ceiling provide good reflections but that the floor is strongly absorptive. For a source at $x = 0$, $y = 3$ m, $z = 2$ m and a listener at $x = 2$ m, $y = 20$ m, $z = 2$ m, calculate the arrival-time sequence of the first several early reflections of an impulsive sound. (*Hints:* Remember the Pythagorean theorem, and imagine that each reflection came from a fictitious image source located behind the corresponding wall.)

12.3. Given a sound energy decaying as $\exp(-2\beta t)$, consider when this energy has decayed to 60 dB below its initial value. Show that this requires t to have the value $6.91/\beta$, and use this to justify Eq. 12.10.

12.4. Suppose for the sake of a simple calculation that the actual sequence of discrete reflections indicated in Fig. 12.1 is replaced by a smoothly decaying sound intensity $I(t) = I_0 \exp(-\gamma t)$. Show that the ratio R of Eq. 12.7 would then be given by $\exp(\gamma\tau) - 1$. For $\tau = 80$ ms, what value of γ would correspond to values $R = 1$ and $R = 2$? Using Eq. 12.23, what are the corresponding values of T_r? Do they seem reasonable?

12.5. For a room with volume 20,000 m^3, what is the total number of normal modes with frequency below **(a)** 200 Hz and **(b)** 2 kHz?

(c, d) How many modes per hertz are there in the vicinity of each of those frequencies?

(e) What is the effective bandwidth of each mode if $T_r = 2.2$ s?

(f) Above what frequency do you expect the room response to be fairly smooth?

(g) Answer the same questions for another room with $V = 200$ m^3 and $T_r = 0.7$ s.

12.6. A small classroom has vinyl tile floors and plaster walls, and 20 percent of the wall area plus the entire ceiling is covered with acoustic tile. Dimensions are $10 \times 6 \times 3$ m, and half the floor area is to seat 30 people. What is T_r at 1 kHz when the room is **(a)** empty, **(b)** occupied? In each case, what SPL_r is produced by an unassisted voice with $P = 100\ \mu\text{W}$?

12.7. Repeat the preceding problem with dimensions changed to $25 \times 15 \times 6$ m and 300 people occupying 300 m^2 of floor area.

12.8. Make an inventory of the major surfaces in your living room and measure their approximate dimensions. Use this information to estimate T_r at 250, 1000, and 4000 Hz. Comment on how much these times might reasonably be changed by altering the room design or furnishings.

12.9. Suppose the interior surfaces of a rectangular auditorium have an average absorption coefficient of 0.30, and dimensions $30 \times 50 \times 15$ m.

(a) What is the reverberation time?

(b) What acoustic power output P is required from a source if it is to produce a steady reverberant SPL of 65 dB?

(c) In order for a person's voice with $P = 50\ \mu\text{W}$ to produce this 65 dB without amplification, to what value would you have to reduce the average absorptivity?

(d) Tell why that would be physically very difficult to achieve, and describe other unwanted consequences that would occur.

12.10. I estimate the volume of my local concert hall to be about 30,000 m^3. If its midrange reverberation time is 2.2 s when set up for orchestra use, what is the effective total absorbing area A_e? But for musical theater shows with amplification, shorter reverberation would be preferred. So high above the seating area there is a catwalk and a heavy curtain that can be deployed (out of a storage bay in the wall) so that it stretches from one side wall to the other, and from catwalk to ceiling. Using curtain dimensions 40×8 m and remembering that both sides are exposed to the sound, estimate how much reduction in T_r may be achieved.

12.11. A small recital hall doubling as a large lecture classroom is roughly rectangular with dimensions $16 \times 25 \times 8$ m. In order to have $T_r = 0.9$ s for speech, what would be the total A_e and the average $\bar{\alpha}$? If a total area of 200 m^2 on the rear and side walls consists of panels whose absorptivity is 0.7, does that still leave a reasonable A_e and $\bar{\alpha}$ for the remaining surfaces? If those panels are reversible to give the room variable acoustics, and the other side has $\alpha = 0.1$, how long a T_r can you achieve for music by turning that side out?

12.12. A reverberation chamber has volume 60 m^3 and reverberation time 2.50 s when empty. Suppose that careful measurements of T_r are found to have an uncertainty of ± 0.05 s. Two panels are made with different materials, each with area 2 m^2, and mounted (one at a time) in the chamber with only one side exposed. Find both the absorption coefficient α and its uncertainty $\Delta\alpha$ in each case, if the observed values of T_r are **(a)** 1.75 s and **(b)** 2.32 s. How much would it increase the precision if you could test panels with twice as much surface area?

12.13. Consider an isotropic source emitting power P into a room with volume V and reverberation time T_r. **(a)** Show that the distance R from the source, beyond which the reverberant sound is stronger than the direct sound, is $R \simeq 0.1\sqrt{V/\pi T_r}$. **(b)** What is R for a $3 \times 4 \times 5$ m room with $T_r = 0.5$ s? **(c)** For $20 \times 30 \times 40$ m with $T = 2.2$ s? **(d)** Would you say that the reverberant sound dominates throughout most of the volume of most rooms?

12.14. In each of the rooms in the preceding problem, what reverberant sound level would be generated by a flute with $P = 2$ mW? How much from a trombone with $P = 2$ W?

12.15. If all inner surfaces of a hollow cube 20 m on a side have $\alpha = 0.06$, what T_r is predicted by Sabine's formula? What if $\alpha = 0.60$ everywhere? What if two opposite faces have $\alpha = 0.60$ while the other four have $\alpha = 0.06$? In the last case, what important feature of the reverberation remains unaccounted for by that formula?

12.16. Discuss the difference between the Sabine and the Eyring-Norris values for T_r in rooms with **(a)** $V = 40{,}000$ m^3, $A = 10{,}000$ m^2, $\bar{\alpha} = 0.32$; **(b)** $V = 4000$ m^3, $A = 2000$ m^2, $\bar{\alpha} = 0.20$; **(c)** $V = 400$ m^3, $A = 400$ m^2, $\bar{\alpha} = 0.12$.

12.17. It is not entirely true that sound is absorbed only during surface reflections; there is also a slight heating of the air through which it passes. This can be approximately allowed for by extending the definition of A_e to include an additional term proportional to the room volume:

$$A_e = \alpha_1 A_1 + \cdots + (0.1V/h)(f/1000)^{1.7}$$

Here h is the relative humidity in percent, and the formula is only intended to be used for $20 < h < 70$ and $f > 1000$ Hz. When $h = 40$, how much does the additional term alter estimates of T_r at 2, 4, and 8 kHz for rooms with **(a)** $V = 1000$ m^3 and A_e (of surfaces alone) $= 160$ m^2 (regardless of frequency); **(b)** $V = 24{,}000$ m^3 and A_e (surfaces) $= 1600$ m^2.

12.18. Suppose a listener's head is modeled as a sphere of diameter 15 cm, with the ears on opposite sides. What is the difference in sound path lengths to the two ears from a distant source located **(a)** directly ahead, **(b)** directly behind, **(c)** straight overhead, **(d)** at 90° off to one side, **(e)** 45° to one side? (*Hint:* Three answers are the same and one of the others is more than 15 cm.)

12.19. Consider sound arriving from a source to one side of a human head. Of the extreme treble and extreme bass, which will arrive with practically equal intensity at both ears, and why? Taking the head diameter as 15 cm, estimate the range of frequency over which you expect intensity differences to provide a useful clue about source direction.

12.20. An audience area located about 42 m from the primary sound source is to be served by a loudspeaker 8 m away, as in Fig. 12.10. What delay-time setting would you choose? In view of the comments in Sec. 12.2 about smoothness, how much leeway might you have in the amount of delay without sacrificing sound quality? How many words ought to be available in the memory of a digital delay unit, if the analog-to-digital conversion is carried out at 20,000 samples per second? In a magnetic-tape delay unit with the tape running at 38 cm/s, how far apart would the recording and playback heads be located for this situation?

LABORATORY EXERCISE

12.A. With sound-level meter, storage oscilloscope, chart recorder, or whatever other equipment is available, make as many diagnostic measurements as you can on some local classroom or auditorium. Possibilities could include sound level from a source of known strength, uniformity of that sound in space, uniformity of frequency response with a pink-noise source, reverberation time, and evidence of echoes. Try to relate those measurements to the perceived acoustical properties of the room.

Chapter 13

Environmental Noise

In the preceding chapter we considered ways of enhancing the production and reception of information-carrying sounds. But in real life our concern is often to eliminate, or at least reduce, annoying extraneous sounds. The word noise usually carries this implication of undesirability. In many cases it is also noise in the technical sense of having a continuous spectrum covering a substantial range of frequencies. But there are also times when we must speak of noise with pure-tone components—for instance, when a table saw radiates sound concentrated at harmonics of the frequency at which the teeth strike the wood, or when neighbors are disturbed by the whine of a gasoline-powered leaf blower.

Our concern with noise control may extend to both indoor (office or factory) and outdoor (highway or amphitheater) environments. In this very brief and incomplete introduction to the subject, we first discuss appropriate ways of measuring and describing noise strength. Then we consider the effectiveness of various barriers in excluding noise, and ways of predicting or alleviating noise levels in living or working spaces. In the final section, we present some elementary information about transportation-system noise.

13.1 NOISE MEASUREMENTS AND STANDARDS

The most complete description of any particular noise would provide a history of how its frequency spectrum changed with time. In especially important cases, such analysis might actually be carried out in attempting to understand the noise source very thoroughly. But most often we prefer a simpler summary that does not overwhelm us with so much information; we do not really want a $20,000 solution to a $200 problem.

In the case of continuing noise that does not change in time, it becomes more attractive and useful to study its frequency spectrum. The appropriate language and units have been introduced in Chap. 3. Practical work is very likely to be carried out with octave or third-octave band analysis, characterizing the noise by its pressure band levels, rather than by high-resolution analysis trying to determine the entire continuous spectrum function $S(f)$. In discussing such spectrum measurements, it is good to always state clearly whether or not the direct physical results have been adjusted (with the A-weighting scale, for instance) to better indicate their perceptual impact on people. Spectral information may sometimes save us from adopting noise-reduction measures that would be most effective above 1 kHz, for example, when the problem at hand is actually worst in the 500-Hz octave band.

In the case of unsteady or intermittent noise we might integrate over frequency rather than time and study the statistics of the time variation of the overall SPL. Finally, we often want to average over both frequency and time in order to present a single number describing the total noise experienced without regard to the details of how it was distributed. We need to give definitions here of some terminology that is widely used in describing such averages.

First, please recall that there is more than one way to form an average of several numbers. As a simple example, what is the average of 2 and 8? The answer depends on whether you have in mind an arithmetic mean $(2 + 8)/2 = 5$, a geometric mean $\sqrt{2 \times 8} = 4$, a root mean square $\sqrt{(4 + 64)/2} = 5.8$, or some other kind. Consistent with the way we began in Chap. 2, we specify that all average noise figures will be based on a simple arithmetic mean of energy or intensity:

$$\langle I \rangle = (1/T) \int_0^T I(t)\, dt. \tag{13.1}$$

This says the same total amount of energy was delivered by the fluctuating sound as if it had been a perfectly steady sound of intensity $\langle I \rangle$ continuing throughout the time T. Since I is proportional to p^2, this means that whenever we want to characterize the sound by its pressure we must deal with rms averages:

$$p_{\text{rms}}^2 = \langle p^2 \rangle = (1/T) \int_0^T p^2(t)\, dt. \tag{13.2}$$

So, finally, if data are recorded in terms of sound pressure levels in decibels, the proper way to combine them is with an exponential average:

$$\boxed{\langle \text{SPL} \rangle = 10 \log \left[(1/T) \int_0^T 10^{\text{SPL}(t)/10}\, dt \right].} \tag{13.3}$$

Take, for example, a situation where the sound level is 80 dB when a machine is running and the background noise is 60 dB with the machine off. If during some length of time T the machine was on during half that time, what was the average sound level? It was *not* (80 + 60)/2 or 70 dB. Rather, it was

$$\langle \text{SPL} \rangle = 10 \log \{(1/T)[(T/2)10^8 + (T/2)10^6]\}$$
$$= 10 \log (50.5 \times 10^6) = 77 \text{ dB}.$$

It is seldom necessary today to take extensive series of SPL readings and calculate their average by hand. This task is better handled by a small microcomputer (such as the one shown in Fig. 13.1), which can be left unattended to take sample readings at regular time intervals (such as 1 s). Each reading is immediately used to update a running average, and after the desired period (such as 24 h) the sampling will stop automatically and the final result will be printed out on command. Other statistical information about how the sound levels varied is also available from such an analyzer.

Since for a given SPL low-frequency noise is generally less annoying to people than middle and high frequencies, it has become widespread practice to use A-weighted levels. (Information on A weighting can be found in Fig. 2.5 and Table 3.1.) Nearly all local noise ordinances are stated in terms of dBA, and you should expect generally to use A-weighted measurements in all community noise work except when there is some specific reason to do otherwise. The average discussed thus far is generally called the **equivalent level** L_{eq}; so the basic average that serves as a departure point is

$$L_{eq,A} = \langle \text{SPL(A)} \rangle. \tag{13.4}$$

A common variation on this is the **day-night average** $L_{dn,A}$. This is a 24-hour average formed like (13.3) except that all sounds between 10 p.m. and 7 a.m. are penalized by using SPL + 10 in place of the actual SPL. If in addition a 5-dB penalty is added in the evening hours of 7 to 10 p.m. we have what is called **community noise equivalent level,** or CNEL.

If the time T = 24 h in the last example and the on and off times were distributed uniformly through the day and night, we would pretend that

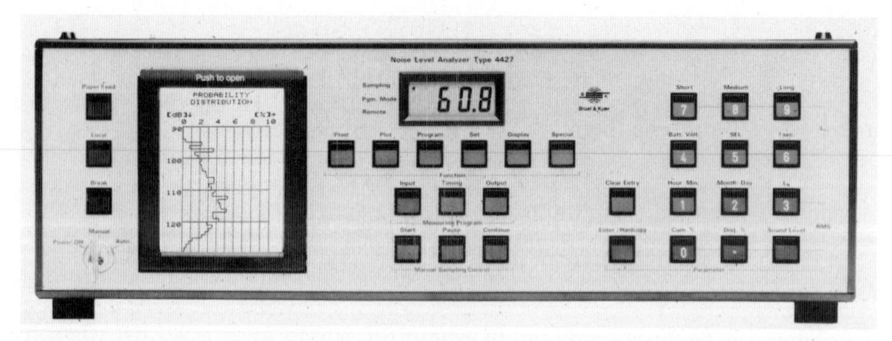

Figure 13.1 A noise-level analyzer, which will automatically accumulate statistical information about noise that varies in time. (Courtesy of Bruel and Kjaer Instruments, Inc., Marlborough, MA.)

the SPL had been 90 dB during half the nine nighttime hours and 70 dB for the other half. Then

$$L_{dn} = 10 \log \{(1/24)[(15/2)10^8 + (15/2)10^6 + (9/2)10^9 + (9/2)10^7]\}$$
$$= 10 \log (221 \times 10^6) = 83 \text{ dB}.$$

In similar fashion, the additional evening penalty would give CNEL = 84 dB.

A single average figure may be inadequate to judge the importance of a noise, because some degree of acclimatization can occur with steady sound whereas intermittent sounds are more difficult to ignore. Consider the two cases represented in Fig. 13.2: Y and Z both have the same average, L_{eq} = 60 dB, but Z involves much greater fluctuations. Such information is often presented in terms of a cumulative distribution, as in Fig. 13.3; for each sound level the curve tells what fraction of the time the actual readings were that much or higher. If you proceed from right to left, you can view the curves of Fig. 13.3 as being the integrals of those in Fig. 13.2.

In interpreting such distributions, it is common to refer to some specific level L_x, the **level exceeded** x percent of the time. L_{50}, for instance, is a level such that half the readings are higher and half lower, which is also called the median. Cases Y and Z above have median levels L_{50} = 60 and 54 dB, respectively, illustrating that L_{50} can be quite different from L_{eq}. The sound level spends 90 percent of the time above L_{90} and 10 percent of the time above L_{10}, and so 80

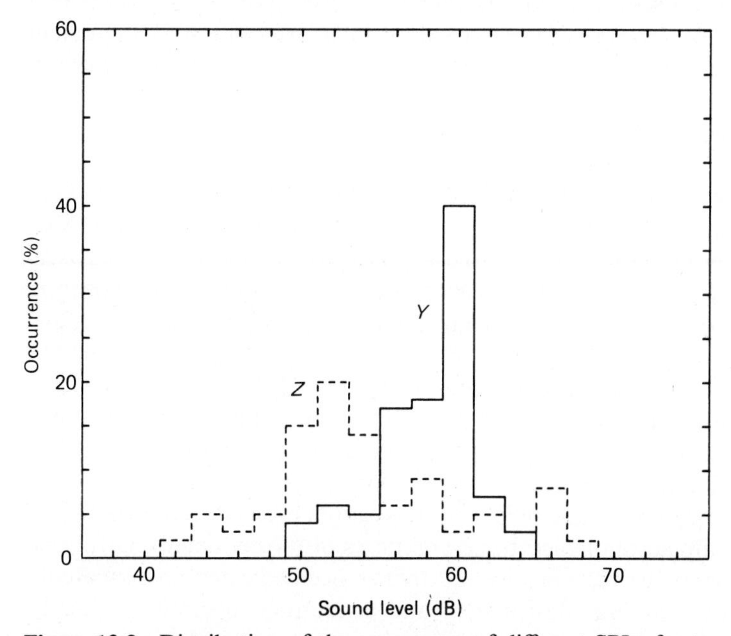

Figure 13.2 Distribution of the occurrence of different SPLs for two different sound histories. Vertical axis is the percent of total time during which SPL fell *within* each range.

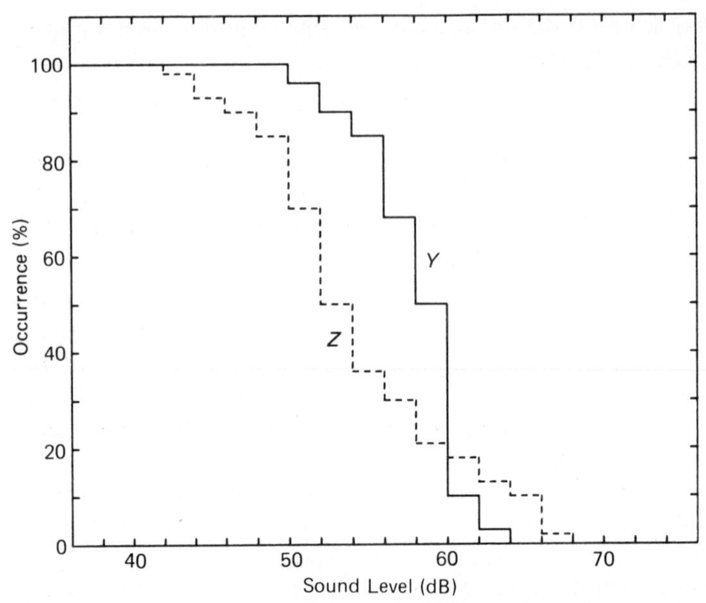

Figure 13.3 Cumulative distribution for the same sounds as in Fig. 13.2. Vertical axis now is the percent of total time during which SPL fell *above* each level.

percent in the range between the two. For case Y this range is only 54 to 62 dB while for Z it is 48 to 66 dB. Note that limits imposed in terms of L_{10} instead of L_{eq} will discriminate more harshly against highly variable sounds as compared with steady ones. An example of a statute written in these terms is the requirement in federal highway construction that protective measures such as noise barriers or buffer zones are to be used wherever L_{10} would exceed 70 dB on adjoining residential property.

Actual community noise ordinances may have rather complex requirements, in their attempt to deal realistically with long-established activities. They do well to emphasize cooperation toward alleviation of problems, reserving fines or other penalties for a last resort in persistent cases of violation. Consider as an example the noise standards adopted by Sacramento County, summarized in Table 13.1. The daytime exterior standard could also be described by saying that none of the limits $L_{50} = 55$, $L_{25} = 60$, $L_{8.3} = 65$, $L_{1.7} = 70$, or $L_0 = 75$ may be exceeded. Note the penalties in recognition that tonal, impulsive, and information-carrying sounds are all more annoying than steady broadband noise. (Because of its insistent stimulation of mental activity, music from an open-air concert several kilometers away may cause complaints even when it is weaker than ambient steady broadband noise and thus cannot even be measured.) These basic standards are accompanied by several pages of further details, largely exemptions for cases such as school athletic activities, licensed entertainment events, emergency work, building construction, and agricultural operations. Clearly, any work with community noise problems requires that you obtain and study carefully the relevant local requirements.

TABLE 13.1 NOISE STANDARDS FOR SACRAMENTO COUNTY, CALIFORNIA

Interior noise standard (limits on noise caused inside any neighboring unit of an apartment, duplex, or condominium. This standard applies only to nighttime hours, 10 p.m. to 7 a.m.)

55 dBA	At any time
50	More than 1 min in any hour
45	More than 5 min in any hour

Exterior noise standard (limits measured 1 ft inside property line in residential areas on noise from any neighboring property)

75 dBA	At any time
70	More than 1 min in any hour
65	More than 5 min in any hour
60	More than 15 min in any hour
55	More than 30 min in any hour

Limits as stated are for 7 a.m. to 10 p.m. and are all reduced 5 dB during nighttime hours. There is also a 5-dB reduction for sounds consisting principally of:

Pure tones
Impulsive noises
Speech or music

TABLE 13.2 OSHA GUIDELINE LIMITS ON PERMISSIBLE DAILY NOISE EXPOSURE IN WORKPLACES, AND MORE CONSERVATIVE LIMITS SUGGESTED FOR AVOIDABLE EXPOSURE

Sound level (dBA)	Maximum 24-hour exposure	
	Occupational	Nonoccupational
80		4 h
85		2 h
90	8 h	1 h
95	4 h	30 min
100	2 h	15 min
105	1 h	8 min
110	30 min	4 min
115	15 min	2 min
120	0 min	0 min

Another kind of standard is involved if we are concerned with protection against hearing loss rather than annoyance. The U.S. Occupational Safety and Health Administration has given guidelines as shown in Table 13.2. These are neither absolute legal requirements nor absolute guarantees of adequate ear protection. This is partly because it is difficult to establish just how much noise exposure will produce a specified amount of damage (such as 25 dB or more reduction in ear sensitivity in 5 percent of the individuals exposed over a working lifetime of 30 years). Note that the halving of exposure time for each 5-dB increase in sound level represents a hypothetical trade-off in which you could tolerate 3.16 times as much intensity for half as much time, or 1.58 times as much total energy deposited in the short time as in the long time. This implies that a given amount of energy causes somewhat less damage if concentrated in short bursts so that the ears can rest in between. Whether that is actually true is still a matter of debate and research. Some European countries have established standards based on the assumption that the total energy determines the damage regardless of how it is spread over time, so that they require a halving of exposure for every 3-dB increase in sound level.

Several forms of ear protection are available. Simple swimming earplugs can provide attenuation ranging from practically nil up to 15 dB or more if they are properly sized and carefully inserted. Earmuff protectors can also give 15 to 20 dB reduction. As much as 25 to 30 dB attenuation can be achieved with individually molded earplugs, which are of course the most expensive as well as most comfortable option. It is also possible to buy miniature ear-insert low-pass filters, which can block much of the noise above 1 kHz while interfering very little with conversation.

13.2 NOISE IN LIVING AND WORKING SPACES

While noise levels generally under 55 dB in daytime or 45 dB at night may be needed for a residential neighborhood to be pleasant, higher levels will be common in offices or workshops. Even without reaching such high levels as to cause long-term damage to the ears, there may be problems in carrying on needed communication among workers. Empirical studies have found that the difficulty may be described very roughly in terms of

$$L_A + 20 \log (r/1 \text{ m}), \tag{13.5}$$

where L_A is the A-weighted background noise level and r is the distance between speaker and listener. If this quantity is less than about 65, communication will be easy with a normal speaking voice. As it rises to 75, extra effort will be required to speak very loudly, and above 90 accurate communication becomes impossible even with shouting. Such a situation may both lower efficiency and create a need for alternate signals, such as danger warning lights.

At a distance $r = 18$ ft $= 5.5$ m, for example,

$$L_A + 20 \log (5.5) = L_A + 15.$$

Then casual speech will be successful only if the background noise is less than about 50 dB. If L_A is less than 60 dB, it will still be possible to converse by speaking loudly, and up to about 75 dB increasingly limited communication could take place.

How may high noise levels be remedied? For noise originating within the room, the best solution would be to eliminate or at least reduce the original noise emission. For noise coming in from outside, we desire ways of achieving noise insulation to prevent its entry. In both cases we also consider further attenuation through use of noise-absorbing materials. Let us consider each of these approaches, starting with the last one.

If a workspace is highly reverberant, any noise created persists for a long time, and the cumulative level is raised. We showed in the preceding chapter that sound sources of total power P in a room with total effective absorption A_e will produce a reverberant sound level

$$\mathrm{SPL}_r = 10 \log (P/A_e) + 126 \text{ dB}. \qquad (12.20)$$

Adding absorbing material, for example, acoustic tile on a ceiling, will increase A_e. But rarely will there be opportunity to decrease SPL_r as much as 10 dB merely through an increase in average absorptivity; that would correspond to converting a very live room into a very dead one. If substantial noise reduction has to be done through absorption, introduction of additional surfaces can be helpful, as illustrated in Fig. 13.4.

Stopping the noise at the source is usually a preferable solution whenever possible. Careful choice of original equipment and good maintenance of machinery should not be overlooked; worn bearings, unbalanced rotating shafts, and especially loose cover panels may all contribute unnecessarily to noise output. Vibration transmitted through the floor to the whole building structure may not only be unpleasant in itself but also radiate sound into the air both in the original room and in others, and may call for carefully designed resilient mounting pads under the offending source. Machines with large vibrating surface areas can sometimes be quieted by making the panels more massive or covering them with absorbent material. This is an extensive subject with a lore of its own that we will not attempt to survey here.

It may occur to you that an attractive way to prevent noise from one source creating problems throughout a large room would be to completely enclose the source (Fig. 13.5). This is correct as long as you understand and avoid one basic pitfall: a simple wall, even if quite heavy, will give no net reduction of sound

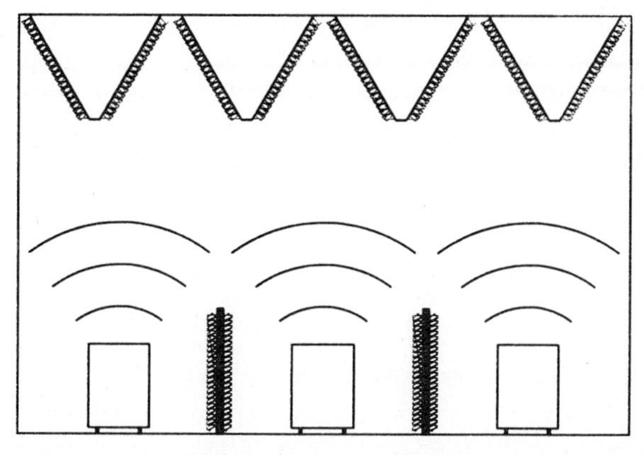

Figure 13.4 Additive of extra surface area can help increase total absorption in a noisy room. Here baffle panels covered with soft absorptive batting material are suggested both between machines on a shop floor and in free space below the ceiling.

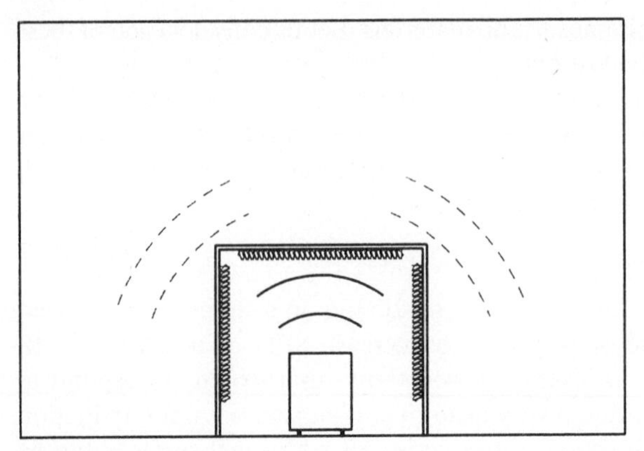

Figure 13.5 Isolation of a noise source with a small enclosure still requires absorbing material to reduce sound levels in the outer room.

level if it has no way to absorb the sound. In Chap. 12 we used terms like "effective absorption" rather indiscriminately to mean anything that removed sound energy from the room in question. But now it is important to distinguish between actual absorption (conversion of sound into some other form of energy) and transmission (sound continuing through a wall into another room). Any such enclosure needs to be *both* massive enough to prevent immediate transmission of sound through its walls *and* absorptive enough to kill the major portion of the sound before it escapes by gradual leakage.

Suppose, for example, a machine generates acoustic power $P = 0.1$ W in a room with $V = 1000\ \text{m}^3$ and effective total absorption $A_e = 100\ \text{m}^2$, thus causing a background noise level

$$\text{SPL}_r = 10 \log (0.1/100) + 126 = 96 \text{ dB}.$$

Let it now be enclosed within a 50 m^3 volume by 60 m^2 of hard walls that transmit only 1 percent of the incident sound. You might at first think this would reduce the sound level outside from 96 to 76 dB. But that is not true, because the other 99 percent of reflected sound keeps hitting the walls again and again until it finally all leaks out. In fact, the room-within-a-room has effective absorbing area of only

$$A_e = 0.01 \times 60\ \text{m}^2 = 0.6\ \text{m}^2,$$

so that the sound level inside it builds up to

$$\text{SPL}_r(\text{in}) = 10 \log (0.1/0.6) + 126 = 118 \text{ dB}.$$

But now the inner walls of the enclosure are receiving an average energy flux

$$P/A_e = (0.1 \text{ W})/(0.6\ \text{m}^2) = 0.16\ \text{W/m}^2,$$

or 10 W total over the whole 60 m^2. One percent of this escaping means the small room acts as a 0.1-W source in the larger room, just as the machine did when unenclosed, and SPL remains 96 dB outside the enclosure.

Only if sound energy can be literally dissipated into heat in the enclosure will the outside sound level be reduced. Suppose, for instance, 30 m^2 of effective absorption can be attained by lining the inner walls with material of absorptivity $\alpha = 0.5$. Then the sound level inside will be reduced to

$$\text{SPL}_r(\text{in}) = 10 \log (0.1/30) + 126 = 101 \text{ dB}.$$

Corresponding to this 17-dB decrease, the outer room now will get a similar benefit:

$$\text{SPL}_r(\text{out}) = 96 - 17 = 79 \text{ dB}.$$

How can we best characterize sound transmission through different types of wall construction? The first and simplest way is to imagine the wall occupying an entire infinite plane, with a plane wave directly incident upon it. Then the ratio of the transmitted and incident wave intensities would be a measure of the attenuation due to the wall, which is called **transmission loss** (or sound-reduction index):

$$\boxed{\text{TL} = 10 \log (I_{\text{inc}}/I_{\text{trans}}).} \qquad (13.6)$$

For simple types of walls, measurements of TL often bear at least some resemblance to Fig. 13.6. Toward the low-frequency end there may be one or more resonances, meaning very poor sound insulation since the wall vibrates readily. These occur because the connection of the wall to its surroundings (foundation, adjoining walls, etc.) provides a certain stiffness against which the mass of the wall must move. There may be additional resonance (usually at higher frequencies, especially for lightweight walls) at the so-called **coincidence frequency.** This

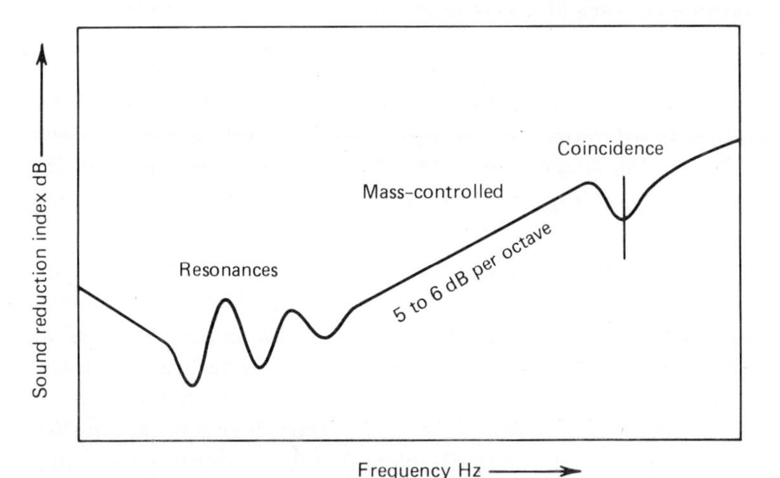

Figure 13.6 Idealized dependence of wall transmission loss on frequency. Wall resonance gives minimum attenuation and often occurs at low audio frequencies, making mass-limited motion (upward slope toward the right) relevant in the midrange. (From Parkin et al., *Acoustics, Noise and Buildings,* by permission of Faber and Faber.)

represents resonant excitation of surface waves traveling along the wall by sound waves approaching at such an angle that their crest-to-crest spacings match.

In between, there may be a significant midfrequency range in which the attenuation increases about 6 dB/octave. This can be interpreted as due to mass-controlled motion. Just as for the simple harmonic oscillator, driving this system above resonance means the stiffness has very little effect on the response. We can easily make a rough estimate of attenuation due to wall mass if we suppose the force needed to launch the weak transmitted wave is negligible compared with that used to move the wall itself. For wave amplitude p_i incident on a heavy wall the reflected wave will have nearly the same amplitude; so a total pressure $2p_i$ acts on the mass per unit area σ. Setting $\sigma\, dv/dt = 2p_i \exp(j\omega t)$, we see that the wall should move with velocity $v = (2p_i/j\omega\sigma)\exp(j\omega t)$. But the immediately adjacent air moves along with the same velocity. According to (10.14), the transmitted sound wave having this velocity must have pressure amplitude $p_t = z_c v = (2z_c/\omega\sigma)p_i$. So finally

$$\mathrm{TL} = 20 \log (p_i/p_t) \sim 20 \log (\omega\sigma/2z_c). \tag{13.7}$$

Writing $\omega = 2\pi f$ and using $z_c = 415$ kg/m²s, this may be put in the practical form

$$\mathrm{TL} \lesssim 20 \log (f\sigma) - 42 \text{ dB}. \tag{13.8a}$$

More complicated calculations averaging over all angles of incidence suggest that for reverberant sound fields the number 42 should be replaced by 48, so that

$$\boxed{\mathrm{TL} \lesssim 20 \log (f\sigma) - 48 \text{ dB}.} \tag{13.8b}$$

Half-inch gypsum wallboard, for instance, with $\sigma = 10$ kg/m², has a predicted transmission loss at 1 kHz of about

$$\mathrm{TL} \simeq 20 \times 4 - 48 = 32 \text{ dB}.$$

Four-inch brick has $\sigma = 200$ kg/m², for which the corresponding predicted TL would be 58 dB, though in practice it may not be that good. In each case the attenuation would improve by 20 log (2) = 6 dB/octave for increasing frequency, as long as it is within the mass-controlled range.

Besides increasing effectiveness toward higher frequency, (13.8) also indicates a 6-dB improvement in attenuation for every doubling of wall mass. Where attenuations of 50 dB or more are required at low frequencies the amount of mass needed would become prohibitive, but there is a better way. Each air-solid interface constitutes an impedance mismatch which tends to reflect waves; so a given total mass divided into two sheets with air space between can improve on the simple wall by another 10 or 15 dB. Note that the proper way to build a good sound-insulating double-leaf partition is to fasten each wall layer to an independent set of studs so that there is no path for vibrations to pass from one sheet to the other except through the enclosed air layer. For details about sound transmission through walls with specific designs and construction materials, a good architectural acoustics handbook (such as Cyril Harris's) should be consulted.

Unfortunately, all the preceding discussion applies to the idealized situation where the transmitted wave simply leaves the scene forever. If it actually finds itself in a reverberant room, it may make further contributions to the measured sound level there. Sometimes what you really want to know for practical purposes is the **noise reduction** NR, defined by

$$\boxed{\text{NR} = 20 \log (p_1/p_2) = \text{SPL}_1 - \text{SPL}_2.} \tag{13.9}$$

Here SPL_1 and SPL_2 refer to sound levels in the source room and receiving room, respectively, and since both are supposed to be reverberant levels they need not be measured close to the common wall. In fact this definition can still apply to more widely separated rooms that have no common wall.

While (13.9) superficially resembles (13.6), they are in fact quite different. If we consider the special case where there is a common wall of area A_c and the amount of sound returning from the second room to the first is relatively negligible, we can relate one to the other. According to (12.14) and (12.18), the energy flux upon the common wall from a source of power P_1 in the first room is

$$P_c = (E_1 c/4V_1)A_c = (P_1/A_{e1})A_c. \tag{13.10}$$

Eq. 13.6 then gives the portion of this that is transmitted,

$$P_2 = 10^{-(\text{TL}/10)} P_1 A_c/A_{e1}. \tag{13.11}$$

This acts as a source in the second room to build up a reverberant sound level as in (12.20):

$$\begin{aligned} \text{SPL}_2 &= 10 \log (P_2/A_{e2}) + 126 \\ &= -\text{TL} + 10 \log (P_1 A_c/A_{e1} A_{e2}) + 126 \\ &= -\text{TL} + 10 \log (A_c/A_{e2}) + \text{SPL}_1. \end{aligned} \tag{13.12}$$

Thus in such a case

$$\boxed{\text{NR} = \text{TL} - 10 \log (A_c/A_{e2}).} \tag{13.13}$$

For instance, a partition with $A_c = 20\ \text{m}^2$ and TL = 32 dB letting sound into a rather live small room with A_{e2} of only 10 m^2 will only have

$$\text{NR} = 32 - 10 \log (20/10) = 29\ \text{dB}.$$

Increasing A_{e2} to 60 m^2 with absorbent materials could increase NR to 37 dB even without any change in the partition.

It would be a mistake to devote our total attention to the details of solid common-wall construction. In practice the insulating potential of a good wall is all too easily wasted by overlooking other ways in which sound can travel from the source to the receiving room. These **flanking paths** (Fig. 13.7) are like parallel branches in an electric circuit; the path of least resistance is the one where the greatest current flow will occur. When you consider that a TL of 40 dB means that no more than 0.01 percent of incident sound energy can be transmitted, you can see how a very small hole in a wall could let this much through regardless of how heavy and solid the wall is otherwise. This is why building-construction codes often prohibit electrical outlet sockets in walls dividing adjoining units in a condominium.

Wherever a door is located, it almost invariably is the determining factor in the transmission of sound through the wall in which it is mounted. Where high attenuation is required, doors must be heavy and closely fit in their frames, preferably with a felt or rubber sealing strip so that there is no place for air to move freely around the edges. Reasonable TL figures might be 15 dB for a well-fitted but unsealed ordinary door, with an increase to 20 or 25 dB from adding good seals. It takes massive special-design doors with detailed sealing to exceed TL = 30 dB.

Privacy between adjoining offices is often limited by an easiest transmission path through a common ventilating-system duct or through the space above a suspended ceiling. Measures that can reduce such problems include routing ducts so that each office is on a separate branch, increasing suspended-ceiling mass, spraying the plenum above the ceiling with acoustic absorbing material, and extending walls all the way to the true ceiling. To give an idea of possible numbers in a medium case, 3/4-in fiber tile (with $\sigma = 6$ kg/m^2) suspended in a lay-in grid with a 60-cm utility space above will leave at best 25 to 30 dB of noise reduction between rooms connected through that space. We might use (13.13) to crudely estimate maximum noise reduction when there is a duct connection as

$$\mathrm{NR}_{\max} = \mathrm{TL}_{\mathrm{duct}} - 10 \log (A_{\mathrm{duct}}/A_{e2}). \tag{13.14}$$

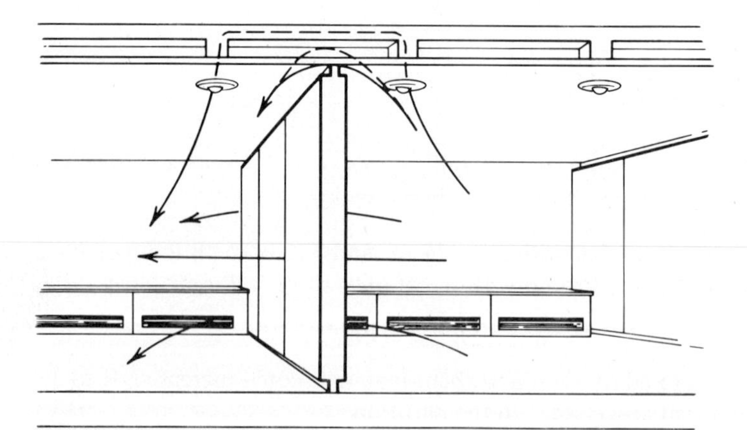

Figure 13.7 A sketch of various flanking paths for sound transmission that would defeat the sound insulating properties of an interior wall. (From Parkin et al., *Acoustics, Noise and Buildings,* by permission of Faber and Faber.)

For instance, the presence of a duct with area 0.06 m^2 leading to a room with $A_{e2} = 10\ m^2$ would keep NR between that room and any other connected to the same duct down to about

$$-10 \log (0.06/10) = 22 \text{ dB}$$

unless there were measures ensuring some TL within the duct.

13.3 TRANSPORTATION NOISE

One of the major sources of noise in modern industrial societies is our systems for moving people and goods. Buildings located close to airports, railroad lines, and main highways are inevitably subject to considerable noise. If everything could be planned with perfect foresight, such locations would be uniformly zoned for industrial use only. But of course in reality many people find themselves in homes (or offices or stores or schools) with significant transportation-noise problems.

It is a considerable task to develop adequate descriptions and measurements for aircraft noise. Each flyover forms a separate sound event, with relatively quiet times in between. Comparisons need to be made in terms of peak level reached, effective duration of levels near that peak, number of events per day, and whether they extend into the night. The most severe noise problems occur for takeoffs, with the plane still close to the ground and using full power to climb. A rough idea of the nature of this noise can be gained from Figs. 13.8 and 13.9. Large commercial airliners present problems different from those of private airplanes not only because of their greater total noise output but because a higher proportion of the noise from their jet engines is in the high-frequency bands where it is most annoying. Total sound power outputs range up to the order of a megawatt for a supersonic transport. The sound levels received depend strongly on the type and size of airplane (Fig. 13.10), location directly under or away from the flight path, and weather conditions. A site with high sound levels yesterday may have no problem today simply because wind shifts have moved the takeoffs to a different runway. Detailed noise surveys near airports must be compared with federal, state, and local requirements.

Highway noise is more often amenable to simple description as a relatively steady noise due to the combined effect of many vehicles in a steady stream of traffic. Let us look at a simple model of this noise, beginning with a single car. Suppose we observe from a distance r, far enough away to treat the car as a point source, and assume the total radiated sound power P goes uniformly in all directions. Since the roadway provides a hard reflecting surface underneath, this sound is limited to a half space and has intensity

$$I(r) = P/2\pi r^2. \tag{13.15}$$

But the car is moving with speed v; let us suppose for simplicity that it is on a straight and level road and that we let $t = 0$ be the time it is closest to a fixed observer a distance d from the road (Fig. 13.11). Since

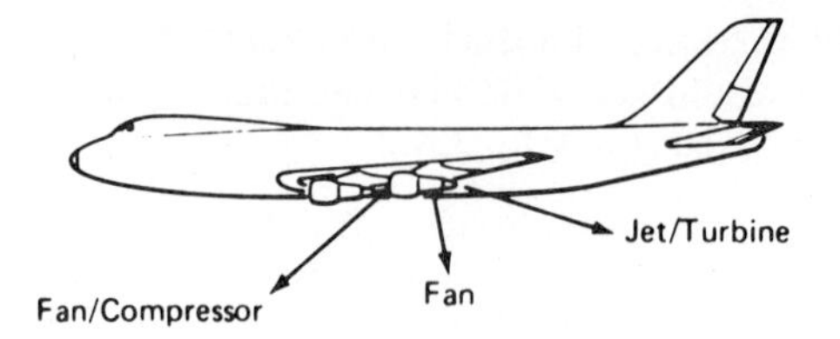

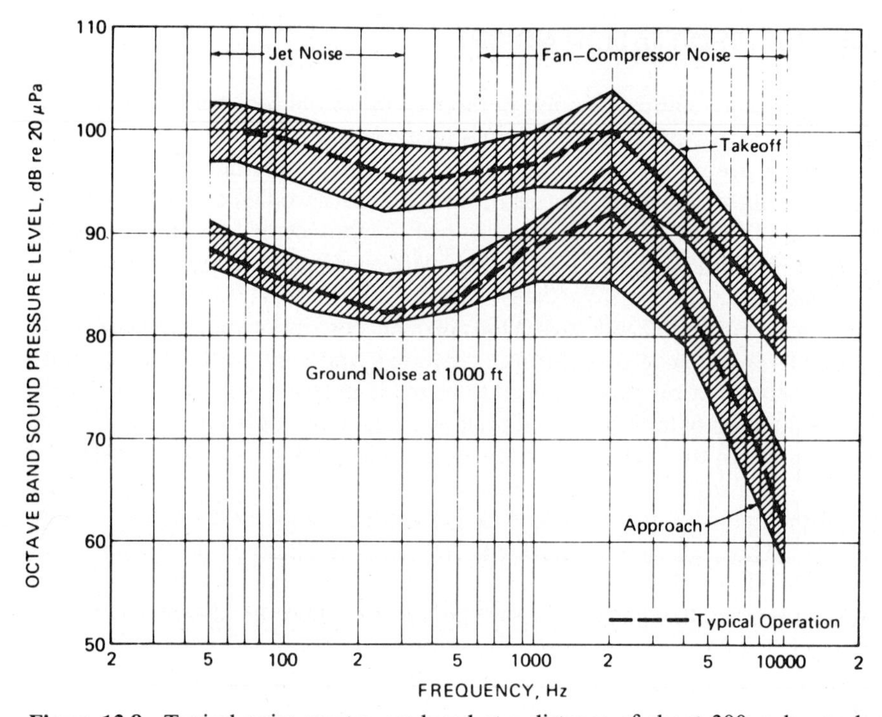

Figure 13.8 Typical noise spectra produced at a distance of about 300 m by modern wide-body fanjet airliners. (From D. N. May, editor, *Handbook of Noise Assessment,* by permission of Van Nostrand Reinhold.)

$$r^2 = d^2 + (vt)^2, \tag{13.16}$$

we can write

$$\begin{aligned} \text{SPL}(d, t) &= 10 \log (p^2/p_{\text{ref}}^2) = 10 \log (\rho c I/p_{\text{ref}}^2) \\ &= 10 \log [P/(d^2 + v^2t^2)] + 112 \text{ dB}, \end{aligned} \tag{13.17}$$

where the last form requires that P be in watts and d in meters. Figure 13.12 illustrates this with $P = 0.01$ W and several values of d. It should be no surprise that each doubling of d reduces the predicted peak SPL by 6 dB. But note also that the effective duration T of the peak increases with d. In fact, if we define T as the time during which SPL is within 3 dB of its maximum, it begins and ends when $r^2 = 2d^2$; thus

$$\boxed{T = 2d/v.} \tag{13.18}$$

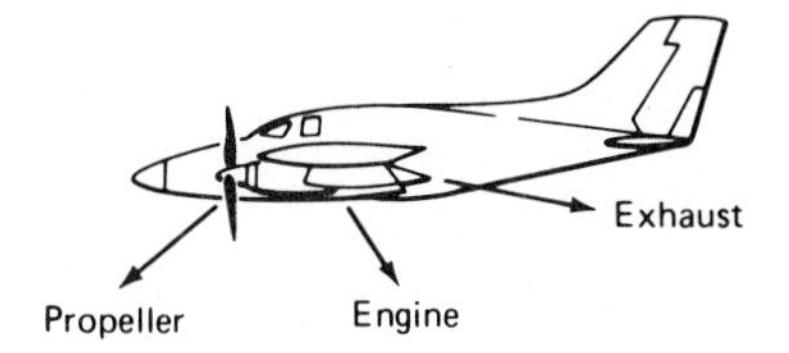

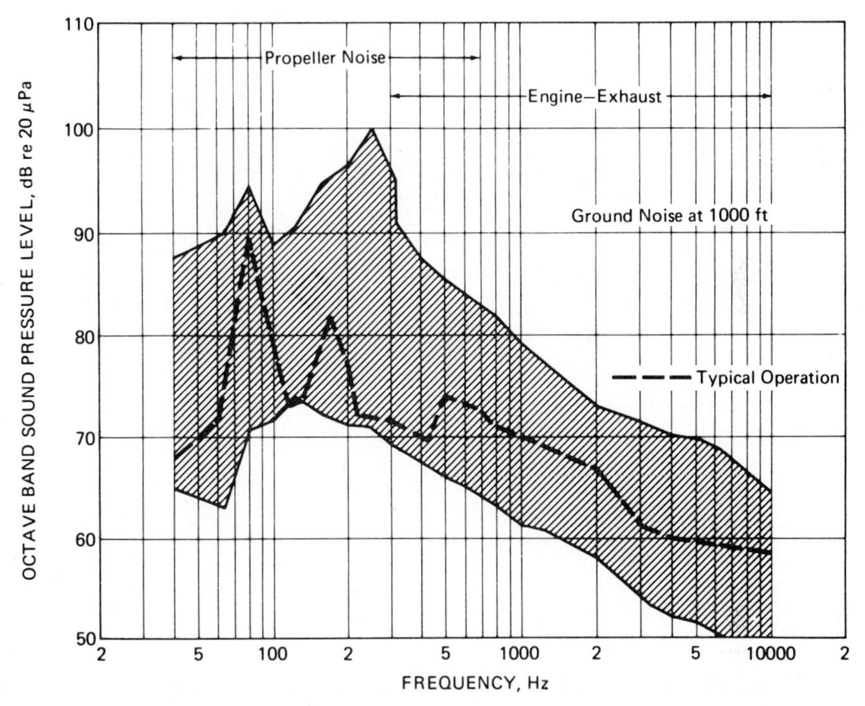

Figure 13.9 Typical noise spectra from small propeller-driven aircraft. (From D. N. May, editor, *Handbook of Noise Assessment,* by permission of Van Nostrand Reinhold.)

When many cars are on the road, we have two possible extremes: either we hear the sound rise and fall for each car before the next comes along or we hear a steady sound due to their combined effect. We could estimate which would be the better description from (13.18), for if the spacing from one car to the next is less than $2d$ the increase in SPL from one approaching will largely compensate the decrease from the other receding. The usual way to describe traffic density is with N, the number of cars passing per unit time. The reciprocal of this would be the average time from one car's arrival to the next, and so v/N is the typical distance from one car to the next. Thus we can predict the criterion

$$\begin{array}{ll} v/N \ll 2d: & \text{steady noise,} \\ v/N \gg 2d: & \text{individual peaks.} \end{array} \tag{13.19}$$

This says that for a given position d, heavier traffic will produce louder and

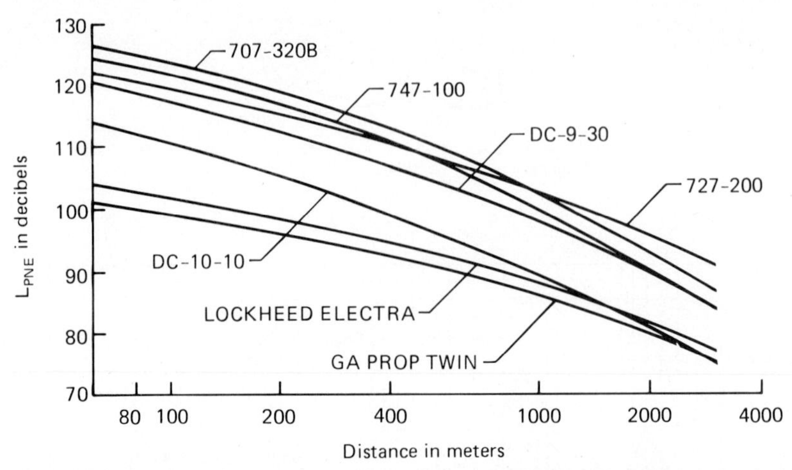

Figure 13.10 Noise level as a function of distance during takeoff for several commercial aircraft. L_{PNE} is another frequency-weighted quantity, the "perceived noise equivalent level." (From C. N. Harris, editor, *Handbook of Noise Control,* by permission of McGraw-Hill Book Co.)

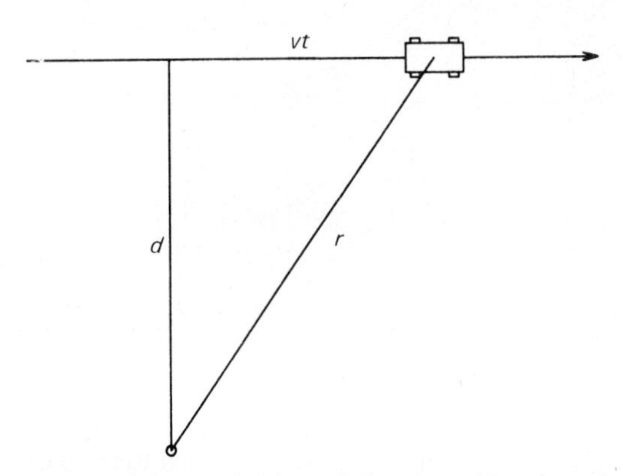

Figure 13.11 Definition of distances d from road and r from car to observer.

steadier noise. But notice that it also says a given stream of traffic will sound more and more steady as you move farther away from it, as illustrated in Fig. 13.13.

When N is large enough to speak of a steady noise level, we could think of the road as a line source radiating power per unit length $P/(v/N)$. This would produce intensity

$$I = (NP/v)/\pi d \tag{13.20}$$

and sound level

$$\boxed{\text{SPL}(d) = 10 \log (\text{NP}/vd) + 115 \text{ dB.}} \tag{13.21}$$

To justify intuitively why this drops only 3 dB for each doubling of d when each car's individual peak contribution falls 6 dB, remember that all cars within a

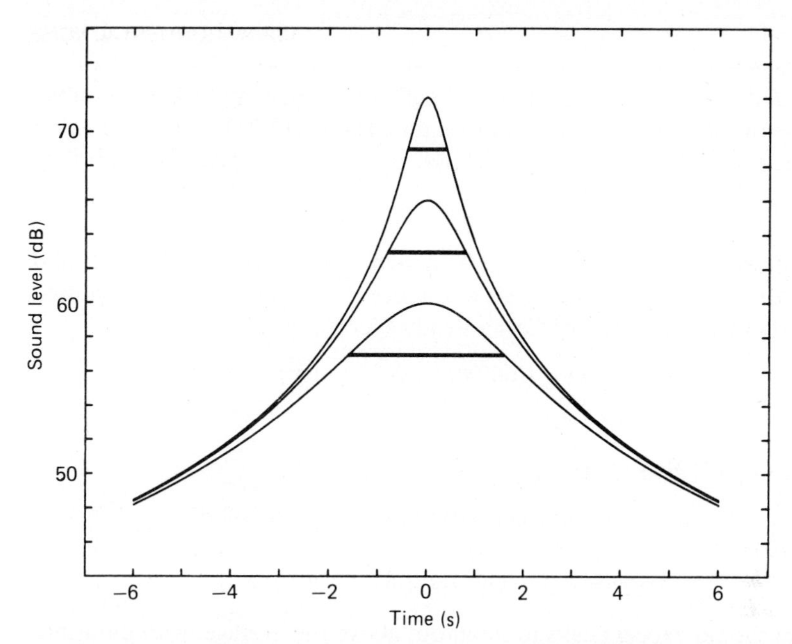

Figure 13.12 Sound levels predicted from (13.17) for a 0.01-W source traveling along a straight road with speed v = 25 m/s (90 km/h). Curves represent, from top to bottom, distances d = 10 m, 20 m, and 40 m from road to observer. Bars indicate durations T = 0.8 s, 1.6 s, and 3.2 s during which sound is within 3 dB of its peak value.

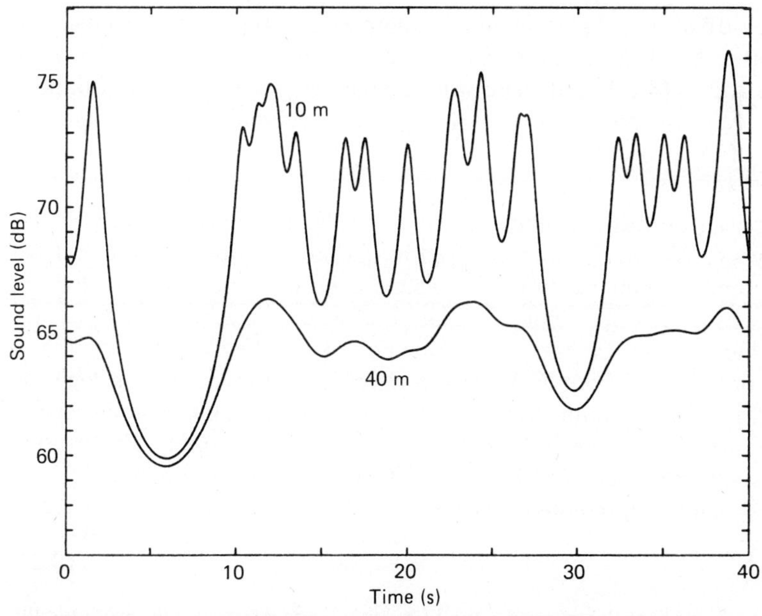

Figure 13.13 Sound levels for a simulated stream of traffic with P = 0.01 W, v = 25 m/s, and average N = 0.6 cars/s but irregular spacing. Upper and lower curves are for distances d = 10 m and 40 m from the road, so that one is smaller and the other larger than $v/2N$ = 21 m.

distance $2d$ along the road contribute strongly to the total, and when d doubles so does this number of important contributors. Using (13.21) even when v/N is not less than $2d$ will nevertheless be useful, because it will give the equivalent average L_{eq} of the fluctuating sound.

For illustration, consider $N = 600$ cars/h moving at $v = 72$ km/h and radiating $P = 0.01$ W apiece. At a distance $d = 15$ m (which is often used as a standard), (13.21) predicts a steady average of

$$\langle \text{SPL} \rangle = 10 \log [(600)(0.01)/(72 \times 10^3)(15)] + 115$$
$$= 62 \text{ dB},$$

while (13.17) gives a peak value of

$$\text{SPL}_{max} = 10 \log [(0.01)/(15)^2] + 112 = 68 \text{ dB}.$$

Here

$$v/N = (72 \text{ km/h})/(600/\text{h}) = 120 \text{ m} \gg d,$$

so we should expect peaks to stand out above the average, with durations on the order of

$$T = (30 \text{ m})/(72 \text{ km/h}) = 1.5 \text{ s}$$

and average separation

$$1/N = (3600 \text{ s/h})/(600/\text{h}) = 6 \text{ s}.$$

The 62-dB average figure should be interpreted as the L_{eq} of this noise. On the other hand, at distance $d = 150$ m the maximum expected from a single car is 48 dB, and there will be relatively little variation around a steady value of about 52 dB. This level is generated mainly by the combined effect of the two or three nearest cars.

The power output P of each vehicle depends, of course, on both vehicle type and speed. Without getting involved in questions of individual driver behavior (especially at intersections) or vehicle condition (such as a rusted-out muffler), let us try to offer "typical" values that would be reasonable to use for average traffic. For automobiles, the engine noise is muffled well enough that it dominates only below about 70 km/h (45 mi/h). It is drowned out at highway speeds by tire noise. With increasing speed v, the tire treads hit the road surface both harder and more often; so the noise output rises very rapidly. For cruising at any constant speed between about 15 and 75 mi/h the combined noise from both sources can be represented roughly by

$$P(v) \simeq (0.02 \text{ W})(v/88)^3, \tag{13.22}$$

where 88 km/h = 55 mi/h represents the U.S. legal limit. Motorcycles are typically worse, with

$$P(v) \simeq (0.1 \text{ W})(v/88)^{2.5}. \tag{13.23}$$

For heavy trucks, empirical measurements are fit better by a pair of formulas,

$$P(v) \simeq (1\ \text{W})(v/88)^2 \qquad (v > 50\ \text{km/h})$$

but

$$\simeq (0.3\ \text{W}) \qquad (v < 50\ \text{km/h}). \tag{13.24}$$

Figure 13.14 gives some idea of the frequency spectrum of traffic noise. Remember that A weighting discriminates against low frequencies, so that the 500- or 1000-Hz bands are the largest contributors to the A-weighted average in all four cases shown.

Serious work with traffic noise will consider several further effects, such as "excess attenuation" when sound travels over grassland or shrubbery instead of a hard flat surface. Sound walls, with noise strength determined by diffraction over the top of the barrier, require development of the theory we introduced late in Chap. 4. Also of interest are trapping of sound in city streets faced by tall buildings, stop-and-go traffic at intersections, type of road surface, and road gradient. Hill slopes of 10 percent, for instance, have been found to increase sound levels by amounts on the order of 3 dB.

Here, as in the earlier sections of this chapter, we must conclude that environmental noise work requires that our understanding of simple basic underlying physics (such as the inverse-square law) be extended by a great deal of empirical data to cover real-life situations. Careful attention to practical measurements and to the exact language of local legal requirements is indispensable in developing workable solutions to noise problems.

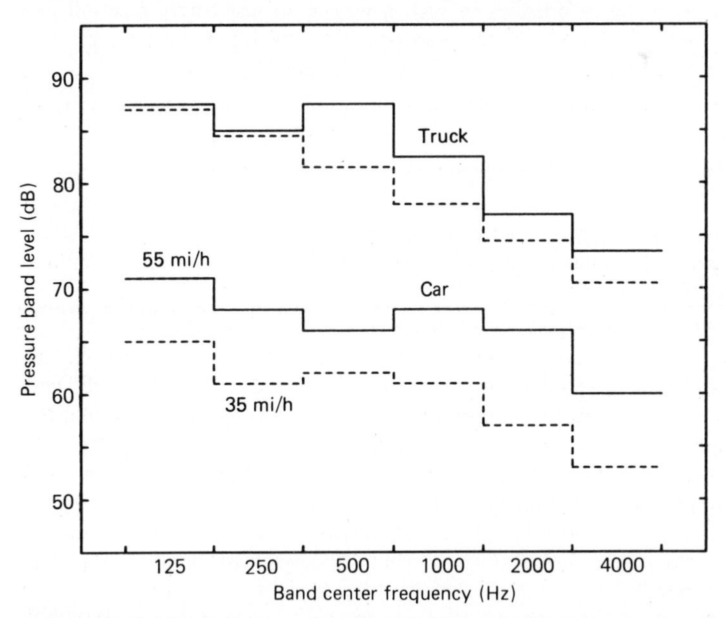

Figure 13.14 Typical octave-band levels for trucks (upper curves) and cars (lower) measured at $d = 15$ m. These are for steady speeds of 88 km/h (55 mi/h, solid curves) and 56 km/h (35 mi/h, dashed) on a level road.

PROBLEMS

13.1. Suppose readings taken in a 1-h survey fell in each range listed below for the number of minutes shown:

70–72 dB	6 min
72–74 dB	10 min
74–76 dB	14 min
76–78 dB	12 min
78–80 dB	12 min
80–82 dB	4 min
82–84 dB	2 min

(a) What are L_{10}, L_{50}, and L_{90} for this distribution? **(b)** What is L_{eq}?

13.2. Suppose a certain air-conditioning compressor unit runs 60 percent of the time from noon to 10 p.m. and 30 percent of the time from 10 p.m. to the following noon. At a neighboring window the A-weighted noise level is 63 dB with the unit on, and relatively negligible when it is off.

(a) What is the ordinary average L_{eq}?

(b) What are the weighted averages L_{dn} and CNEL?

(c) Is this unit in violation of the Sacramento County ordinance described in the text?

(d) Would it violate the simpler requirement that L_{10} not exceed 65 dB in daytime or 55 dB at night? Or that L_{dn} not exceed 55 dB?

(e) How much attenuation is needed in order to obey each of these limits?

13.3. Suppose a noise source operates in such a way as to just barely meet all five criteria of the Sacramento exterior noise standard. Sketch its cumulative level distribution, as in Fig. 13.3. What are its L_{50}, L_{10}, and L_{eq}?

13.4. Discuss how far apart two people can still communicate in the presence of background noise with level L_A of **(a)** 50 dB, **(b)** 70 dB, **(c)** 90 dB. Consider both extremes of barely communicating and carrying on a reasonable conversation.

13.5. A wall is needed to protect nearby apartment windows from the sound of a cooling-system compressor, by blocking the line of sight.

(a) What minimum mass density would you specify if the needed attenuation is 18 dB at 1 kHz and 14 dB at 500 Hz?

(b) Why is this alone not enough to guarantee a satisfactory result?

13.6. Laboratory tests show that a certain type of wall has transmission loss 25, 35, and 40 dB for frequencies 250, 500, and 1000 Hz, respectively. Let such a wall, with area 40 m^2, separate two identical rooms of volume 500 m^3 each, in which the reverberation time is 1.5, 1.2, and 1.0 s for those frequencies.

(a) What noise reduction NR do you predict at each frequency?

(b) Why in practice may it turn out to be somewhat less?

(c) If a source in one room emits 10 mW of acoustic power in each of three octave bands at the frequencies mentioned, what is the weighted sound level SPL(A) in each room?

13.7. One wall of a 5000 m^3 auditorium with $T_r = 2$ s faces a street with traffic noise levels on the order of $L_{10} = 80$ dBA.

(a) If it is desired that the background noise level inside the auditorium remain less than 30 dBA and the wall has area 500 m^2, what transmission loss does it need to provide?

(b) What further information about the noise is needed to design the wall most effectively?

(c) If the greatest problem is identified as being in the frequency range around 250 Hz, what more would you suggest about what type of wall would be best?

13.8. A house is located 80 m from a highway, on level ground with no sound barriers. Suppose **(a)** rush-hour traffic has 7200 cars/h at 72 km/h and 0.01 W apiece, while **(b)** late at night there are only 200 cars/h. Describe the expected noise in both cases.

13.9. Suppose a stream of traffic moving at 80 km/h consists of 1800 cars and 100 heavy trucks per hour, with $P = 0.01$ W for a car and 0.5 W for a truck.

(a) How far from the roadway do you need to be to treat the cars as a continuous stream? How far for the trucks?

(b) At $d = 50$ m, how would you describe the noise levels?

13.10. **(a)** A freeway sometimes carries light traffic consisting of 500 cars and 20 trucks per hour at 100 km/h. **(b)** When the load increases to 4000 cars and 50 trucks the average speed decreases to 80 km/h, and **(c)** sometimes a rush-hour jam may involve 16,000 cars and 60 trucks per hour at only 40 km/h. Estimate the noise level at $d = 100$ m in each case.

13.11. Discuss the noise that might be experienced in a residence near enough to the end of a major airport runway that jet airliners pass 1 km directly overhead during takeoff. Make a reasonable estimate of the duration of each noise peak, how long an interval there would be from one to another, and both peak and average outdoor noise levels on the ground. Then consider, with some thought about the frequency spectrum, how much less the noise might be indoors and whether these would be satisfactory living conditions. Can you obtain information for the major airport nearest you as to the distance along the ground from the runway end to the nearest houses, and the corresponding altitude of the airplanes as they pass over that point?

LABORATORY EXERCISES

13.A. Take some simple readings with a sound-level meter of noise levels in one or more workplaces. Discuss such things as their spatial variation, time variation, degree of annoyance or hazard, sources, and possible measures for noise reduction.

13.B. Take simple sound-level-meter readings of traffic noise at several sites. Make note of your distance from the roadway, rate of traffic flow, type of vehicles, etc. Discuss the results, and particularly how much time variation in sound level occurs in each case.

13.C. For one example of noise (such as those in the two preceding exercises), carry out a more careful and detailed analysis. Obtain both a time-averaged frequency spectrum and information about the statistics of time variation of the overall level. The latter can be done to some extent from a strip-chart recording or even from a series of measurements recorded by hand, but automatic analysis with a microcomputer is highly preferable if you have access to such equipment.

Chapter 14

Pipes and Resonators

In the applications we have considered so far, sound waves travel more or less freely in three dimensions. But there are other important situations where sound is confined within relatively small spaces. Examples include long narrow pipes (musical instruments or ventilation ducts), cavities (speaker cabinets or car exhaust mufflers), and gradually flaring conduits (horn loudspeakers). This chapter introduces some of the techniques that are useful in analyzing such confined waves. In particular, we develop the concept of acoustic impedance at greater length to get an extended analogy to electric-circuit theory just as we did for mechanical systems.

14.1 GUIDED WAVES

Consider the propagation of sound waves through a long narrow pipe with rigid walls and uniform cross section (Fig. 14.1). Under what circumstances might we be able to treat this as a one-dimensional problem? That is, when would it be valid to suppose that all the air in any given cross section of the pipe moves in the same way? Let us see now that this requires the pipe to be neither too large nor too small.

First remember the intuitive picture that bits of air within about a quarter wavelength of each other act as if they are in communication and can arrive at an agreement on how to move together. But air parcels a half wavelength or more apart may be moving quite differently. This suggests that the greatest chord D that can be drawn across the pipe (which is simply the diameter in the case of a circular pipe) should not exceed $\lambda/4$. But since $\lambda/4 = c/4f$, this means it will be waves below a critical frequency that can be treated as one-dimensional:

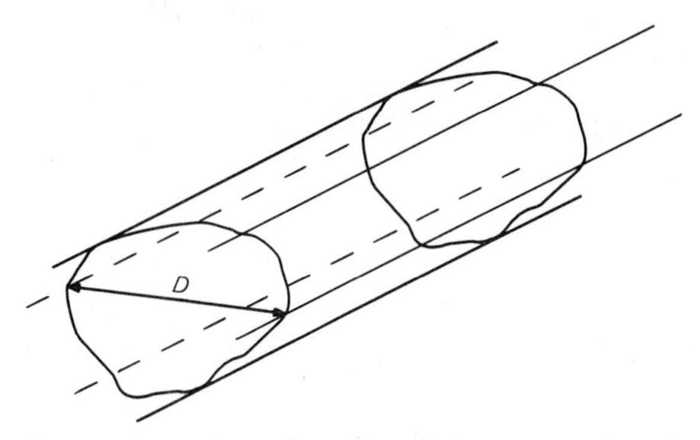

Figure 14.1 A long pipe with uniform cross section of maximum diameter D.

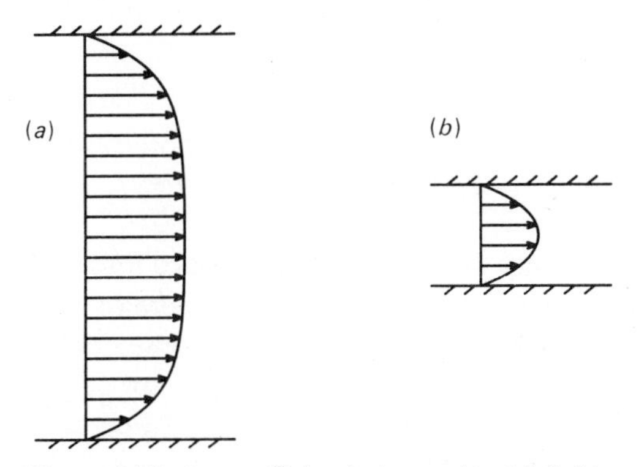

Figure 14.2 In a sufficiently large pipe (a) fluid motion can be uniform over almost the entire cross section, but in a small pipe (b) the viscous boundary layer is important.

$$\boxed{f < f_{cr} \sim c/4D.} \tag{14.1}$$

Below this frequency it is correct to picture all air motion as being parallel to the pipe axis, but at higher frequencies we would also have to consider waves that zigzag across the pipe as they go, reflecting back and forth from opposite sides. The exact theory (Pierce, Chap. 7) shows that our argument is overly conservative: the critical frequency for a circular pipe with diameter D is actually $c/1.7D$.

At the other extreme, we would expect friction at the walls to be important in very small pipes. A thin layer of fluid adjacent to the wall cannot move freely as can that in the middle of the pipe. If this boundary layer controlled by viscosity is much thinner than the pipe diameter, we may be able to ignore it (Fig. 14.2a) and suppose that the fluid velocity is nearly the same over the entire cross section. But if the boundary layer fills the pipe (Fig. 14.2b), the entire flow is dominated by viscosity, and sound waves in the sense we have been studying cannot propagate at all. On dimensional grounds alone we can make a rough estimate of the boundary-layer thickness h. It is physically determined only by the fluid's intrinsic properties of density ρ (kg/m^3) and viscosity η (kg/m-s), and by the rate at which the layer must reverse its flow as expressed in the frequency $f(1/s)$. The only combination of these three quantities that has the units of a length is $h \sim \sqrt{\eta/f\rho}$. Then the requirement that this boundary-layer thickness h be much less than the pipe size D limits the frequency to

$$\boxed{f \gg \eta/\rho D^2.} \tag{14.2}$$

Take, for instance, an air-filled pipe with diameter $D = 10$ cm. We must consider only frequencies

$$f < f_{cr} = (340 \text{ m/s})/(0.17 \text{ m}) = 2 \text{ kHz}.$$

Since air has density $\rho = 1.2$ kg/m^3 and viscosity $\eta = 1.8 \times 10^{-5}$ kg/m-s, we also require

$$f \gg (1.8 \times 10^{-5})/(1.2)(0.01) = 1.5 \times 10^{-3} \text{ Hz}.$$

Only for capillary tubes with diameters under a millimeter would we be forced to give up the picture of Fig. 14.2*a* at audible frequencies.

Subject to these limitations, we can use the expressions developed for plane waves in Chap. 9 to represent the guided waves in a pipe. The exact shape of its cross section will not matter. For sine waves, in particular, a pressure

$$p(x, t) = p_i e^{j(\omega t - kx)} \tag{14.3}$$

will be accompanied by a velocity

$$v(x, t) = (p_i/z_c)e^{j(\omega t - kx)}, \tag{14.4}$$

where the characteristic impedance z_c is $\rho_0 c$. For many purposes the total volume of air moving in or out of a section of pipe is of more interest, and we multiply by the cross-sectional area A to get the volume flow rate

$$U(x, t) = Av = (Ap_i/z_c)\, e^{j(\omega t - kx)}. \tag{14.5}$$

Since the wave covers only a finite area, it can appropriately be described by the total power it carries. This is

$$\boxed{P = IA = \langle pv \rangle A = \langle pU \rangle = Ap_i^2/2z_c.} \tag{14.6}$$

This picture of guided plane waves is useful even in the case of curving pipes, such as you see in a trumpet or a French horn. As long as the radius of curvature of the bend is larger than a wavelength, the waves will follow the bend as easily as a car rounds a curve on a well-banked highway.

14.2 ACOUSTIC IMPEDANCE

We have already suggested in Chap. 11 that it is useful to consider the ratio of acoustic pressure p to the volume velocity U associated with it. In the case of the guided traveling wave, Eqs. 14.3 and 14.5 give

$$Z_a = p/U = z_c/A = \rho_0 c/A. \tag{14.7}$$

We would like to extend this concept into a full-fledged analogy to electrical and mechanical impedance, including the representation of many sound-wave problems by equivalent acoustic circuits.

A basic point we must understand before proceeding very far is the distinction between **lumped** and **distributed** impedance. When we refer to the current entering an inductor, we imply that exactly that same current flows in every loop of the coil; similarly, the voltage across a capacitor is not even well defined unless that same voltage exists on every part of the conducting plate. We treat these as pointlike devices called lumped circuit elements, and find from Kirchhoff's laws that circuits constructed from such elements obey ordinary differential equations. A length of coaxial cable, on the other hand, may have different values of current and voltage at different places; so we must speak of it as having distributed inductance and capacitance. The corresponding transmission-line theory involves partial differential equations.

Similarly, when every part of a small rigid body moves in exactly the same way it may be treated as a point mass and the analogous circuit element is a lumped inductance. But a string has its mass distributed so that different parts may move in different ways; so our theory of its motion (Chap. 7) had to be like that of a transmission line. The key condition that must be satisfied in order to treat any mechanical impedance as lumped is that every part of the object move in the same way. But this is equivalent to saying that the dimensions of the object must be less than about a quarter wavelength of any waves that are going to be considered. Just as with (14.1), this means that for disturbances with frequency less than $c/4L$ it will be adequate to treat something with length L as lumped, but for higher frequencies it must be considered distributed. In the case of a string, c represents the speed $\sqrt{T/\mu}$ of its transverse waves, while for electric circuits c is the speed of light. By taking c as the speed of sound, we may state a similar criterion for acoustic-circuit analysis: for the air contained in any region whose greatest dimension is D, its motion may be summarized with a lumped impedance for frequencies below $c/4D$, but for higher frequencies the impedance must be viewed as distributed.

For instance, a 10-m length of coaxial cable could be treated as a simple lumped impedance only for frequencies well below $(3 \times 10^8 \text{ m/s})/(40 \text{ m}) = 7.5$ MHz. Similarly, the air in an empty soft-drink bottle 20 cm high presents a simple lumped impedance to a sound wave trying to enter the bottle if that wave has $f < (340 \text{ m/s})/(0.8 \text{ m}) = 425$ Hz.

Subject to these limitations, let us write out some details of the analogy between mechanical and acoustic impedance. In the mechanical case, we take total force F as the measure of how much push occurs and ordinary velocity v as the measure of how much motion results from it. In the acoustic case, we take pressure p and resulting volume velocity U as more appropriate. But whenever we deal with a flow that has uniform velocity v over a cross-sectional area A, we have the connection $U = vA$. Similarly, if the pressure p has the same value everywhere on this surface, $F = pA$. Then the two types of impedance will be related by

$$\boxed{Z_a = p/U = (F/A)/(vA) = F/vA^2 = Z_m/A^2.} \tag{14.8}$$

When we split each type of impedance into its resistive and reactive components, $Z = R + jX$, (14.8) will apply to each part individually. In particular, a slug of air with length Δx and cross section A (Fig. 14.3a) has mass $m = \rho_0 A \times \Delta x$ and mechanical impedance $j\omega m$. Suppose we want to represent the analogous acoustic impedance as $j\omega M$, where M is an "acoustic mass." This is also sometimes called an **inertance,** since it expresses how the inertia of the air limits its response to the applied pressure. Assuming that Δx and $\sqrt{A}$ are both small compared with $c/4f$, we must have the relation

$$\boxed{M = m/A^2 = \rho_0 A \ \Delta x/A^2 = \rho_0 \ \Delta x/A.} \tag{14.9}$$

It seems obvious that greater density or length mean greater inertance, but you may need to think carefully about how greater area actually makes it easier to move the air: Greater A means that a given $U = vA$ can be achieved with a smaller v, which does not require as much pressure applied.

Similarly, the springiness of a confined volume of air V being compressed (Fig. 14.3b) must be represented by an acoustic impedance $S/j\omega$ analogous to the mechanical impedance $s/j\omega$, with $S = s/A^2$. Since the spring stiffness s is

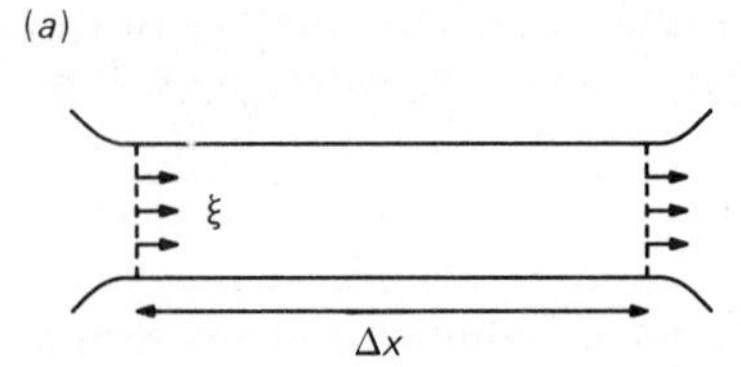

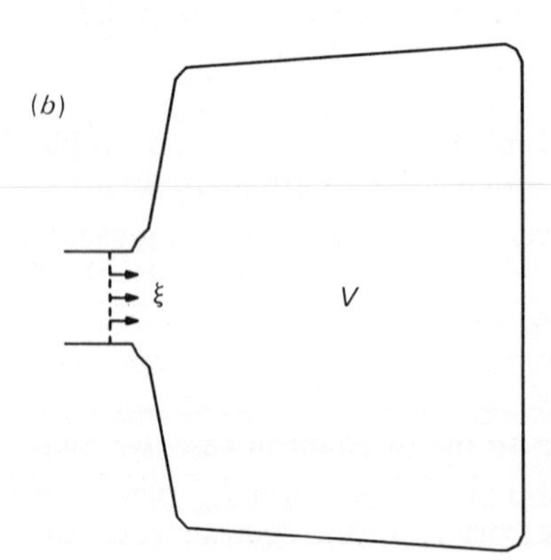

Figure 14.3 (a) A plug of air located between x and $x + \Delta x$, all experiencing about the same displacement ξ, has inertance $M = \rho_0 \Delta x/A$. (b) A container of volume V with air entering through a neck of area A has compliance $C = V/\rho_0 c^2$.

defined as $\Delta F/\Delta\xi$ and a pressure change Δp corresponds to an additional force $\Delta F = A\ \Delta p$, we have $S = \Delta p/A\ \Delta\xi$. But displacement of an area A through a distance $\Delta\xi$ reduces the enclosed volume by $\Delta V = -A\ \Delta\xi$; so the definition of acoustic stiffness must be $S = -\Delta p/\Delta V$. When V is small enough to treat this as a lumped circuit element, the increase in density is uniform throughout the volume V, and $(\Delta\rho/\rho) = -(\Delta V/V)$. This can be used to write S as $\rho\ \Delta p/V\ \Delta\rho$. But $\Delta p/\Delta\rho$ is c^2; so $S = \rho c^2/V$. It is more usual to express this in terms of the inverse stiffness $C_a = 1/S$, called **acoustic compliance:**

$$\boxed{C_a = V/\rho_0 c^2.} \tag{14.10}$$

This compliance is directly analogous to electrical capacitance and says simply that the larger the volume V the easier it will be to squeeze more air into it.

To illustrate these definitions, consider the air in an empty half-liter wine bottle. It has compliance

$$C = (5 \times 10^{-4}\text{m}^3)/(415\ \text{kg/m}^2\text{s})(340\ \text{m/s})$$
$$= 3.5 \times 10^{-9}\ \text{m}^4\text{s}^2/\text{kg},$$

where the last units are the same as m^3/Pa. If the neck of this bottle is well described as a tube of length 6 cm and diameter 2 cm, the air in it has mass

$$m = (1.2\ \text{kg/m}^3)(0.06\ \text{m})(\pi \times 10^{-4}\ \text{m}^2) = 23\ \text{mg}$$

and inertance

$$M = (1.2\ \text{kg/m}^3)(0.06\ \text{m})/(\pi \times 10^{-4}\ \text{m}^2) = 230\ \text{kg/m}^4.$$

In Table 14.1 we list again the analogies we gave in Chap. 6 for electric and mechanical circuits and add the acoustic case. The symbols sometimes used to represent lumped acoustic elements are intended to suggest their function: Constrictions with small cross section contribute inertance, loose material partially filling the pipe presents resistance to airflow, and an opening into a larger volume offers compliance. (One potentially confusing part of the analogy you will find in other books is use of the word ohm for all three kinds of impedance. That is, an mks acoustic ohm is defined to be 1 $\text{kg/m}^4\text{s}$, which, however, is quite different from the mks mechanical ohm of 1 kg/s. Furthermore, some sources use cgs acoustic ohms, 1 $\text{g/cm}^4\text{s} = 10^5$ mks ohm, or cgs mechanical ohms, 1 $\text{g/s} = 10^{-3}$ mks ohm. I prefer to write out these units.)

14.3 THE HELMHOLTZ RESONATOR

The simplest example of acoustic circuit analysis is also of great practical importance. Suppose that air can only flow in or out of an enclosed volume V through a relatively narrow neck or opening that has length l and cross-sectional

TABLE 14.1 ANALOGIES AMONG ACOUSTIC, MECHANICAL, AND ELECTRICAL SYSTEMS*

Acoustic		Mechanical		Electrical	
Volume change	ΔV	Displacement	x	Charge	q
Volume velocity	U	Velocity	v	Current	i
Inertance	M	Mass	m	Inductance	L
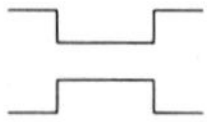		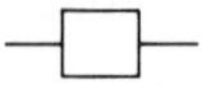		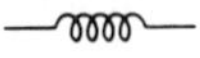	
Resistance	R_a	Resistance	R_m	Resistance	R_e
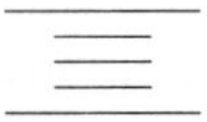		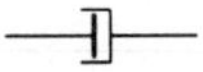			
Compliance	$C_a = 1/S$	Compliance	$C_m = 1/s$	Capacitance	C_e
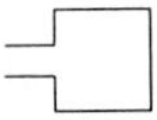		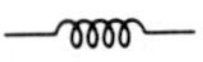		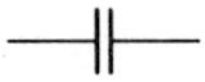	
Pressure	p	Force	F	emf	$\mathcal{E}$
Energy and power:					
$MU^2/2 = E_k$		$mv^2/2 = E_k$		$Li^2/2 = U_B$	
$(\Delta V)^2/2C_a = E_p$		$sx^2/2 = E_p$		$q^2/2C_e = U_E$	
$R_aU^2 = P_R$		$R_mv^2 = P_R$		$R_ei^2 = P_R$	
$pU = P$		$Fv = P$		$\mathcal{E}i = P$	
Impedance and reactance:					
$Z_a = R_a + jX_a$		$Z_m = R_m + jX_m$		$Z_e = R_e + jX_e$	
$j\omega M =$	jX_M	$j\omega m =$	jX_m	$j\omega L =$	jX_L
$1/j\omega C_a =$	jX_C	$s/j\omega =$	jX_s	$1/j\omega C_e =$	jX_C

* Symbols Z, R, X, and C play identical roles in all cases, and subscripts a, m, and e will be used where needed to avoid ambiguity.

area A. Chemists' flasks and cider jugs are good examples, and Fig. 14.4 suggests how we view them. This acoustic system is analogous to the series LRC circuit and to the simple harmonic oscillator, because the volume flow, or "current," that passes through the constriction must be exactly the same as that entering the compliant volume. We can immediately transfer all our knowledge about simple harmonic motion to this case. The following analysis is limited, of course, to those frequencies low enough for all dimensions of the resonator to be less than a quarter wavelength.

First, what about free oscillations? Using the analogy to s/m or $1/LC_e$, we know it will be useful to define a frequency

$$\boxed{\Omega = \sqrt{1/MC_a} = \sqrt{(A/\rho l)(\rho c^2/V)} = c\sqrt{A/lV}.} \tag{14.11}$$

Be sure to note that l and A refer here to the length and cross section of the neck only, not to the main volume V. There will also be a damping constant just like $R_m/2m$ or $R_e/2L$,

$$\beta = R_a/2M = R_aA/2\rho l, \tag{14.12}$$

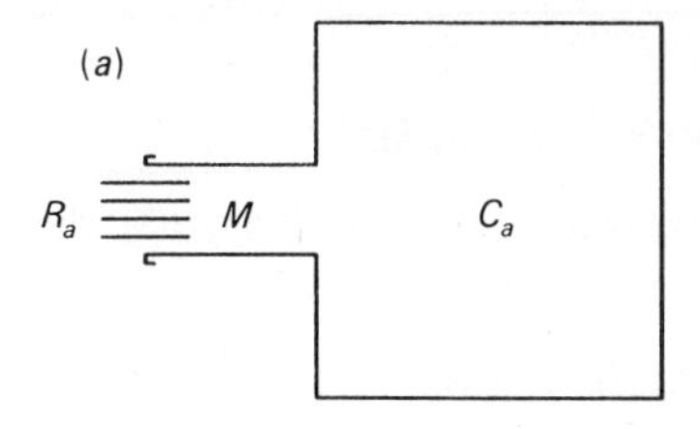

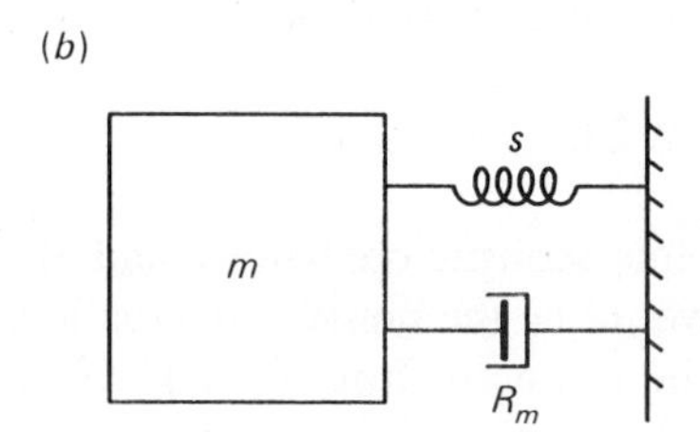

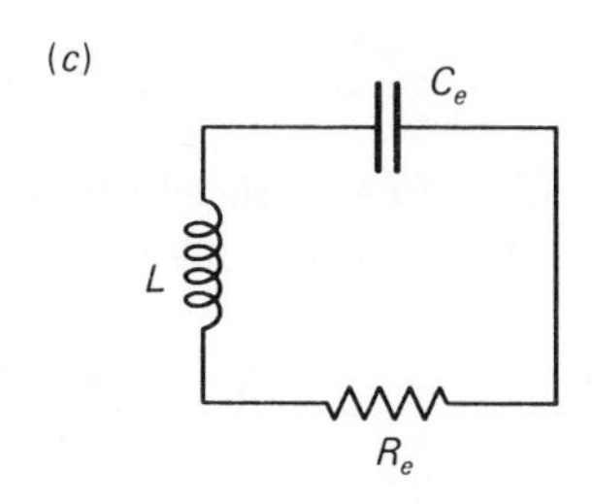

Figure 14.4 (*a*) A narrow-necked flask has inertance and acoustic resistance concentrated in the neck and compliance associated with the enclosed volume. (*b, c*) The analogous mechanical and electrical systems.

and a quality factor

$$\boxed{Q = \Omega/2\beta = \sqrt{M/C_a}/R_a = (\rho c/R_a)\sqrt{l/AV}.} \tag{14.13}$$

Any initial disturbance (such as slapping the mouth of a jug with your hand) will be followed by a damped simple harmonic oscillation. The flow through the neck, for instance, will be

$$U(t) = U_0 e^{-\beta t} e^{j(\Omega_d t - \phi)}, \tag{14.14}$$

where

$$\Omega_d = \sqrt{\Omega^2 - \beta^2}. \tag{14.15}$$

For example, consider a jug whose neck has $l = 5$ cm and $A = 4$ cm^2, and whose interior volume is $V = 4000$ cm^3. The inertance of the air in the neck is

$$M = \rho l/A = (1.2 \text{ kg/m}^3)(0.05 \text{ m})/(0.0004 \text{ m}^2)$$
$$= 150 \text{ kg/m}^4,$$

and the compliance behind it is

$$C_a = V/\rho c^2 = (0.004 \text{ m}^3)/(1.2 \text{ kg/m}^3)(340 \text{ m/s})^2$$
$$= 2.9 \times 10^{-8} \text{ m}^4\text{s}^2/\text{kg}.$$

The undamped natural frequency will be

$$\Omega = (340 \text{ m/s})\sqrt{2/\text{m}^2} = 480/\text{s} \quad \text{or} \quad 76 \text{ Hz}.$$

If a wad of cotton in the neck provides a resistance $R = 3000 \text{ kg/m}^4\text{s}$, the damping rate will be

$$\beta = (3000 \text{ kg/m}^4\text{s})/(300 \text{ kg/m}^4) = 10/\text{s}.$$

Then Ω_d will not be significantly less than Ω, but the oscillations will be damped in 0.1 s to $1/e$ of their original amplitude (an 8.7-dB reduction). This oscillator has

$$Q = 480/20 = 24.$$

Like a mechanical or electrical system, this acoustic oscillator could also be driven. An appropriate driving influence would be the pressure fluctuations of a sound wave incident on the mouth of the resonator from outside. If the driving pressure has rms average p_{out} and varies sinusoidally at frequency ω, it must work against an impedance

$$Z_a = R_a + j(\omega M - 1/\omega C_a). \tag{14.16}$$

(We are proceeding by analogy with Secs. 5.3 and 6.2, and you should think again especially of Figs. 5.3, 5.4, and 6.2 through 6.5.) The airflow allowed by this impedance is

$$U_{\text{rms}} = p_{\text{out}}/|Z_a|. \tag{14.17}$$

It will lag the driving pressure by a phase angle

$$\phi_{Up} = \arctan\,[(\omega M - 1/\omega C_a)/R_a], \tag{14.18}$$

and average power

$$\boxed{P = p_{\text{out}}^2 R_a/|Z_a|^2 = U_{\text{rms}}^2 R_a} \tag{14.19}$$

will be absorbed from the driving wave and dissipated in the resistance. All the phenomena of resonant response occur here just as they did for the simple harmonic oscillator.

In the example of the jug described above, suppose an incident sound wave has SPL = 80 dB so that $p_{\text{out}} = 0.2$ Pa and $f = 85$ Hz. Then $\omega = (6.28)(85) = 534/\text{s}$, and

$$X_a = (534)(150) - 1/(534)(2.9 \times 10^{-8})$$
$$= 1.55 \times 10^4 \text{ kg/m}^4\text{s}.$$

Since $R_a = 0.3 \times 10^4 \text{ kg/m}^4\text{s}$, Z_a has magnitude $1.6 \times 10^4 \text{ kg/m}^4\text{s}$ and phase angle arctan (1.55/0.3) = 79°. The airflow induced in the neck has rms average

$$U = (0.2 \text{ Pa})/(1.6 \times 10^4 \text{ kg/m}^4\text{s}) = 1.25 \times 10^{-5} \text{ m}^3/\text{s},$$

or $$v = U/A = (1.25 \times 10^{-5} \text{ m}^3/\text{s})/(4 \times 10^{-4} \text{ m}^2) = 0.031 \text{ m/s}.$$

The power dissipated in the resonator is

$$P = (1.25 \times 10^{-5}\ \mathrm{m^3/s})^2(3000\ \mathrm{kg/m^4s}) = 0.47\ \mu\mathrm{W}.$$

But if the incident wave has $f = 76$ Hz so that X_a is zero, then

$$U = (0.2\ \mathrm{Pa})/(3000\ \mathrm{kg/m^4s}) = 6.7 \times 10^{-5}\ \mathrm{m^3/s},$$

$v = 0.17$ m/s, and $P = 13\ \mu$W, illustrating how much more strongly the system responds at its own natural frequency. For comparison, the incident plane wave has

$$v = p/\rho c = (0.2\ \mathrm{Pa})/(415\ \mathrm{kg/m^2s}) = 0.48\ \mathrm{mm/s},$$

$$I = p^2/\rho c = 100\ \mu\mathrm{W/m^2},$$

and power 0.04 μW in a 4-cm^2 section. This presents the paradox that the resonator seems to soak up energy from a much larger portion of the wave than just the neck area A. This must be understood in terms of the outflow from the neck extending with significant strength over a region roughly a half wavelength across, so that a portion of the incident wave carrying power $I(\lambda/2)^2$ (about 400 μW in this case) has the opportunity to perform work upon that flow.

Just as the voltage across a capacitor in a resonating circuit may be larger than the driving voltage, so also can the pressure fluctuations inside a Helmholtz resonator greatly exceed those in the externally incident wave. We would associate a voltage $q/C_e = i/\omega C_e$ with the capacitor; so we expect a pressure difference $U/\omega C_a$ to belong to the compliance. In general, U is given by $p_{\mathrm{out}}/|Z_a|$, but this has its maximum value when $\omega = \Omega$. In that case we may have a pressure inside the resonator as large as

$$p_{\mathrm{in}}(\omega = \Omega) = U(\Omega)/\Omega C_a = (p_{\mathrm{out}}/R_a)/\sqrt{C_a/M}. \qquad (14.20a)$$

But $\sqrt{M/C_a}/R_a$ is just the quality factor; so

$$\boxed{p_{\mathrm{in}}(\omega = \Omega) = Q\, p_{\mathrm{out}}.} \qquad (14.20b)$$

This could also be described by saying the sound level inside the resonator can be 20 log Q higher than outside.

In the example above, the driving pressure $p_{\mathrm{out}} = 0.2$ Pa at 85 Hz produces

$$p_{\mathrm{in}} = U/\omega C_a = (1.25 \times 10^{-5}\ \mathrm{m^3/s})/(534/\mathrm{s})(2.9 \times 10^{-8}\ \mathrm{m^4s^2/kg}) = 0.8\ \mathrm{Pa}$$

or SPL = 92 dB. But at 76 Hz the resonant response is

$$p_{\mathrm{in}} = Q\, p_{\mathrm{out}} = (24)(0.2\ \mathrm{Pa}) = 4.8\ \mathrm{Pa},$$

or SPL = 108 dB. This is 20 log (24) = 28 dB above the driving pressure.

What if we leave all damping material out of the neck? It might seem that $R = 0$ would allow undamped free oscillation, or $Q = \infty$ and unlimited response to small driving pressure. And what if there is merely a hole in the side of a thin-walled vessel, with no neck? Putting $l = 0$ would appear to make $M = 0$ and $\Omega = \infty$, making it impossible to talk sensibly about oscillations. Both questions are solved by recognizing that the air in the space surrounding the opening presents a load that must still be moved even when $l = 0$. We have studied the impedance of a similar air load at length in Chap. 11 and know that it always has a resistive as well as a reactive part. Thus neither R nor M will ever be zero for a real Helmholtz resonator.

An equivalent circuit explicitly showing the radiation load is given in Fig. 14.5. In general the total resistance $R_a + R_{a,\mathrm{rad}}$ and total inertance $M + M_{\mathrm{rad}}$ must be used in place of R_a and M in all the preceding equations. We are concerned here only with the case where the neck dimensions are small compared with a wavelength, which means $kb \ll 1$ in Chap. 11. So according to (11.8) and (11.42) we should use

$$\boxed{R_{a,\mathrm{rad}} = \rho c k^2/4\pi} \tag{14.21}$$

for an unflanged neck that allows radiation in all directions, but

$$\boxed{R_{a,\mathrm{rad}} = \rho c k^2/2\pi} \tag{14.22}$$

for a flanged or baffled opening that allows radiation only into a half space. Similarly, identifying $j\omega M_{\mathrm{rad}}$ with $jX_{a,\mathrm{rad}}$, we may write $M_{\mathrm{rad}} = X_{a,\mathrm{rad}}/ck$. The imaginary part of (11.7) suggests that this would be roughly $\rho/4\pi b$ for an unflanged neck, but the actual inertance is considerably higher since this flow does not actually have spherical symmetry. A detailed calculation by H. Levine and J. Schwinger (*Phys. Rev.*, **73,** 383, 1948) shows that $0.61\ \rho/\pi b$ is the correct value for the idealized case in which viscosity and turbulence where the flow rounds the sharp corner are ignored. This can be written as

$$M_{\mathrm{rad}} = (\rho/A)(0.61b). \tag{14.23}$$

For a flanged opening of radius b, (11.43) as corrected by (11.70) suggests

$$M_{\mathrm{rad}} \simeq (\rho/A)(8b/3\pi) = (\rho/A)(0.85b). \tag{14.24}$$

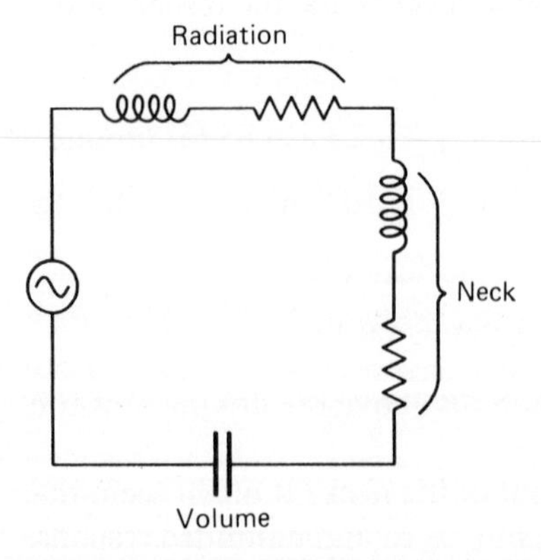

Figure 14.5 Equivalent circuit emphasizing that radiation into the air outside the mouth of a Helmholtz resonator constitutes an additional load that must be added to the impedance of the air actually in the neck.

If the opening is not circular, we may still hope that (14.23) and (14.24) will be correct if we use the value for a circular opening of the same area, that is, define b to be $\sqrt{A/\pi}$.

If we compare these formulas with $M = (\rho/A)l$ for the neck, they suggest that the radiation reactance has all the same effects as if the length of the neck were extended slightly. It is common to refer to this as an **end correction** and to say that we may easily lump M and M_{rad} together by just using an **effective length** l_e:

$$\boxed{M_{\text{tot}} = (\rho/A)l_e,} \tag{14.25}$$

where we must set

$$\boxed{l_e = l + \alpha b.} \tag{14.26}$$

In order to correctly specify α we must make a short digression. Equation 14.24 is really not quite right, because it pertains to air motion near a baffled solid disk. The air near the rim of the opening will actually move a little more readily than that in the center (Fig. 14.6). There are two extreme cases where analytic theory provides an exact solution, and we will only quote the results here. For a flanged end on a semi-infinite pipe the extra inertance is $(\rho/A)\,0.82b$ (L. V. King, *Phil. Mag.*, **21,** 128, 1936). A flanged zero-length pipe (i.e., an opening in a thin solid sheet) has $(\rho/A)(\pi b/4)$ for each side. So we may use

$$\boxed{\begin{array}{ll} \alpha = 0.61 & \text{(unflanged long pipe),} \\ \alpha = 0.82 & \text{(flanged long pipe),} \\ \alpha = 0.79 & \text{(flanged short pipe).} \end{array}} \tag{14.27}$$

For a pipe of finite length, the proper factor is presumably somewhere between 0.79 and 0.82, with the transition taking place when l is comparable with $2b$. Similarly, experiments verify that a flange of finite size must call for an α somewhere between 0.61 and 0.82 (Fig. 14.7).

Consider again the example of the jug whose neck has area $A = 4\ \text{cm}^2$, which would correspond to

$$b = \sqrt{4/\pi} = 1.13\ \text{cm}.$$

This would mean an end correction for the outer end of

$$\alpha b = (0.61)(1.13) = 0.69\ \text{cm},$$

assuming the neck is unflanged, as would be appropriate for most jugs. But the diverging flow at the inner end of the neck also contributes some inertance. The jug itself provides a flange at the inner end; so we estimate an additional correction

$$\alpha b = (0.82)(1.13) = 0.93\ \text{cm}.$$

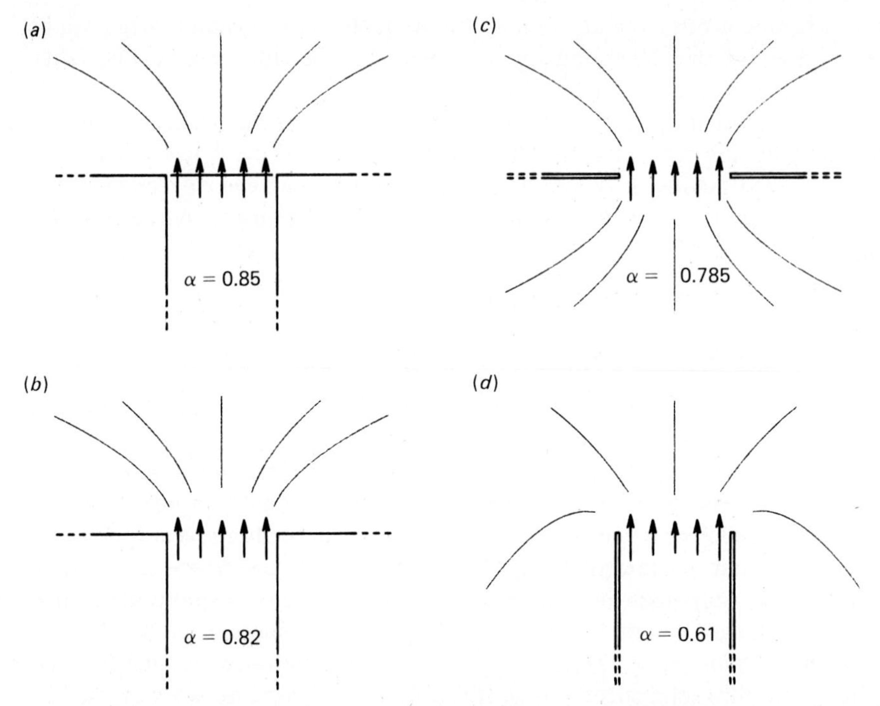

Figure 14.6 Airflow through various openings, and associated end correction coefficients α. (*a*) A light solid disk in the end of a long pipe could force the air to have uniform outward velocity over the entire cross section, so that the theory of Chap. 11 would apply. (*b*) Actual flow in the open end of a long pipe set into an infinite baffle is less constrained, so has a slightly smaller α. (*c*) For pipe length much less than the diameter of the opening, airflow is even less constrained than in (*b*). (*d*) With no flange at all, the end correction has its smallest value.

Instead of recalculating everything with this increased inertance, let us just say that all the results we found before would still apply if the actual neck length were only $l = 3.38$ cm so that the total effective length

$$l_e = 3.38 + 0.69 + 0.93 = 5.0 \text{ cm}.$$

Even if we remove the cotton from the neck, there will still be a radiation resistance

$$R_{a,\text{rad}} = \rho c k^2/4\pi = \pi \rho f^2/c$$
$$= (3.14)(1.2 \text{ kg/m}^3)(76/\text{s})^2/(340 \text{ m/s}) = 64 \text{ kg/m}^4\text{s}$$

at the resonant frequency. Thus the damping rate of free oscillations must be at least

$$\beta_{\min} = R_{a,\text{rad}}/2M$$
$$= (64 \text{ kg/m}^4\text{s})/(300 \text{ kg/m}^4) = 0.21/\text{s},$$

so the $1/e$ damping time scale is no more than $1/\beta = 4.7$ s. Q cannot be

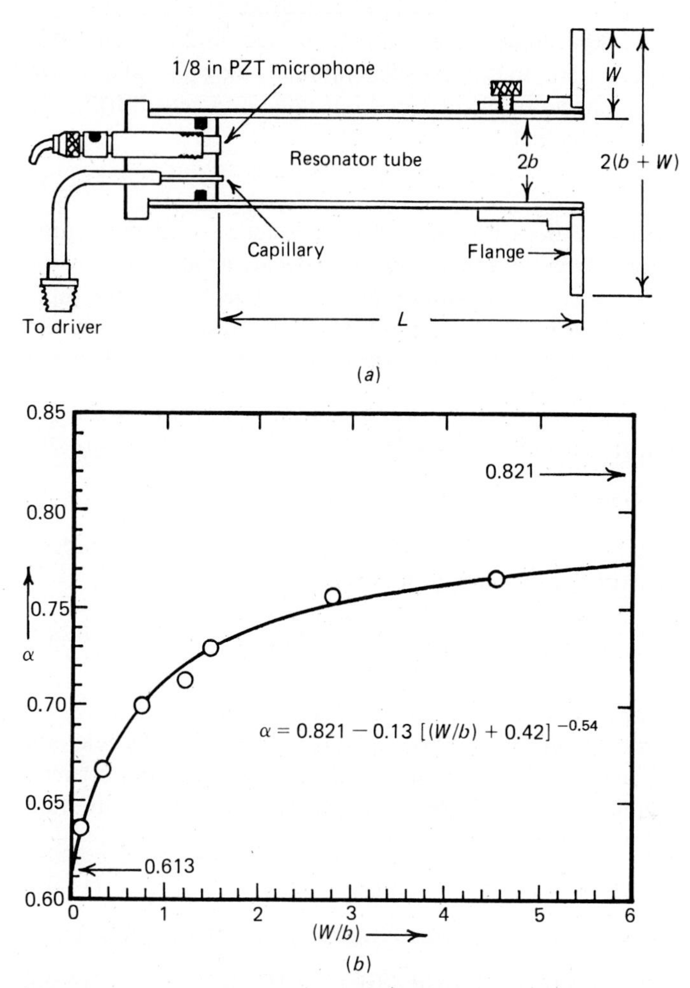

Figure 14.7 (*a*) Apparatus used by Benade and Murday to find effective length with flanges of finite diameter $2W$ by measuring the frequency at which the pipe resonated. (*b*) Resulting values of α, and an empirical equation that fits the data well. (Courtesy of A. H. Benade.)

greater than $\Omega/2\beta = 480/0.43 = 1100$, and p_{in} for the driven resonator cannot be more than 20 log 1100 or 61 dB stronger than p_{out}. Laboratory measurements would find a Q somewhat less than 1100 because of energy losses to viscosity, heat conduction, and turbulent flow at sharp corners.

14.4 REFLECTIONS AND STANDING WAVES

Most interesting acoustic devices involve some distributed impedance. Without doing a complete theory of transmission lines, we would like to ask what the inlet of a long pipe looks like to the lumped elements that drive air into it. That is, if we can describe the **input impedance** of a pipe, we can find out what role it will play in some larger system. In the case of an infinitely long pipe of uniform cross section, waves travel away from the input but none ever return from the other end. Then (14.7) tells us that

$$Z_{\text{in}} = \rho c/A, \tag{14.28}$$

and this acoustic impedance does not depend on frequency. (Let us assume without using the subscript a that all impedances below are acoustic.)

But what if there is some discontinuity a finite distance down the pipe, which reflects waves back toward the input? The total disturbance in the pipe then includes waves traveling in both directions, generally with different amplitudes:

$$p(x, t) = p_i e^{j(\omega t - kx)} + p_r e^{j(\omega t + kx)} \tag{14.29}$$

and

$$U(x, t) = (Ap_i/z_c)e^{j(\omega t - kx)} - (Ap_r/z_c)e^{j(\omega t + kx)}. \tag{14.30}$$

(By letting p_r be complex we can allow for different phases as well as amplitudes.) The acoustic impedance of this combination of waves is

$$Z(x) = p(x, t)/U(x, t) = \frac{z_c}{A}\frac{p_i e^{-jkx} + p_r e^{+jkx}}{p_i e^{-jkx} - p_r e^{+jkx}}. \tag{14.31}$$

Since p_i and p_r have fixed values in each specific situation, the dependence on position x is all shown explicitly in the exponential factors. We may take advantage of this to eliminate the wave amplitudes and get a direct relation between the values of Z at two different points in the tube. Let the two points in question be labeled $x = 0$ and $x = L$; then

$$Z(0) = (z_c/A)(p_i + p_r)/(p_i - p_r). \tag{14.32}$$

This can be solved for p_r to get

$$p_r = p_i[Z(0) - z_c/A]/[Z(0) + z_c/A], \tag{14.33}$$

which can be used to eliminate p_r from (14.31). The result can be put in the form

$$Z(L) = \frac{z_c}{A}\frac{[Z(0)A/z_c] - j\tan kL}{1 - j[Z(0)A/z_c]\tan kL}. \tag{14.34}$$

By algebraically inverting this we just get the same result as if we interchange the roles of $x = 0$ and $x = L$ by changing the sign of kL:

$$\boxed{Z(0) = \frac{z_c}{A}\frac{[Z(L)A/z_c] + j\tan kL}{1 + j[Z(L)A/z_c]\tan kL}.} \tag{14.35}$$

Implicit in the following discussion is the idea that the impedance of the combined (incident and reflected) waves at $x = L$ must be the same as the impedance of the load they are driving there. This is a consequence of two things:

(1) if the moving air in the pipe and the adjoining material (yielding solid wall or air outside) are not to lose contact or to both occupy the same space, they must both have the same velocity at the boundary, and (2) Newton's third law of motion requires that the pressure exerted by the sound waves upon the termination must be precisely equal and opposite to the pressure acting back upon them. Since p and U individually match across the boundary, so must the ratio p/U.

Let us illustrate what these rather complex formulas mean with some simple cases. If the pipe is terminated at $x = L$ with a solid wall at which it is impossible for the flow U to have any value other than zero, then $Z(L) = \infty$ and the input impedance is

$$Z(0) = (z_c/A)(-j \cot kL). \tag{14.36}$$

By comparing (7.47), you may see that this is exactly analogous to the mechanical input impedance of a string whose far end is fixed. It is also the same as the electrical input impedance of a cable whose far end is an open circuit. In each case the zero input impedance for $kL = (2n - 1)\pi/2$ identifies the resonance at each of the normal-mode frequencies $f_n = (2n - 1)c/4L$. This justifies the treatment of a closed end on a pipe in Fig. 9.4.

In the electrical case it is easy to short-circuit the end of a cable; setting $Z(L) = 0$ then gives

$$Z(0) = (z_c/A)(j \tan kL). \tag{14.37}$$

Now the resonances occur for $kL = n\pi$, and the normal-mode frequencies $f_n = nc/2L$ correspond to n half wavelengths matching the length of the line, as was suggested in Fig. 9.3. But this is never quite what really happens in the acoustic case, for when the end of a pipe is left open the pressure there is not quite zero. The air outside the pipe presents a small but finite impedance $Z(L) = Z_{\text{rad}}$. For sufficiently small pipe radius b this radiation impedance is mainly reactive: $Z(L) \simeq j\omega M_{\text{rad}}$, with the equivalent inertance given by (14.23) or (14.24). If we use $\omega = ck$ and let α stand for the end correction coefficient (14.27), we can write

$$Z(L) \simeq j\,(z_c/A)\,\alpha kb. \tag{14.38}$$

Then (14.35) gives

$$Z(0) \simeq (z_c/A)(j\alpha kb + j \tan kL)/(1 - \alpha kb \tan kL), \tag{14.39}$$

which may remind you of the identity $\tan (y + z) = (\tan y + \tan z)/(1 - \tan y \times \tan z)$. Since we have already limited this discussion to $kb < 1$, we may replace each αkb by $\tan (\alpha kb)$ and use the trigonometric identity to write

$$Z(0) \simeq (z_c/A)[j \tan k(L + \alpha b)]. \tag{14.40}$$

Thus the open-ended pipe of any length L acts to a good approximation as if it were a pipe of length $L + \alpha b$ terminated in a true zero impedance. This says the idea of "effective length" is useful under more general conditions than those of the discussion where we introduced it in Sec. 14.3; there we had assumed that L was short compared with λ.

If we take into account the small but finite radiation resistance of an open-ended pipe, we will never have $Z(0)$ exactly equal to zero at any frequency. Using $Z(L) = j\omega M_{\rm rad} + R_{\rm rad}$ with $R_{\rm rad} = z_c k^2/4\pi$ for an unflanged pipe, we would replace (14.39) with

$$Z(0) = \frac{z_c}{A} \frac{k^2b^2/4 + j(\alpha kb + \tan kL)}{(1 - \alpha kb \tan kL) + j(k^2b^2/4) \tan kL}. \tag{14.41}$$

Even when $\alpha kb + \tan kL$ is zero, $Z(0)$ is approximately $(z_c/A)(k^2b^2/4)$, from which a finite wave amplitude and finite power absorbed can be calculated for driving at resonance. The Q of the resonance can also be estimated by asking what small change in k would make $\tan (kL_e)$ as large as $k^2b^2/4$; that would double $|Z|^2$ and reduce the response 3 dB below maximum. Since the slope of the tangent function is unity near any of its zeros, the answer is $\Delta kL \simeq k^2b^2/4$, or

$$\Delta f = (c/2\pi)\, \Delta k \simeq ck^2b^2/8\pi L = \pi f^2 b^2/2cL. \tag{14.42}$$

Then

$$Q = f/2\Delta f \simeq cL/\pi b^2 f, \tag{14.43}$$

and since $f_n \sim nc/2L$ for the nth mode, we can say that

$$Q_n \simeq 2L^2/n\pi b^2 \tag{14.44}$$

gets progressively smaller for higher modes of resonance.

If instead of $kb \ll 1$ we consider $kb \gg 1$, Fig. 11.8 shows that $Z(L)$ is quite close to z_c/A, and then (14.35) says that $Z(0)$ also is nearly equal to z_c/A. That is, for high frequencies the tube shows no strong resonances. By setting $kb \sim 2$ we can estimate a critical frequency $f_{\rm max} = ck_{\rm max}/2\pi$ or $c/\pi b$ above which there are no sharp resonant peaks in pipe response. But since $f_n \simeq nc/2L$, we can say this occurs for all modes numbered above about

$$n_{\rm max} \simeq 2L/\pi b. \tag{14.45}$$

Thus for an organ pipe of a given length, larger diameter will mean fewer harmonics playing a prominent role in the operation of the pipe. Figure 14.8 illustrates these points by plotting the real part of the admittance $Y(0)$ as a function of frequency. Since $P_{\rm in} = p^2(0)\, {\rm Re}[Y(0)]$, the admittance peaks mark the resonances at which the pipe absorbs most power from the driver.

Wave reflections in a tube form the basis for a standard method of determining acoustic impedance of various materials such as noise-absorbing fiber panels. A sample of the material is used to terminate a long tube, and sine waves are sent toward it from a speaker at the other end (Fig. 14.9). A probe microphone measuring the resulting standing waves then gives information about the reflecting properties of the material. If both sound pressure p and flow U were measured at some point in the pipe, their ratio could be used in (14.34) to deduce the impedance at the termination. It is much easier, however, to measure pressure at two different locations and not attempt to measure U.

To see how to interpret such measurements, let us go back to Eq. 14.29 and remove a common factor:

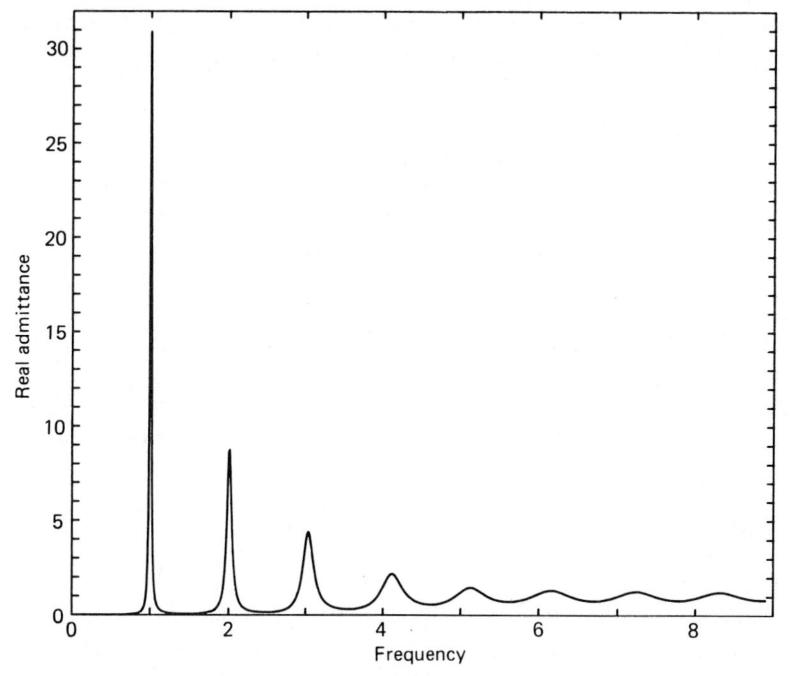

Figure 14.8 Real part of input admittance $Y(0)$ of a pipe with length-to-diameter ratio $(L/2b) = 4$. Y is in units of A/z_c and f in units of $f_1 = c/2L_e$. Calculated from (14.35) in the approximation that R_{rad} in units of z_c/A is $k^2b^2/4$ for $kb < 2$ but 1 for $kb > 2$, and that M_{rad} is $0.6\ \rho b/A$ for $kb < 1$ but $(2/\pi k^2b^2)\rho b/A$ for $kb > 1$. (14.45) predicts disappearance of resonances beyond $n \sim 5$. Note that mode frequencies f_n for larger n are slightly greater than nf_1.

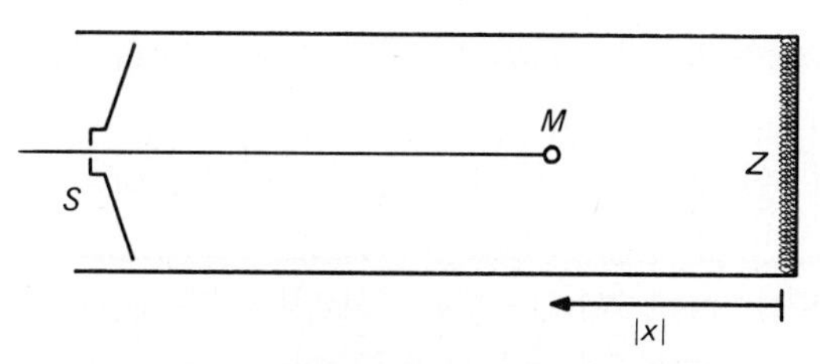

Figure 14.9 Standing-wave tube for measurement of acoustic impedance Z of material sample. Probe microphone M measures amplitude of waves generated by speaker S.

$$p(x, t) = e^{j\omega t}[p_i e^{-jkx} + p_r e^{+jkx}]. \tag{14.46}$$

In the bracketed factor we must interpret p_r as a phasor that contains information about both amplitude and phase of the reflected wave as compared with the incident one. We can show that explicitly by writing

$$p_r/p_i = Be^{j\theta}, \tag{14.47}$$

where B is real and less than 1 and θ represents a phase change caused by the reflection. If we scan along all values of x, the two terms in (14.46) take on all relative phases. In particular, the standing-wave pattern will have antinodes where the combination is constructive and relative nodes (though not actual zeros) where it is destructive:

$$p_{\max} = |p_i| + |p_r| \qquad p_{\min} = |p_i| - |p_r|. \tag{14.48}$$

Though the individual pressure amplitudes would depend on the source, they are all proportional to the driver amplitude. The **standing-wave ratio**

$$\boxed{S = p_{\max}/p_{\min} = (1 + B)/(1 - B)} \tag{14.49}$$

is determined by the terminating impedance alone and is easily measured as the microphone is moved along the tube axis.

We can keep the algebra simpler by labeling the termination as $x = 0$, so that the measurements are all in the region $x < 0$. Then the desired result will be expressed by (14.32):

$$\begin{aligned} Z(0) &= (z_c/A)(p_i + p_r)/(p_i - p_r) \\ &= (z_c/A)(1 + Be^{j\theta})/(1 - Be^{j\theta}). \end{aligned} \tag{14.50}$$

Since the information about the relative phase of p_r and p_i is missing from the standing-wave ratio S, it is also necessary to measure the location of at least one node to supply that information. When (14.47) is substituted in (14.46), it gives

$$p(x, t) = p_i e^{j(\omega t - kx)}[1 + Be^{j(\theta + 2kx)}]. \tag{14.51}$$

The first node of p must occur at the first location $x = x_1$ where the phase of the last term is 180°. Thus

$$\theta = -\pi - 2kx_1 = -\pi + 2k|x_1| \tag{14.52}$$

and

$$e^{j\theta} = -e^{2jk|x_1|} = -\cos 2k|x_1| - j \sin 2k|x_1|. \tag{14.53}$$

Now this can be substituted into (14.50), along with $B = (S - 1)/(S + 1)$, which comes from rearranging (14.49). The final result expresses the desired impedance entirely in terms of measured quantities:

$$\frac{Z(0)}{z_c/A} = \frac{(S + 1) - (S - 1)(\cos 2k|x_1| + j \sin 2k|x_1|)}{(S + 1) + (S - 1)(\cos 2k|x_1| + j \sin 2k|x_1|)}. \tag{14.54}$$

Finally, the complex fraction may be rationalized to show the real and imaginary parts explicitly:

$$\frac{Z(0)}{z_c/A} = \frac{2S - j(S^2 - 1) \sin 2k|x_1|}{(S^2 + 1) + (S^2 - 1) \cos 2k|x_1|}. \tag{14.55}$$

The result is often expressed as a specific acoustic impedance of the material at the termination, relative to that of air, by writing the left-hand side as $(AZ)/z_c = z(\text{sample})/z_c(\text{air})$.

For example, if we observed a maximum difference in SPL of 12 dB between nodes and antinodes, that would mean the standing-wave ratio is

$$S = 10^{12/20} = 4.$$

Then $$B = (4 - 1)/(4 + 1) = 3/5,$$

meaning that the reflected wave carries $(3/5)^2 = 9/25$ or 36 percent of the intensity of the incident wave. Thus we could describe the termination as having an absorption coefficient of 0.64 for normal incidence. (Unfortunately, that is not the same as α averaged over all angles of incidence as we would want for use in Chap. 12.) If the first node was 10 cm from the end for a frequency of 500 Hz, this would make

$$2k|x_1| = 4\pi f\,|x_1|/c$$

$$= (6280/\text{s})(0.1\ \text{m})/(340\ \text{m/s}) = 1.85\ \text{rad} = 106°.$$

Then (14.55) gives z for this material as

$$z/(415\ \text{kg/m}^2\text{s}) = [8 - j15 \sin(1.85)]/[17 + 15 \cos(1.85)]$$

$$= (8 - j14.4)/(12.9) = 0.62 - j1.11,$$

or $z = 260 - j460\ \text{kg/m}^2\text{s}$.

*14.5 JUNCTIONS AND FILTERS

Suppose a guided wave arrives at a point where the tube branches. By passing through an opening on the side, the moving air could enter another long tube, or a Helmholtz resonator, or even escape into the unconfined region outside (Fig. 14.10a). If the junction region is small compared with a quarter wavelength, it is analogous to a point where several lumped impedances are connected together in an ordinary electric circuit (Fig. 14.10b). It is the same pressure

$$p_1 = p_2 = p_3 = \cdots = p_{\text{in}} \tag{14.56}$$

that acts on the entrance to each branch. It is the sum of the volume flows into all the branches that must account for the total flow just upstream on the input side:

$$U_1 + U_2 + U_3 + \cdots = U_{\text{in}}. \tag{14.57}$$

You should recognize these two statements as being equivalent to Kirchhoff's laws for parallel branches in a circuit.

It is convenient to look at the acoustic admittance $Y = U/p$ rather than impedance, for then dividing (14.56) into (14.57) simply gives

$$Y_1 + Y_2 + Y_3 + \cdots = Y_{\text{in}}. \tag{14.58}$$

One way to use this is to take this total Y as $1/Z(L)$ and put it into (14.35) to find out what input impedance there would be at the driving end of the input tube. Another is to find out how the current divides itself among the branches, and that is what we would like to pursue here.

As in an electric circuit, we expect to have the largest current flowing through the branch with the greatest admittance. Each branch will receive a flow

(a)

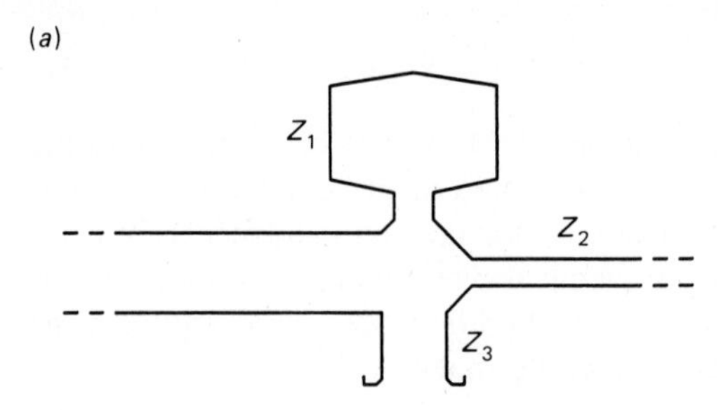

(b)

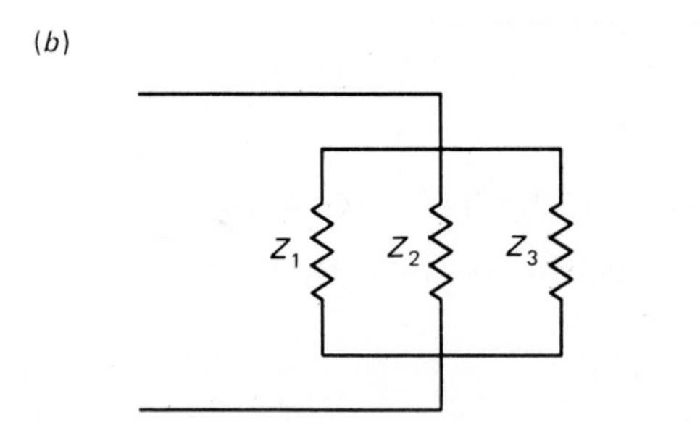

Figure 14.10 An acoustic junction at which air has a choice of several branches into which it may flow, and the analogous circuit.

$$\boxed{U_n = Y_n p_{\text{in}},} \tag{14.59}$$

where we must remember that the admittances may be complex and so the phasors U_n may not all reach maximum at the same time. In many cases we are even more interested in how much power is transmitted into the branch. This will be

$$P_n = \langle p_n U_n \rangle = \text{Re}(Y_n) p_{\text{in}}^2 \tag{14.60}$$

if we take p_{in} to mean an rms average. We are reminded that net power can be transmitted only into a branch that has a nonzero conductance $G = \text{Re}(Y)$.

We would like to compare each P_n with the power P_i in the incident wave to find the fractional transmission of power into that branch,

$$\boxed{T_n = P_n / P_i.} \tag{14.61}$$

But we must remember that both the incident and the reflected waves contribute to the total pressure on the input side. In (14.29) and (14.30) we may take the point we are concerned with as $x = 0$ and write

$$p_{\text{in}} = p_i + p_r, \qquad U_{\text{in}} = Y_0(p_i - p_r), \tag{14.62}$$

where $Y_0 = A/z_c$ is the acoustic admittance of an infinite tube with cross section A. This gives

$$Y_{\text{in}} = Y_0(p_i - p_r)/(p_i + p_r), \tag{14.63}$$

which may be inverted to obtain

$$p_r/p_i = (Y_0 - Y_{in})/(Y_0 + Y_{in}) \tag{14.64}$$

and

$$p_{in}/p_i = 1 + p_r/p_i = 2Y_0/(Y_{in} + Y_0). \tag{14.65}$$

Now we can use $P_i = Y_0 p_i^2$ in (14.61), along with (14.60) and (14.65), to write our final result:

$$\boxed{T_n = [4\ \mathrm{Re}(Y_n)/Y_0]/|1 + Y_{in}/Y_0|^2.} \tag{14.66}$$

Along with this we can write the fraction of power reflected back toward the source:

$$\boxed{R = \frac{P_r}{P_i} = \left|\frac{p_r}{p_i}\right|^2 = \left|\frac{1 - Y_{in}/Y_0}{1 + Y_{in}/Y_0}\right|^2.} \tag{14.67}$$

If all the branch admittances are specified in (14.58), these transmission and reflection coefficients can be calculated. They are in general functions of frequency, which gives us the opportunity to construct various kinds of acoustic filters by properly choosing which elements to connect at a junction.

Take as a simple example an expansion chamber with volume V and very long inlet and outlet pipes of identical area A (Fig. 14.11*a*). This is a very crude model of how an automobile muffler works. Airflow into this junction has two alternatives: it can compress the air in the chamber, or it can continue down the outlet pipe. The total admittance of this parallel combination is

$$Y_{in} = j\omega V/\rho c^2 + A/z_c, \tag{14.68}$$

and when this and $Y_0 = A/z_c$ are used in (14.66) and (14.67) they yield

$$T_{out} = 1/[1 + (\omega V/2cA)^2], \tag{14.69}$$

$$T_V = 0, \tag{14.70}$$

and

$$R = 1/[1 + (2cA/\omega V)^2]. \tag{14.71}$$

At low frequencies there is nearly 100 percent transmission into the outlet, but the high-frequency noise is mostly reflected back toward the source; that is, this is a low-pass filter, as the equivalent circuit (Fig. 14.11*b*) suggests.

Suppose for illustration that $A = 20\ \mathrm{cm}^2$ and the chamber is 40 by 20 by 10 cm so that $V = 0.008\ \mathrm{m}^3$. Then the frequency above which T is less than 50 percent is given by $(\omega V/2cA)^2 = 1$, or

$$f_c = cA/\pi V \simeq (380\ \mathrm{m/s})(0.002\ \mathrm{m}^2)/(3.14)(0.008\ \mathrm{m}^3) = 30\ \mathrm{Hz}.$$

Here we have made some allowance for the raised temperature of exhaust gases by using $c = 380$ m/s. At 3600 rpm, a four-cylinder two-stroke engine is producing 120 explosions per second so that noise is concentrated at

(a)

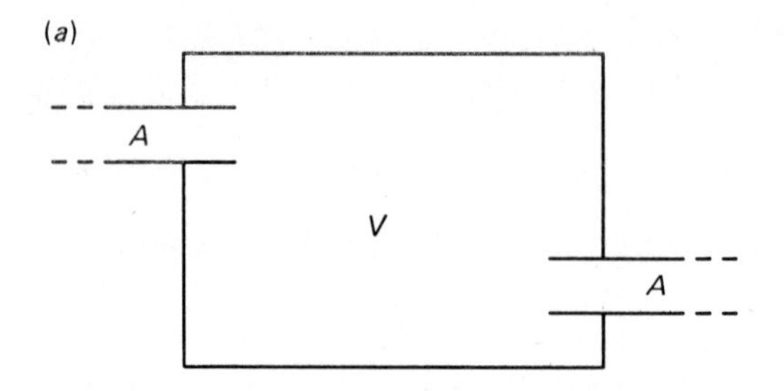

(b)

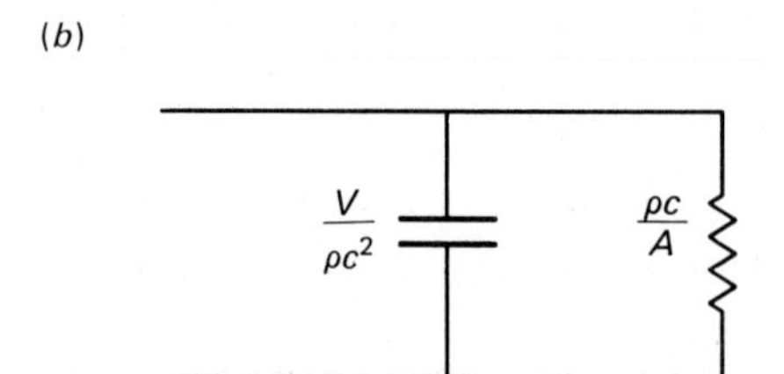

Figure 14.11 A junction consisting of an expansion chamber of volume V and long outlet pipe of area A, and its equivalent circuit.

harmonics of that frequency. According to this simple model, the muffler would have $T < 1/17$ for $f > 120$ Hz; so it would attenuate such frequencies by about 12 dB or more. We must remember, however, that this lumped-element analysis is not valid unless $\lambda/4$ is greater than the maximum dimension of 40 cm. Thus for $f > (380 \text{ m/s})/(1.6 \text{ m}) = 240$ Hz the attenuation is probably not as good as this prediction. At higher frequencies the muffler's performance will depend more on such details as whether the inlet and outlet pipes are in line or offset, whether they extend into the chamber with perforated sides, and whether the chamber is packed with some absorbing material.

A simple hole in the side of a pipe can act as a high-pass filter, and a Helmholtz resonator can function as a band-reject filter. The latter can be useful in reducing noise transmitted through ductwork from building ventilation equipment when that noise is generated mainly at a single frequency. We will leave details of such examples for the problems.

PROBLEMS

14.1. **(a)** For $f = 100$ Hz in air, what is the largest tube diameter for which plane waves are the only propagating mode?
(b) What if $f = 10$ kHz?

14.2. **(a)** For glycerin, what limits would you expect on the frequency range for which ordinary sound waves could propagate along a tube of diameter $D = 10$ cm?
(b) What if $D = 1$ cm?

14.3. A hole with diameter 2 cm allows air to escape from a pipe with diameter 6 cm. For what range of frequencies will it be valid to treat this as an acoustic transmission line with a simple lumped-impedance side branch?

14.4. One type of microphone calibration device, called the pistonphone, has a small chamber of volume V that is enclosed when the microphone is inserted snugly into its sleeve. A battery-operated camshaft moves a small spring-loaded piston of area A in and out with displacement amplitude ξ, thus compressing the enclosed air. Use the language of acoustic impedance to show what the amplitude of the pressure fluctuations will be. For what range of frequencies can you argue that this calibration will be unaffected by whether the batteries are weak and drive the camshaft more slowly? Illustrate the results with the example where $V = 20\ \text{cm}^3$, $A = 7\ \text{mm}^2$, and $\xi = 0.9$ mm.

14.5. Suppose an opening 5 cm across leads through the 2-cm-thick outer shell of a submerged submarine into a ballast tank of volume 0.1 m^3 filled with water. Rapid flow of water past the opening might excite the natural oscillations of this Helmholtz resonator, thus generating sound that an enemy vessel could detect. At what frequency would this occur?

14.6. Design a Helmholtz resonator for air such that its frequency of maximum response will be 1 kHz and its maximum gain for an oncoming plane wave in free space will be 30 dB. Over how wide a band of frequencies is the response strong? Besides the dimensions, give the effective M, R, and C of your resonator. This problem does not have a single unique solution, but do check yours for self-consistency.

14.7. Under the name of "krummhorn," you may find sets of cylindrical resonators on some pipe organs, each driven by a vibrating reed at one end and open at the other (with no flange). If such a pipe has length 40 cm and diameter 2 cm, what is its effective length? If a very high input impedance is needed in order for the resonator to work with the reed, what is the frequency at which this pipe will speak well? By considering the radiation resistance, estimate the maximum input impedance and the Q of the first-mode resonance. Roughly how many modes might resonate strongly?

14.8. If the effective specific acoustic impedance AZ_a of a certain layer of sediment on the ocean floor is $(1.9 - j0.5) \times 10^6\ \text{kg/m}^2\text{s}$, what will be the values of B and θ in (14.47) for a sound wave incident from the water above? How many decibels lower is the SPL of the reflected wave than the incident one? What is the standing-wave ratio, and how many wavelengths above the bottom is the first node located?

14.9. Suppose you measure a sample of acoustic tile in a standing-wave tube and find $S = 9$ and the first node 19 cm from the surface of the sample when $f = 400$ Hz. What is the specific acoustic impedance of the material? To see how sensitive this technique may be to measuring errors, show how much the answer would change if S were changed to 8 or if $|x_1|$ were 18 cm.

14.10. What value(s) of terminating impedance $Z(0)$ in a standing-wave tube would make it possible to get a standing-wave ratio $S = 1$? What $Z(0)$ for $S = \infty$? What kind of $Z(0)$ would it take to make the first node occur at $x_1 = 0$? At $\lambda/4$ away from the sample? At $\lambda/8$? At $3\lambda/8$?

14.11. Prove that our expressions (14.66) and (14.67) have the attractive property that the reflection R at a junction plus the total transmission T_n into all branches adds up to 100 percent.

14.12. A hole of radius b is drilled in the side of a thin-walled pipe of radius a and infinite length. Draw the equivalent circuit for this junction, and explain how it should function at high and low frequencies. Write out the expressions for R, T_{pipe}, and T_{hole} as functions of frequency, and sketch their graphs. Apply them to the case where $a = 2.5$ cm and $b = 1$ cm, pointing out the frequency at which only half the power would be transmitted. Without repeating all the calculations, can you describe roughly how the answer would change if you used a finite wall thickness of 2 mm instead of zero?

14.13. Draw the equivalent circuit for a Helmholtz resonator attached to a hole in the side of a pipe, and explain what kind of filtering it should cause. Make an argument for omitting R_a in this case. Using V for the volume, b and l_e for the neck radius and effective length, and a for the pipe radius, write expressions for the reflection and transmission coefficients R and T as functions of frequency. At what frequencies will you have $T = 0$? $T = 0.5$? Illustrate these formulas with $V = 4000$ cm^3, $l_e = 5$ cm, $b = 1.13$ cm (the example in the text), and $a = 3$ cm. What actual neck length l would be needed to make $l_e = 5$ cm?

14.14. In order to make a highly absorptive wall, a perforated metal screen is sometimes placed a distance D out from a hard backing surface. If the perforations form a square lattice with nearest neighbors spaced a distance H apart, what portion of the enclosed volume behind the screen belongs to each hole? In order to preferentially absorb at $f = 2$ kHz with $D = 3$ cm and $H = 1$ cm, what diameter $2b$ should the holes have? First get an approximate answer by pretending the metal screen has zero thickness; then correct this to allow for a thickness of 0.8 mm. What difference might it make whether a soft batting material is placed between the screen and backing?

LABORATORY EXERCISES

14.A. Choose two or three beverage bottles or laboratory flasks of different sizes and study their Helmholtz resonances; that is, measure their dimensions and compare predicted with observed frequencies. Excite them by placing their mouths near a loudspeaker driven by a sine-wave generator. Find a small enough microphone to place inside the resonator, and estimate its Q both from p_{in}/p_{out} and from width Δf of the resonant response curve. Try changing the Q with some sound-absorbing material.

14.B. Place a loudspeaker near one end of a piece of PVC water pipe and a microphone at the other, but without blocking these too much to be "open" ends. A length on the order of 50 or 60 cm may be convenient. Carefully measure the frequencies of the first several natural modes in the pipe by watching for maxima in the microphone signal on an oscilloscope while slowly tuning the sine-wave generator. Verify in a couple of cases that you are correctly identifying the mode numbers, by dangling a small microphone down through the tube and comparing its output with Fig. 9.3. Do all modes agree on the value of $l_e = nc/2f_n$? Compare the actual and effective lengths to see whether the end correction is as predicted, but note that you cannot expect this to work unless you have checked the temperature in your lab and used the corresponding value of c. Do this for several different diameters of tubing. Investigate the effect of a hole in the side of the tube (for instance, 10 cm away from one end) on the effective length.

Chapter 15

The Dynamic Loudspeaker: A Case Study

We mentioned in Chap. 2 that there are several types of electroacoustic transducers. That is, there are several different physical effects that can be used in order to convert electrical signals into sound waves. In many ways the dynamic transducer might be considered of special interest; in particular, it finds very wide use in nearly all home-entertainment equipment as well as in sound-amplification systems for auditoriums, churches, and offices. In this chapter we will use the dynamic loudspeaker to illustrate how underlying physical principles are involved in understanding the design of a good speaker. We will not provide as much detail about other speaker types but will indicate briefly in Chap. 18 some ways in which the theory should be strongly parallel for all types.

15.1 THE BASIC MECHANISM

The fundamental requirement before us is to take an alternating electrical signal and get it to move some air. But air is an extremely poor conductor of electric current. Unless we consider exotic schemes like ionizing some of the air molecules, we have an electrically neutral fluid that is almost totally unresponsive to electric or magnetic forces. So we use a two-step process: first, send electric current through a solid material that is a good conductor so that magnetic forces can be exerted on it; then let the motion of this material be passed on to the adjoining air. The conducting material here can take various forms, such as long strips deposited on a light membrane to make a panel-type speaker. But far the most common form is a compact coil of wire, which can more easily be placed in a region of strong magnetic field.

A basic problem here is that a coil of copper wire has several thousand times greater mass density than air. The bare coil can move easily through the air but carries very little air along with it, and thus is very inefficient in producing sound. There are two fundamentally different solutions to this problem. One is to confine the air inside a narrow tube so it cannot easily move aside to let the coil slip past; this idea leads to horn loudspeakers, which we will discuss in the next chapter. The other solution, which we examine here, is to attach a light membrane with a relatively large area to the coil. For a tweeter, a simple cap or shallow dome with the coil at the edge may provide sufficient area to radiate the high-frequency sound, but a woofer generally needs a radiating area much larger than the coil cross section. A shallow cone works better than a flat disk because it is inherently stiffer even when made of very thin material (commonly plastic or heavy paper) and will move rigidly along with the coil instead of flopping back and forth. Since in most applications we are concerned with wavelengths larger than the cone diameter, however, we will continue to use the theory for the flat disk as a good approximation for predicting the radiation from a shallow cone.

Thus we arrive at the basic idea indicated in Fig. 15.1. The steady magnetic field could be supplied by a separate coil carrying a large direct current, but it usually comes instead from a permanent magnet; that is why dynamic speakers are heavy to lift. This magnet is shaped so that the field lines crossing the gap between its north and south poles will point radially outward and cut across every part of the coil in the same way. Then every part of the coil experiences force in the same (axial) direction and the total force is simply

$$\boxed{F_{\text{mag}} = Bli,} \tag{15.1}$$

where F will be in newtons if B is the magnetic field strength in tesla (1 T = 10^4 gauss), i is the alternating current in amperes, and l is the total length of wire in the coil in meters. In order to avoid nonlinear response (harmonic distortion) in the speaker, the magnet must be designed to produce a uniform field. One of the main limits on the power-handling capability of the speaker will be the signal distortion that occurs if the amplitude of coil motion becomes so great that it enters a region where the magnetic field strength is different.

Some further mechanical details are necessary. In order to allow the coil to move easily along its axis and yet not vibrate sideways and bump against the pole pieces, it needs a guiding connection with the magnet that will allow motion only in one direction. And even along the axis, some restoring force should be present to prevent the coil from moving entirely out of the gap when it carries dc or very-low-frequency currents. This can be done with various arrangements of light springy material called the **suspension,** which connects the small end of the cone and the coil to the frame that holds the magnet.

There is also an elementary acoustic problem. A bare cone speaker as shown thus far is almost never used because it has extremely weak bass response. This inefficiency is simply because it will act as a dipole source for all wavelengths greater than about its own diameter. To get the greater efficiency of a monopole

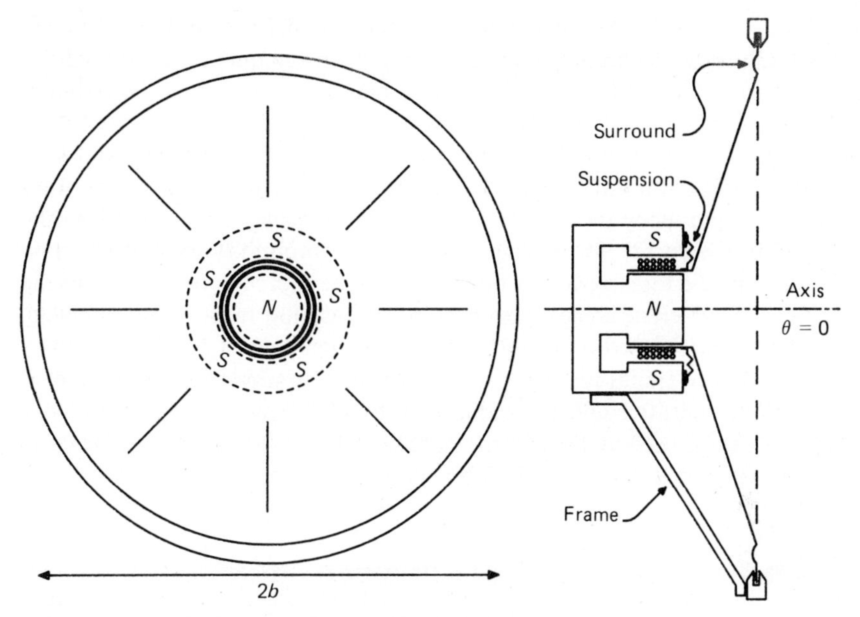

Figure 15.1 Basic features of a cone dynamic loudspeaker.

source we must mount the speaker in some sort of cabinet or baffle so that only the front side of the cone can radiate into the space where the sound is wanted. In order to have an airtight connection between the moving cone and the stationary rim where it is attached to the cabinet, the cone will have a **surround** in the form of a corrugated or half-rolled ring of soft foam plastic or rubberized material.

What should we have in mind as desirable goals in speaker design? Usually one of the first considerations is whether it has flat **frequency response;** that is, accurate reproduction of music or speech requires that the ratio of acoustic output to electrical input should be independent of frequency. Another way to judge a speaker is by its **efficiency;** especially if we need to fill a large auditorium with sound, it will be important whether the device can successfully convert a large percentage of its input power from an amplifier into useful acoustic output. Finally, we should ask about the **spatial distribution** of sound created. In some public-address applications we may want a fairly narrow beam of sound to be directed at a certain part of the audience, but in other cases (including most home hi-fi situations) we want the speaker to distribute the sound quite evenly in all directions.

Where we briefly defined speaker output before as

$$S = p(d)/V_{\text{in}}, \tag{2.2}$$

we might now show explicitly how it varies with frequency and direction by writing

$$S = S_0(d, f_0)\, g_0(f)\, h(\theta, \phi; f). \tag{15.2}$$

S_0 gives the response at distance d along the speaker axis at some chosen reference frequency f_0, while the dimensionless function g_0 is defined to be unity when $f = f_0$ and tells how the on-axis response changes with frequency. Similarly, h is defined as unity for $\theta = 0$ and tells for each frequency how much the response drops in other directions. g_0 may be measured with the loudspeaker in an anechoic chamber by slowly sweeping the frequency of a sinusoidal test signal, and h by putting the speaker on a turntable. These would be the most appropriate ways to describe output when the application involves primarily direct sound (in the sense used in Chap. 12), for instance, in outdoor amphitheaters. But in a reverberant room, the total power output $P(f)$ may be more appropriate since the room itself effectively averages h over all directions, and most of the following discussion will be focused on how to make P independent of f.

We will keep things relatively simple for now by supposing the speaker is mounted in an infinite baffle and radiating into a half space, so that the theory of Chap. 11 can be used to describe the radiation. In Chap. 16 we consider some differences associated with realistic cabinets.

15.2 SOME GENERAL DESIGN CONSIDERATIONS

We might best see how to construct a theory for the dynamic loudspeaker by considering Fig. 15.2. The coil has an electrical resistance R_e and self-inductance L, and if this alone is connected to a source of emf (possibly with some internal impedance Z_{amp}), it makes a simple series circuit that looks as if it should be easy to study. The most important thing to consider is the competition between the resistance R_e and the reactance $2\pi fL$ in limiting the current flow. We must expect different behavior according to whether we consider frequencies above or below

$$\boxed{f_{LR} = R_e/2\pi L.} \tag{15.3}$$

Similarly, the coil and cone viewed as a purely mechanical system have a total mass m, the mounting provides some stiffness s, and the flexing of material

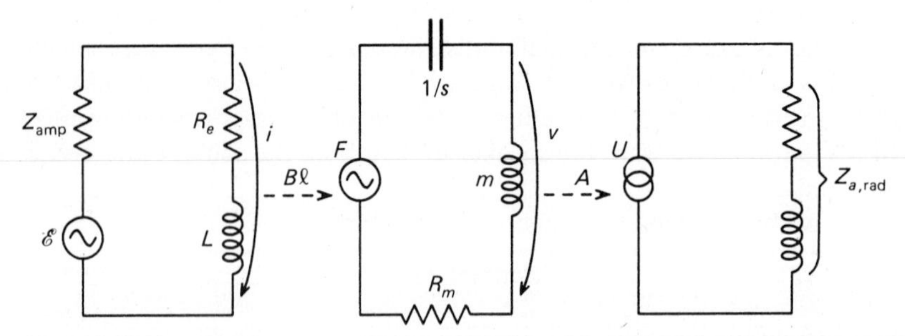

Figure 15.2 Conceptual operation of a loudspeaker system. The amplifier supplies emf $\mathcal{E}$ which drives current i through the coil. This causes force $F = Bli$ on the cone and coil, with resulting motion v. Then for cone area A the volume flow $U = vA$ works against the radiation impedance $Z_{a,\text{rad}}$ of the surrounding air to generate sound pressure p.

may necessitate including a resistance R_m. This is just a simple harmonic oscillator, and we would like to think of it as driven by a force that can be considered given once the electric-circuit problem has been solved. Again, the key to understanding this motion will lie in considering whether it is the reactance of the mass or of the spring that limits the coil motion. We look for different behavior according to whether the frequency is above or below the frequency of mechanical resonance,

$$\boxed{f_0 = \Omega/2\pi = \sqrt{s/m}/2\pi.} \tag{15.4}$$

Finally, once the motion in the mechanical circuit has been found, its velocity multiplied by the effective cone area A tells how much air is moving. If this is taken as given, we can analyze the very simple acoustic circuit on the right, where the assumed velocity U acts as an ideal current source. The radiation impedance of the circular piston serves to determine what pressure p and radiated acoustic power P_a must be associated with the volume flow U. Again, we know that the key parameter determining the radiation process is $kb = 2\pi fb/c$. Remembering Fig. 11.8, we must expect different types of speaker behavior according to whether kb is greater or less than roughly 2, that is, for frequencies above or below

$$\boxed{f_b = c/\pi b.} \tag{15.5}$$

Now, for a specified amplitude of voltage input, can we determine the electric current i, force F, motion v or U, and acoustic power P_a as functions of frequency? Let us develop an equation that will suggest how that can be achieved. The result will tell whether the speaker has the desired flat frequency response. Starting from the acoustic side of Fig. 15.2, we know that the power output can be stated in terms of the volume flow and radiation resistance:

$$P_a = R_{a,\text{rad}} U^2. \tag{15.6}$$

(Throughout this chapter, we assume that all amplitude symbols, such as this U, are actually rms averages.) Working our way back one step at a time, $U = vA$ and $v = F/|Z_m|$, so

$$P_a = R_{a,\text{rad}} A^2 F^2/|Z_m|^2. \tag{15.7}$$

But $F = Bli$ and $i = \mathcal{E}/|Z_e|$, so

$$\boxed{P_a = (AlB\mathcal{E})^2 R_{a,\text{rad}}/|Z_e|^2|Z_m|^2.} \tag{15.8}$$

In this equation, $R_{a,\text{rad}}$, Z_e, and Z_m all depend on frequency; our task is to look for some way to ensure that these dependences cancel out. Consider the possibilities: (1) $|Z_e|^2$ might be dominated by L and thus proportional to f^2, or dominated by R_e and independent of f; this will depend on whether f is above or below f_{LR}. (2) $|Z_m|^2$ might be dominated by m and proportional to f^2, or dominated by s and proportional to f^{-2}, according to whether f is above or below the resonant frequency f_0. (The third possibility of dominant R_m and no dependence on f is an unlikely choice since it would require the mechanical system to be heavily damped and thus to absorb most of the input power, giving a very inefficient speaker.) (3) $R_{a,\text{rad}}$ can be either constant or proportional to f^2, depending on whether kb is greater or less than about 2, that is, f greater or less than f_b.

Out of eight possible combinations for these three factors, only two result in a P_a that is independent of frequency. The first way is to have $|Z_e|^2 \propto f^2$ (which requires $f > f_{LR}$), $|Z_m|^2 \propto f^{-2}$ ($f < f_0$), and $R_{a,\text{rad}}$ constant ($f > f_b$). This can be done self-consistently if $\sqrt{s/m}$ is much greater than either R_e/L or $2c/b$. For instance, a coil with $R_e = 2\ \Omega$ and $L = 0.2$ mH mounted with $b = 5$ cm, $m = 0.1$ g, and $s = 10^6$ N/m could satisfy these conditions in the range from about 2 to 16 kHz. Such a design might be considered for a tweeter, but since $kb > 2$ its radiation would be confined more or less near the axis; an external diffuser could be added to broaden the final radiation pattern. In an even higher frequency range this scheme might offer an alternative to the more common piezoelectric ultrasonic projector. But for lower audio frequencies this would encounter serious difficulties: maintaining the condition $kb > 2$ would require large radius b, but the cone mass increases at least as fast as b^2 (or possibly even more in order to maintain adequate cone stiffness). Then the mounting stiffness s must remain quite large, and the amplitude of motion will be very limited even though larger cone excursions are needed in order to get the same output power at low frequencies. Furthermore, in ordinary audio applications of cone speakers it is usually desired to have $kb < 2$ in order to get broad dispersion.

Therefore, most of the speakers you encounter will be based on the other design possibility: The electric circuit is to be dominated by resistance (Z_e constant, $f < f_{LR}$) and the mechanical circuit by mass ($|Z_m|^2 \propto f^2, f > f_0$), and the wavelength is to be larger than the cone ($R_{a,\text{rad}} \propto f^2, f < f_b$). This can be done consistently if the mechanical resonant frequency $\Omega = \sqrt{s/m}$ satisfies both conditions

$$\Omega \ll R_e/L \quad \text{and} \quad \Omega \ll 2c/b. \tag{15.9}$$

For instance, $R_e = 8\ \Omega$ and $L = 0.5$ mH would mean

$$f_{LR} = 8/(6.3)(0.0005) = 2.6 \text{ kHz}.$$

If this speaker has $b = 12$ cm, then

$$f_b = 340/(3.14)(0.12) = 900 \text{ Hz}.$$

If $m = 16$ g and $s = 1200$ N/m, the speaker resonance is expected at about

$$f_0 = \sqrt{1200/0.016}/6.3 = 44 \text{ Hz}.$$

These parameters could work together to make a bass speaker operating in the range from about 50 Hz to 1 kHz.

On either side of this range the response is expected to fall off as shown in Fig. 15.3. Below the mechanical resonance at f_0, $P \propto f^4$ and there is a very rapid (12 dB/octave) bass rolloff. In order not to have a prominent hump at f_0 where

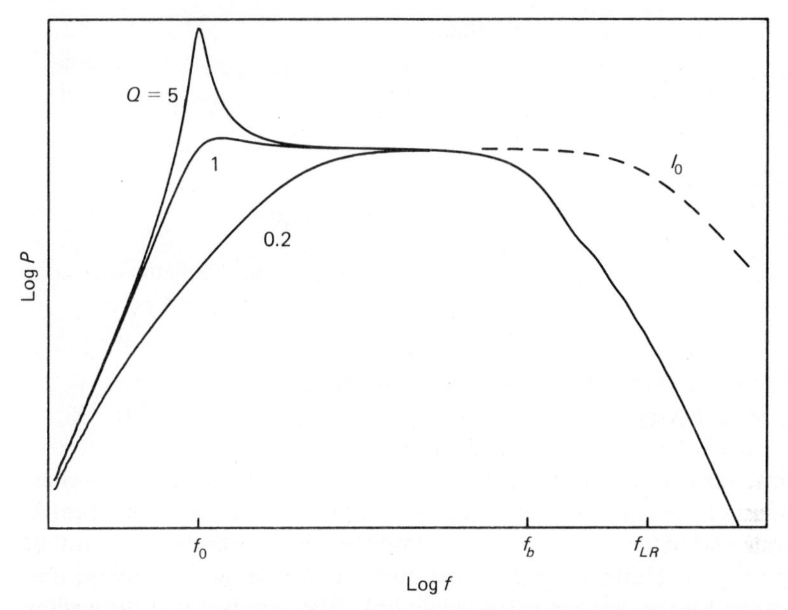

Figure 15.3 Approximate speaker behavior expected on the basis of Fig. 15.2. Operating range with flat response is limited to $f_0 < f < f_b$, with 12 dB/octave bass rolloff below the mechanical resonance frequency. Power output rolls off 6 dB/octave at the treble end above f_b even though on-axis intensity does not, and going beyond f_{LR} increases the slope another 6 dB/octave.

the speaker makes one particular bass note boom out above all others, we can see that we would want the mechanical system to include enough damping R_m that its Q_m will not be much greater than unity. For $f > c/\pi b$ the total power output $P \propto f^{-2}$ (treble rolloff 6 dB/octave), although the on-axis intensity I_0 remains constant (see Eq. 11.63). If f is also above f_{LR}, an additional 6 dB/octave drop will set in.

In reality, things are not quite so simple. In the following section we consider the need for a more sophisticated theory.

Nevertheless, let us go ahead and use (15.8) to make a crude estimate of the power output in the example just described. Suppose the coil has length $l = 10$ m, the field strength is $B = 0.5$ T, and the driving voltage $\mathcal{E} = 5$ V. Then

$$A = \pi b^2 = (3.14)(0.12 \text{ m})^2 = 0.045 \text{ m}^2$$

and

$$\begin{aligned} AlB\mathcal{E} &= (0.045 \text{ m}^2)(10 \text{ m})(0.5 \text{ T})(5 \text{ V}) \\ &= 1.13 \text{ N m}^2 \,\Omega. \end{aligned}$$

$R_{a,\text{rad}}$ is given by (11.42) as

$$\begin{aligned} 2\pi f^2 \rho/c &= (6.28) f^2 (1.2 \text{ kg/m}^3)/(340 \text{ m/s}) \\ &= (0.022 \text{ kg-s/m}^4) f^2, \end{aligned}$$

$$|Z_m| \sim 2\pi f m = (0.10 \text{ kg}) f,$$

and Z_e is simply 8 Ω. So we estimate

$$P_a = (1.13 \text{ N m}^2\ \Omega)^2(0.022 \text{ kg-s/m}^4)/(64\ \Omega^2)(0.010 \text{ kg}^2)$$

$$= 0.043 \text{ W}.$$

On the electrical side we have input power

$$\mathcal{E}^2/R = (25 \text{ V}^2)/(8\ \Omega) = 3 \text{ W},$$

so the efficiency is only about 1.4 percent. It is generally difficult with cone dynamic speakers to achieve efficiencies of more than 3 or 4 percent.

15.3 DETAILED THEORY OF THE DYNAMIC LOUDSPEAKER

There are two serious flaws in the preceding discussion, both of which represent the same kind of oversight. In going from left to right in Fig. 15.2, we assumed twice that one part of the system exercises complete control over the next without feeling any reciprocal influence acting back upon it. We can most easily see that this is incorrect for the mechanical/acoustic link: The very fact that the airflow U generates a pressure p ensures that this pressure is pushing back against the cone. There will be a reaction force F_{rad} acting on the coil along with the magnetic force we considered before. By ignoring this, we effectively studied how the coil and cone would move in a vacuum chamber. Only if their mass is much larger than the air mass being carried along would they continue to move in nearly the same way when the air load is added. In fact, the stronger the radiation produced, the greater the back reaction and the more the radiation load tends to take over as the limiting factor in the mechanical circuit. We would actually like this to be so for efficiency's sake: only if the radiation load takes a large share of the total magnetic force input can it convert a large share of the power input into acoustic power output.

Since the back-reaction force is $F_{\text{rad}} = -pA$ and $p = Z_{a,\text{rad}}U$, the proper equation for the mechanical circuit must be

$$F_{\text{mag}} + F_{\text{rad}} = Z_m v \qquad (15.10)$$

or

$$F_{\text{mag}} - AZ_{a,\text{rad}}(Av) = Z_m v. \qquad (15.11)$$

We could move the middle term to the right side and rewrite this as

$$F_{\text{mag}} = (Z_m + A^2 Z_{a,\text{rad}})v. \qquad (15.12)$$

This suggests that the acoustical load can be fully incorporated into the mechanical circuit by adding another series element that is the mechanical equivalent of the acoustic impedance (Fig. 15.4). If we do this, the back reaction is automatically accounted for and we can pretend that the only external force driving the system is the magnetic force. Let us introduce an additional subscript to represent the total mechanical impedance including the acoustic load:

$$\boxed{Z_{m,t} = Z_m + A^2 Z_{a,\text{rad}}.} \qquad (15.13)$$

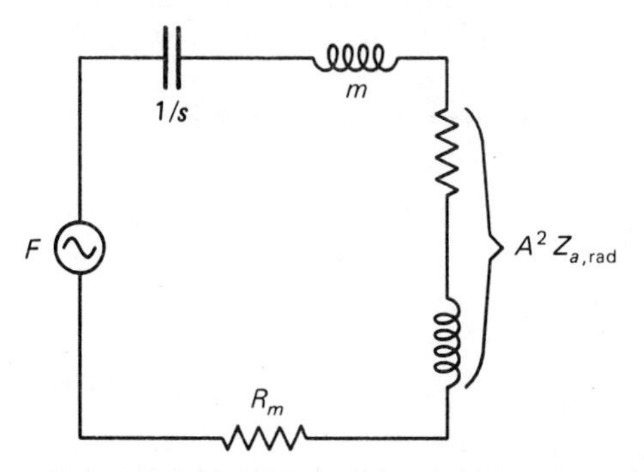

Figure 15.4 Addition of the mechanical equivalent of the radiation impedance to give the correct equivalent circuit for the mechanical system.

In the frequency range where $kb < 1$, Sec. 11.5 suggests that the additional impedance could be approximated by

$$A^2 Z_{a,\text{rad}} \simeq \rho c \pi b^2[(kb)^2/2 + j(8/3\pi)kb]. \qquad (15.14)$$

Since $kb < 1$, the imaginary part is predominant over most of the frequency range of interest. Using $ck = \omega$, we can conveniently write this reactance as $j\omega m_{\text{rad}}$ by defining

$$\boxed{m_{\text{rad}} = (8/3)\rho b^3.} \qquad (15.15)$$

In most cases, the cone must provide kinetic energy to an air load on the back side as well as on the front; so we double (15.15) before adding it to the cone mass.

In the bass speaker example above with $b = 12$ cm, we would have

$$m_{\text{rad}} = (8/3)(1.2\ \text{kg/m}^3)(0.12\ \text{m})^3 = 5.5\ \text{g}.$$

Adding twice this to the coil and cone mass $m = 16$ g means there is really a total mass load of 27 g. Thus we should revise our estimate of the resonant frequency f_0 downward from 44 Hz to

$$\sqrt{(1200\ \text{N/m})/(0.027\ \text{kg})}/6.28 = 34\ \text{Hz}.$$

This would also reduce our estimate of the power output by a factor $(16/27)^2$, from 0.043 W down to 0.015 W, for only 0.5 percent efficiency.

Note that there is a potentially misleading thing about the equivalent circuit picture of Fig. 15.4: the mere introduction of names or symbols like air mass m_{rad} or air resistance $R_{m,\text{rad}}$ does *not* necessarily mean that this circuit element has a simple constant numerical value. While m_{rad} above is constant for $kb < 1$, it becomes strongly frequency-dependent for $kb > 1$. The opposite is true of $R_{m,\text{rad}}$.

Now that we have remedied the original lack of radiation reaction in the mechanical system, consider how a similar omission occurred when we supposed that the current in the electric circuit was unaffected by anything happening in the mechanical circuit. Just as our original picture of the cone motion applied only in a vacuum, so our original remarks about the electric circuit would be true only if the coil and cone were firmly clamped so they could not actually move. That situation, called a "blocked speaker," corresponds to a very great value of the stiffness s. In reality, when the speaker does move, a coil of wire is cutting across magnetic field lines. According to Faraday's law this produces an **induced emf** proportional to the field strength and to the rate of motion:

$$\boxed{\mathscr{E}_{\text{ind}} = -Blv.} \tag{15.16}$$

The minus sign here is the result of two successive applications of right-hand rules and reminds us that when a current i produces a force F and resulting velocity v, the consequent induced emf must be in such a direction as to oppose the original current flow.

It is no accident that it is the same multiplier Bl that occurs in both (15.1) and (15.16). When systems like these interact with each other, there are fundamental principles of thermodynamics that require the action of X upon Y to be balanced by a reaction of Y upon X that has in some appropriate sense an equal strength. (I like to call it "poetic justice.") This requirement of reciprocity will receive further attention in Chap. 18.

Including the induced emf means the equation governing the electric circuit is now written as

$$\mathscr{E} + \mathscr{E}_{\text{ind}} = Z_e i \tag{15.17}$$

or

$$\mathscr{E} - Blv = Z_e i. \tag{15.18}$$

But $v = F/Z_{m,t}$ and $F = Bli$; so

$$\mathscr{E} - (Bl)^2 i/Z_{m,t} = Z_e i. \tag{15.19}$$

Just as with (15.11), the middle term is proportional to i, so profitably can be moved to the right side:

$$\mathscr{E} = [Z_e + (Bl)^2/Z_{m,t}]i. \tag{15.20}$$

Again we find that the simplicity of treating the electric circuit alone can be recovered by saying that there is an effective total electrical impedance

$$\boxed{Z_{e,t} = Z_e + (Bl)^2/Z_{m,t}.} \tag{15.21}$$

The added term is referred to as **motional impedance,** Z_{mot}. It is an effective electrical impedance representing the load upon the circuit because of the mechanical system it is driving. The lighter that mechanical load (i.e., the smaller the Z_m), the greater will be its effect in the electric circuit.

In the bass speaker example, Bl = (0.5 T)(10 m) = 5 T-m, and in most of the frequency range of interest $Z_{m,t}$ is approximately $j\omega m_{\text{tot}}$ or $j(6.28)f(0.027$ kg). So the motional impedance is about

$$Z_{\text{mot}} = (25\ \text{T}^2\ \text{m}^2)/j(0.17\ \text{kg})f = -j(150\ \Omega/\text{s})/f,$$

that is, equivalent to a capacitance of $1/(2\pi)(150) = 1060\ \mu F$. At the upper end of the range 50 Hz $< f <$ 600 Hz, this impedance is only a fraction of an ohm and thus negligible in comparison with $R_e = 8\ \Omega$. But at the lower end of the range it is several ohms and has a significant effect on how much current will flow. In fact, if we go on down to the resonant frequency f_0, the total mechanical reactance is close to zero. At that frequency the motional impedance would have its maximum value of $(Bl)^2/R_m$. Suppose, for instance, that $R_{m,t}$ for this speaker is 1.8 kg/s. (The radiation resistance at this low frequency would contribute only $2\pi^3 \rho b^4 f^2/c = 0.16$ kg/s to this total.) The motional impedance would then be a resistance of $25/1.8 = 14\ \Omega$. The mechanical Q in this case is

$$\sqrt{sm}/R_m = \sqrt{(1200\ \text{N/m})(0.027\ \text{kg})}/(1.8\ \text{kg/s}) = 3.2,$$

which sounds a little high. But the sharp peak this would tend to create in the frequency response is somewhat reduced by its contribution to the motional impedance (Fig. 15.5). Thus mechanical Q's somewhat greater than unity can be artfully used to produce acceptable frequency response around and even a little below f_0, if the magnetic coupling is chosen so that $(Bl)^2$ is comparable with $(Q - 1)R_e R_m$.

Just as we incorporated the radiation load into the mechanical circuit in Fig. 15.4, can we give a more specific picture of standard circuit elements that would be equivalent to the motional impedance? One answer to this is simply

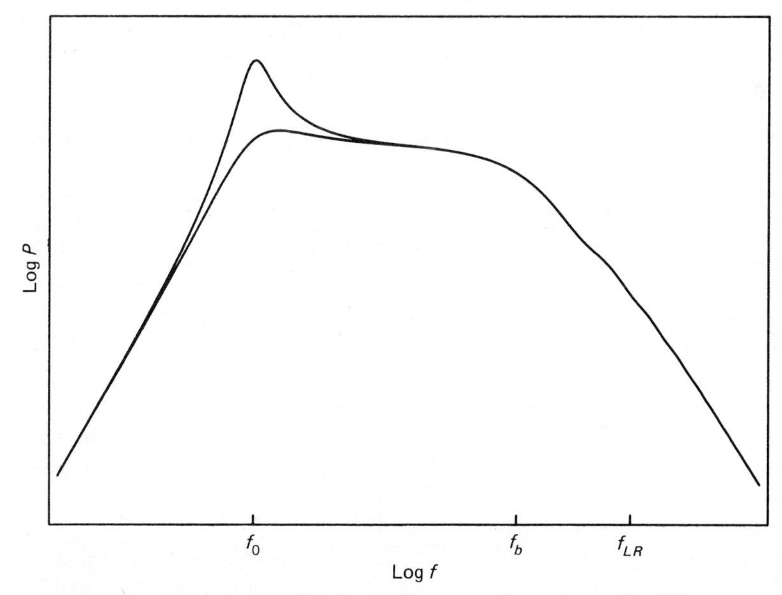

Figure 15.5 Acoustic power output for the example discussed in the text, with the moderately high mechanical Q of 3.2. Upper curve: motional emf omitted. Lower curve: motional emf included, with $(Bl)^2/R_e R_m \simeq 1.9$.

to calculate $(Bl)^2/Z_{m,t}$ and set it equal to $R_{\text{mot}} + jX_{\text{mot}}$. Since the mechanical reactance is positive (masslike) over most of the range of interest, X_{mot} will be negative, which suggests representing it as a capacitance (Fig. 15.6*b*). Although this will work at any one frequency, it gives a somewhat misleading picture because the corresponding R_{mot} and capacitance C_{mot} are frequency-dependent (see Prob. 15.7).

A more attractive alternative is to look at the motional admittance, $Y_{\text{mot}} = Z_{m,t}/(Bl)^2$. Since the mechanical circuit elements are in series, their impedances are additive, and that means each can individually be thought of as making an additive contribution to Y_{mot}. But for admittances to add, the corresponding circuit elements must be in parallel, hence the picture of Fig. 15.6*c*. Here the equivalent inductance comes entirely from the mechanical stiffness: $(1/j\omega L_p) = (s/j\omega)/(Bl)^2$, or

$$L_p = (Bl)^2/s. \tag{15.22}$$

Similarly, the equivalent capacitance comes from the mass: $j\omega C_p = [j\omega(m + m_{\text{rad}})]/(Bl)^2$, or

$$C_p = (m + m_{\text{rad}})/(Bl)^2. \tag{15.23}$$

Finally, the resistances are related by $(1/R_p) = R_{m,t}/(Bl)^2$, or

$$R_p = (Bl)^2/(R_m + R_{m,\text{rad}}). \tag{15.24}$$

These have the distinct advantage that (aside from the small radiation resistance, or from the dropoff in m_{rad} beyond $kb = 2$) they have set numerical values that remain nearly constant over a wide range of frequencies.

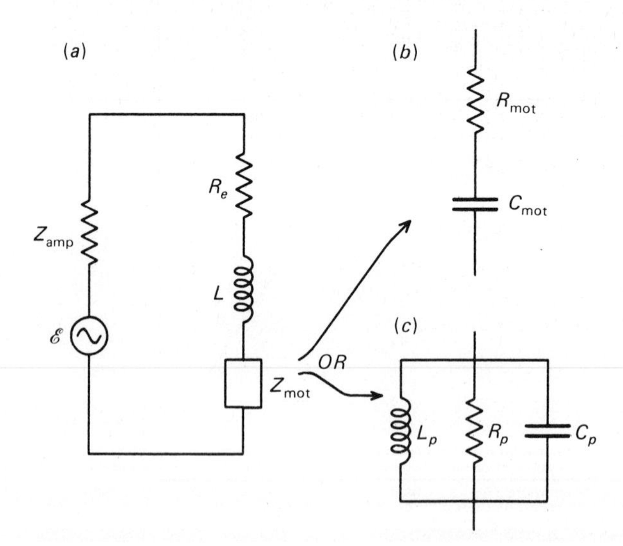

Figure 15.6 (*a*) Equivalent electric circuit with motional impedance included, and representation of that impedance with series (*b*) or parallel (*c*) simple circuit elements.

In the ongoing bass speaker example, they would be

$$L_p = (5 \text{ T-m})^2/(1200 \text{ N/m}) = 21 \text{ mH},$$

$$C_p = (0.027 \text{ kg})/(5 \text{ T-m})^2 = 1080 \ \mu\text{F},$$

and

$$R_p = (5 \text{ T-m})^2/(1.8 \text{ kg/s}) = 14 \ \Omega.$$

Since the mass is the limiting mechanical impedance over the operating range of frequencies, it is the capacitive branch that must have the greatest admittance and carry most of the current in the equivalent circuit.

The final conclusion to this discussion is that Eq. 15.8 for the acoustic power output will be correct *if* we put into it the *total* impedances (15.13) and (15.21). It would be possible to combine all these into a single grand equation, but one that still depends on several parameters (R_e, L, s, m, R_m, b, Bl) as well as on frequency and is too complex for its meaning to be seen easily. Practical speaker design must build on the concepts presented here by carrying out extensive numerical calculations to show the total effect of each possible change in parameters upon the ultimate performance. Problem 15.9 will suggest some sample calculations of this type.

Let us give a short illustration of such a calculation while making one further point about the fundamental equations. Figure 15.7 shows the result of this procedure for the example discussed above. That is, we have taken R_e =

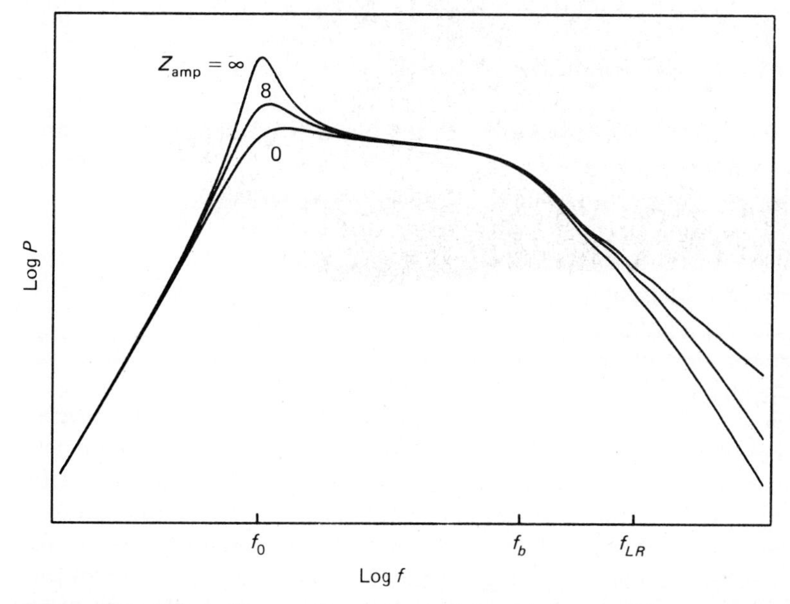

Figure 15.7 Effect of internal impedance of the driving amplifier upon speaker output for the example described in the text. Bottom curve for an ideal voltage source repeats the lower curve of Fig. 15.5. Top curve is for an ideal current source, and middle curve for $Z_{amp} = 8 \ \Omega$.

8 Ω, L = 0.5 mH, s = 1200 N/m, m = 16 g, R_m = 1.8 kg/s, b = 12 cm, and Bl = 5 T-m, used these to calculate first the radiation impedance (using results from Chap. 11 in order to be correct for f above as well as below f_b), then used that in calculating Z_m, that in turn in finding Z_e, and all of them to get P_a. The results should be compared with our original expectations as shown in Fig. 15.3.

The three curves represent different assumptions about Z_{amp}. If the amplifier that drives the speaker is an ideal voltage source, we are entitled to suppose that it provides a fixed emf $\mathscr{E}$ regardless of the current drawn by its load. This case is represented by setting $\mathscr{E}$ = 5 V and Z_{amp} = 0. At the opposite extreme we might imagine an ideal current source, which always provides a specified current regardless of how much emf is required to drive it through the load. Predicted output in this case was calculated by simply replacing $\mathscr{E}/Z_{e,t}$ by i = (5 V)/(8 Ω) = 0.625 A and decreeing that i was independent of frequency. (You may choose to think of this as the result of $\mathscr{E}$ and Z_{amp} both being infinite.) Any real amplifier must have a finite internal impedance characteristic of its particular design, and that impedance may depend on frequency. The third curve in Fig. 15.7 is calculated on the rather artificial assumption of a constant real impedance Z_{amp} = 8 Ω. Here the emf $\mathscr{E}$ has been set to 10 V, so that the output terminal voltage of the amplifier operating into a nominal 8-Ω load would be the same 5 V as in the first case. The point is to remind you that a speaker becomes part of a total system and that totally accurate speaker design can only be done in a framework including some knowledge of how the speaker will be driven.

15.4 EVALUATING REAL SPEAKERS

Relatively few readers are going to be in the business of using the foregoing theory to design speakers at the drawing board. More often than not, your interest is likely to be in trying to understand an already existing speaker and interpret its measured performance. Frequency-response curves may give a few clues when interpreted in terms of Fig. 15.3, but you may wish for more specific information about each individual parameter. One way to get this is through destructive testing: if you are willing to cut the device apart, you can directly measure the cone-coil mass, unwind the coil and measure its length, measure the magnetic field in the gap, and so on.

But how much of this could you find out without damaging the device? Let us propose a series of measurements, primarily electrical, that would determine the speaker parameters. The most simple and obvious place to start is with an ohmmeter. A dc measurement will be sure to give you R_e with no complications either from the coil inductance or from the motional impedance. By going to the other extreme of very high frequency (perhaps even ultrasonic) you could measure an input impedance $2\pi fL$ and thus determine L. This might be done either with a standard impedance bridge or by putting a known resistance R_0 in series with the speaker and measuring ac voltages across each (Fig. 15.8). The impedance is then $V_{\text{sp}}/i = V_{\text{sp}}/(V_{\text{res}}/R_0)$, which will be especially convenient to read if R_0 is chosen to be 1 Ω and V_{res} is adjusted to 1 V.

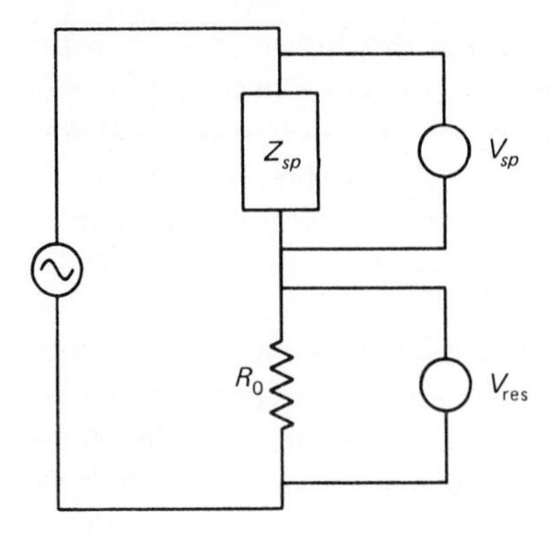

Figure 15.8 Measurement of speaker impedance. If R_0 is much larger than the speaker impedance, current will remain nearly constant during frequency sweep with a constant-voltage generator, making V_{sp} directly proportional to the magnitude of $Z_{e,sp}$. A more sophisticated way to achieve this is with a feedback loop which adjusts the generator output to keep V_{res} constant.

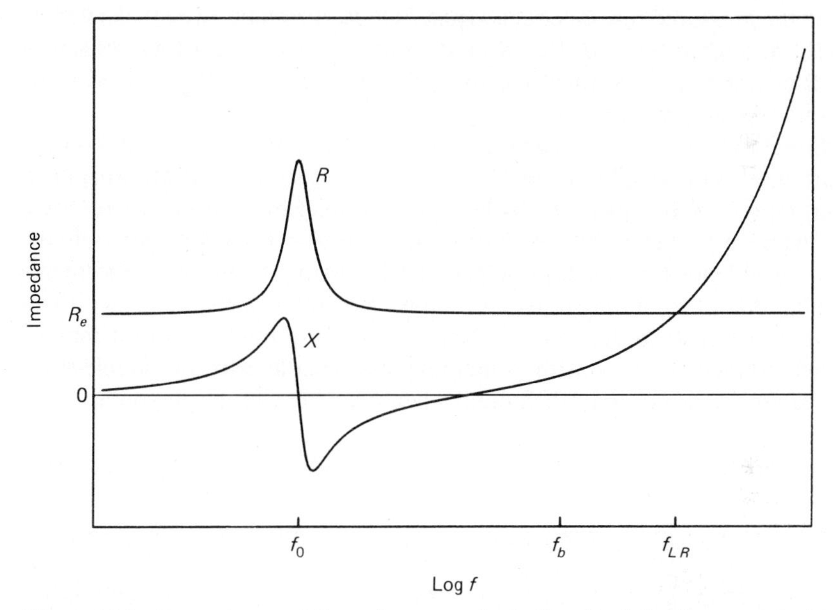

Figure 15.9 Dependence of Z_e on frequency (including Z_{mot}) for the example described in the text.

Even more information would come from measuring the input impedance of the speaker as a function of frequency. Again the circuit of Fig. 15.8 could be used repeatedly at different frequencies. If you record only the magnitude of the impedance, the main features you expect to see are the rise of $2\pi fL$ above R_e at the high-frequency end, and a peak at about f_0 due to the motional impedance. If you measure the resistance and reactance separately, you will have somewhat more detailed information (Fig. 15.9; compare Fig. 6.9). How could that be done? One way is to display V_{sp} and V_{res} together on a dual-trace oscilloscope, where the fraction of a cycle by which the speaker voltage leads the current tells the phase angle of Z as a complex number. Or if you have a lock-in amplifier available, it can be set to pick out either the in-phase or the out-of-phase component of V_{sp} using V_{res} as a reference signal. Once $Z(f)$ is determined, an alternative form in which some people like to plot this information is in the complex impedance plane, where each mechanical resonance of the system will be represented by more or less of a loop. Figure 15.10 may best be compared with Fig. 6.12 since the motional electrical impedance is proportional to the mechanical admittance. If you do this with a real speaker, you may find small secondary loops at higher frequencies, indicating that the cone can resonate in other modes for which it does not remain rigid.

It is likely to be easier to identify the resonant frequency f_0 from such impedance measurements than from the acoustic frequency-response curve, because the very fact that the electrical impedance has a peak tends to flatten the acoustic response (Fig. 15.5). Since the motional reactance vanishes near resonance, and since this is generally at a low enough frequency for L to be unimportant, the measured resistance at f_0 is just $R_e + R_p$. From this we can determine $(Bl)^2/R_m$, though still not B, l, or R_m individually. Measuring a total resistance of 12 Ω at resonance, for instance, when R_e was previously determined to be 7 Ω, would mean $(Bl)^2/R_m = 5$ Ω.

Simple observation of f_0 determines $s/m = (2\pi f_0)^2$, but how can you determine either s or m individually? It would be most straightforward to apply a known static force (for instance, by laying the speaker on its back and resting a 100-g mass on its face) and measuring the deflection it causes; then s is the applied force divided by the displacement. Rather than measuring displacements of a millimeter or so, it may be easier to make a *dynamic* measurement with an extra mass attached to the cone. As little as 5 or 10 g may do, and it may be attached nondestructively with something like candle wax or double-sided gummed tape so as not to rattle around when the cone vibrates. If you find that

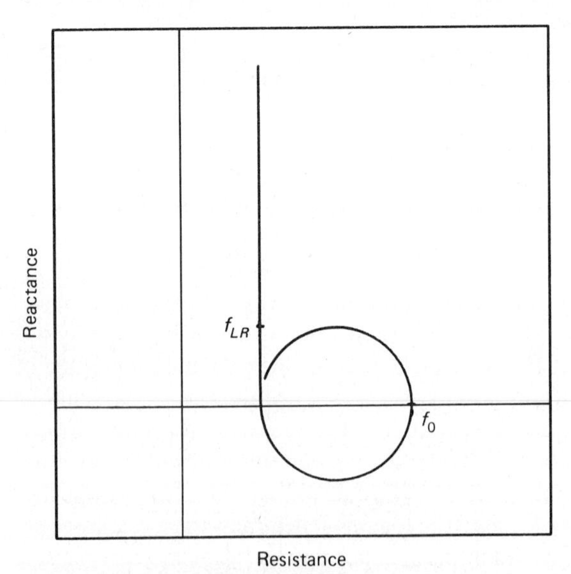

Figure 15.10 Same information as in Fig. 15.9, presented as a trace in the complex impedance plane.

the resonant frequency f_0 for the speaker alone is shifted downward to f_1 when a mass m_1 is added, you have $2\pi f_1 = \sqrt{s/(m + m_1)}$. Substituting $s = m(2\pi f_0)^2$ and rearranging gives

$$m = m_1 f_1^2/(f_0^2 - f_1^2). \tag{15.25}$$

Once m is known, s immediately follows from f_0.

For example, observation of $f_0 = 80$ Hz and $f_1 = 70$ Hz with $m_1 = 5$ g would give

$$m = (5\text{ g})(4900)/(6400 - 4900) = 16\text{ g}$$

and

$$s = (16\text{ g})(500/\text{s})^2 = 4000\text{ N/m}.$$

These, however, do not determine L_p and C_p as well until $(Bl)^2$ is known.

In the preceding paragraph m must actually include the radiation mass load. In order to subtract that off and obtain the cone/coil mass alone, you might try estimating m_{rad} with (15.15). Unfortunately, any real mounting is at best a rough approximation of an infinite baffle, so the correctness of $2(8/3)\rho b^3$ will be in doubt. Specifically, f_0 is usually low enough that a modest mounting cabinet would be smaller than a wavelength, casting doubt on whether $(8/3)\rho b^3$ for the outside air might be reduced to as little as $(\pi/4)\rho b^3$ as it would be for an unbaffled monopole source. Worse yet, the air enclosed in a cabinet contributes a stiffness as well as a mass load, which tends to obscure the whole procedure by causing the resonant frequency to increase rather than decrease.

It may occur to you that it would be simpler to make all these measurements with the speaker removed from its cabinet. It is tempting to suppose that the unbaffled speaker, acting as a dipole source, may have such a small radiation load that the measured m will be essentially that of the cone and coil alone. This, however, is not true: even though the real part of the impedance and corresponding radiated power are much smaller at f_0 for the dipole, the imaginary part and corresponding m_{rad} remain on the order of two-thirds as much as for the baffled speaker. But there is another way to add or subtract m_{rad} at will, so that it can play the role of m_1 and leave no need to attach other solid masses to the cone. That is to measure f_0 once in a vacuum chamber and then again with an air load. This can easily be done for a small bare speaker in a modest Bell jar, but clearly may be impractical for a larger speaker, especially for the more interesting case when it is mounted for actual use in a cabinet. All this is likely to force you back to adding a separate mass m_1 after all.

Consider finally two possible ways of determining Bl and thus knowing essentially everything that matters. (Note that B and l individually do not matter.) First we might make an acoustic measurement. If we determined the total output P_a for a known driving voltage $\mathscr{E}$ at a frequency well above resonance and yet well below f_{LR}, we would be able to use $Z_e \simeq R_e$ and $Z_m \simeq 2\pi fm$ in (15.8) to determine Bl. But we would probably prefer to avoid both the inconvenience of adding measurements of intensity I in many different directions to get total power, and doubts about whether the actual mounting simulates an infinite baffle well enough. This can be accomplished if we measure instead just the on-axis intensity I_0 at a distance r and a frequency high enough to make kb greater than about 4 (if this can be done without getting above f_{LR}). Then the speaker radiates a directional pattern without depending much on how it is baffled, and (11.63) can be combined with (15.8) to show that Bl should be given by

$$(Bl)^2 \simeq (4R_e^2 m^2 c r^2 I_0 / \mathcal{E}^2 \rho b^4)[1 - J_1(2kb)/kb]. \qquad (15.26)$$

The other way we might be able to extract information is by looking at more detail on the input impedance peak. If this peak is sufficiently prominent, we can measure something like the Q of the resonance. Specifically, say you subtract off R_e and then make note of the frequencies f_- and f_+ at which the motional impedance has either 45° phase angle or magnitude $1/\sqrt{2}$ times R_p (these being the half-power or 3-dB down points of the mechanical resonance). Since Q equals both $f_0/(f_+ - f_-)$ and $\sqrt{sm}/R_{m,t}$, this determines $R_{m,t}$ if s and m are already known. $(Bl)^2$ is then just $R_{m,t}R_p$.

For example, if in the case above with $R_p = 5\ \Omega$, $f_0 = 80$ Hz, $m = 16$ g, and $s = 4000$ N/m we found peak width $f_+ - f_- = 8$ Hz, that would give

$$R_{m,t} = (8/80)\sqrt{(4000)(0.016)} = 0.8 \text{ kg/s}$$

and

$$Bl = \sqrt{(0.8)(5)} = 2 \text{ T-m}.$$

If the speaker is well baffled with cone radius $b = 10$ cm, its air load would be contributing

$$2(8/3)(1.2 \text{ kg/m}^3)(0.001 \text{ m}^3) = 6.4 \text{ g}$$

to the total mass and only

$$\rho c\, \pi b^2\, (kb)^2/2 = 0.14 \text{ kg/s}$$

to the resistance at f_0.

PROBLEMS

15.1. If a speaker is going to have radius $b = 5$ cm and mass $m = 3$ g (cone and coil alone), and you want the lower end of its range to be at 300 Hz, what mounting stiffness s does it need?

15.2. One way to get more power from a speaker is simply to drive it with a larger voltage. Unfortunately, heat is deposited in the coil because of its resistance, and too much heat can melt the lacquer insulation coating that separates adjacent strands of conductor; temperatures above 250°C generally must be avoided. If the total mass of copper in a voice coil is 2 g and its specific heat capacity is 0.39 J/g-K, make a rough estimate of its maximum steady power-handling capability. In order to do this, make an intuitive guess about the time scale over which the coil can pass heat on to its surroundings; that is, if you have been running the speaker hot and suddenly turn off the signal, will it cool off significantly in 1, 10, 100, or 1000 s?

15.3. It would appear from (15.8) that the efficiency of a speaker could be made arbitrarily large (even larger than 100 percent??) simply by increasing B or l or A. Discuss the problems that will be encountered if you try to pursue each of these possibilities.

15.4. A small intercom speaker has $m = 4$ g, $s = 6 \times 10^4$ N/m, $R_m = 8$ kg/s, $b = 6$ cm, $Bl = 8$ T-m, $R_e = 2\ \Omega$, and $L = 0.03$ mH. Using appropriate approximations to keep the discussion primarily physical rather than mathematical, tell as much as you can about the performance to be expected. Consider frequency range, directional radiation, efficiency, reasonable maximum power output, input impedance, etc.

15.5. Suppose you have a spool of No. 34 copper wire whose resistance is 0.0086 Ω/cm. Its diameter is 0.16 mm, and the density of copper is 8.9 g/cm³. You also have a magnet from an old speaker whose voice-coil gap has diameter 3 cm and field strength 0.7 T. How many turns of wire would you wind to make a coil with nominal impedance 6 Ω? What is this coil's mass? It is difficult to write an exact expression for the self-inductance of a coil, but a rough approximation is $L \sim N^2\mu_0 a \ln(a/p)$ where $\mu_0 = 4\pi \times 10^{-7}$ H/m, N is the number of turns in the coil of radius a, and $p \sim (d/2)\sqrt{N}$ is the approximate radius of a bundle of N wires each with diameter d. What estimate does this give for the inductance of your coil, and what corresponding limiting frequency f_{LR}?

15.6. For an infinite-baffle speaker with $b = 10$ cm, what is the radiation impedance at $f_1 = 200$ Hz and at $f_2 = 1$ kHz? If it has $m = 8$ g, $s = 2000$ N/m, $R_m = 0.5$ kg/s, and $Bl = 6$ T-m, what is its total mechanical impedance at each of these frequencies? What is the motional electrical impedance at each frequency? What are the values of the equivalent circuit elements in Fig. 15.6*b* and *c*?

15.7. For the series-equivalent circuit of Fig. 15.6*b* show that the motional resistance is $(Bl)^2 R_{m,t}/|Z_{m,t}|^2$. What is the corresponding expression for C_{mot}? Discuss how these depend on frequency.

15.8. Let a voltage signal with square waveform of frequency f be sent to a loudspeaker. Discuss qualitatively, using the concepts of spectral analysis, how nearly square the waveform of the sound output will be in the cases **(a)** f a little larger than f_0, **(b)** f a little smaller that f_b, and **(c)** f well below f_0. (In order to compute quantitatively exact waveforms, you would need information on what happens to phases as well as amplitudes for different frequency components, and only the latter has been spelled out in this chapter.)

15.9. Using whatever computer or programmable calculator you have available, write a program that will calculate acoustic output P_a for a series of different frequencies when all the parameters defining a speaker are given. Include the radiation load in the total mechanical impedance, and the motional component of the total electrical impedance, but for simplicity let $Z_{\text{amp}} = 0$. If possible, let the results be provided in the form of graphic output by the computer. In order to make comparisons with Fig. 15.7, take as a baseline case the parameters of the example discussed in Sec. 15.3. Then take one parameter at a time (m, for instance) and produce graphs showing the effect of making this parameter smaller or larger while keeping all the others fixed. Try to give a qualitative physical explanation for the changes you find.

LABORATORY EXERCISES

15.A. Obtain an expendable speaker, perhaps the cheapest available at a local electronic supply store. Measure its frequency response with whatever techniques you have available, with the speaker mounted in some approximation to an infinite baffle. Make some kind of measurement of its directional radiation pattern at several frequencies. Then dismantle the speaker and make direct measurements of as many of its parameters as possible. Use this information to compare the measured performance with theoretical expectations.

15.B. Using a reasonably good speaker, find out as much as you can about it with nondestructive techniques. In particular, use measurements of its input impedance as a function of frequency. As in 15.A, compare theoretical predictions based on m, s, b, R_e, etc., with actual acoustic performance.

Chapter 16

Loudspeaker Mounting

We have tried to make it as simple as possible to understand the operation of a loudspeaker by supposing it to be mounted in an infinite baffle. But most practical applications require the use of a finite baffle or cabinet. We want to ask how that affects (1) the total power radiated for a given input signal, that is, the efficiency; and (2) the directional distribution of that radiation. We find that judicious choice of cabinet design can extend the range of high-quality response of a speaker. After considering the effects of simple baffles and boxlike enclosures, we take a brief look at baffles folded forward to form a horn in front of the moving diaphragm and find that these offer useful options in directivity and high efficiency. The final section points out the need for circuits that will distribute input signals properly among systems of speakers working together to provide full coverage of the audible frequency range.

16.1 OPEN BAFFLES

You may have been concerned while studying the preceding chapter about our neglect of what happens on the back side of a speaker diaphragm. Must it not move just as much air out of its way on the back side as on the front? Even in the case of a practically infinite baffle such as a mounting in the ceiling of an office or restaurant, should we not ask about radiation on both sides? We may simply be unconcerned with the sound going into the back-side space, as long as that space is totally separated from the front-side space. The only modification needed to give a correct theory may then be a doubling of the radiation impedance load in the mechanical-circuit analysis, but that has rather little effect other than close to the speaker's resonant frequency anyhow.

More interesting questions arise with finite baffles. Consider first a speaker cone of radius b mounted in the center of a plane disk-shaped baffle with radius B (Fig. 16.1). At the high-frequency extreme where $kb > 4$, that is, $f > 2c/\pi b$, Eq. 11.64 says that the speaker radiates mainly into angles $\theta <$ arcsin $(3.83/kb) < 90°$. Then the size or even the mere presence of the baffle has very little effect. One might say that the speaker in this case is big enough to act as its own baffle. At the low-frequency end, for f less than the resonant frequency $f_0 = \sqrt{s/m}/2\pi$, the speaker output is very weak regardless of any baffle properties.

Between these two extremes lies the range where the speaker would ordinarily be used. In this range kb is small enough that radiation goes out tangent to the baffle, at $\theta = 90°$, with strength not much less than on axis at $\theta = 0°$. Thus for $r > B$ radiation from the front and from the back will have the opportunity to merge. But these two signals are 180° out of phase with each other since the cone is always pulling away from the air on one side just when it is pushing the air on the other side. So there will be cancellation, and the net intensity $I(90°)$ should be very small. There are two different pictures of how this occurs; let us consider each in turn.

If $kB \gg 1$, or $\lambda \ll 2\pi B$, the radiation is already effectively free from its source by the time it passes the edge of the baffle. From the discussion of anisotropic radiation in the first part of Sec. 10.4, we expect that sound moving out near the axis (small θ) must continue onward and be hardly affected at all by what is happening in the plane of the baffle. That is, the on-axis intensity I_0, the total power P, the efficiency, and the radiation impedance will all be nearly the same as for an infinite baffle. Only in a relatively narrow range of angles near 90° is there reduced output. By asking how large an angle is subtended by a quarter wavelength at a distance B away we might even guess that the width of that range could be as small as $(\pi/2)/(kB)$ radians.

On the other hand, for $kB \ll 1$ the edge of the baffle is still in the induction zone. This means that the distance $2B$ from the front side of the cone around the edge of the baffle to the back side is less than $2/k = \lambda/\pi$. Then the air motion is slow enough that there is plenty of time for the front-side and back-side air

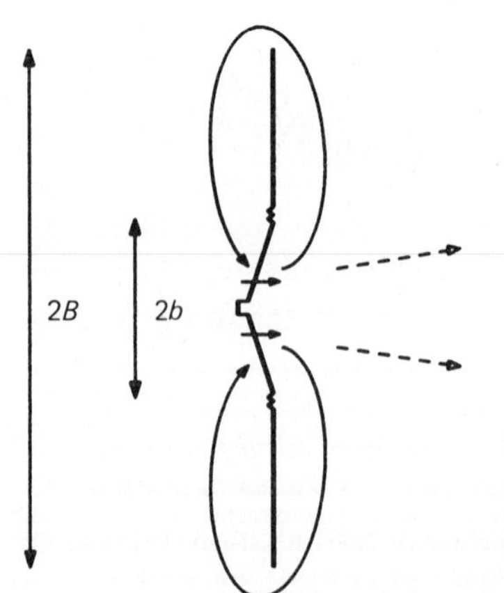

Figure 16.1 A speaker of diameter $2b$ mounted in a baffle of diameter $2B$. For $kB \gg 1$ cone motion will radiate effectively (dashed lines) but for $kB < 1$ it is easier for air to flow around the baffle edges and the speaker acts as a dipole source.

to coordinate their motion: air being pushed away on one side can find its way around the baffle to fill in the void being left on the other side. This easy alternative reduces the effectiveness of this air in moving radially outward to launch a sound wave (Fig. 16.1). In more sophisticated language, it is now acting as a dipole source and is correspondingly inefficient. Comparing (11.15) with (11.37) shows four powers of k for the dipole case instead of two; so the acoustic power output for a given volume of air moved by the cone is less for the dipole than for the monopole by a factor of about $(kB)^2$ or $(2\pi B/c)^2 f^2$. This causes a 6-dB/octave dropoff in bass response (Fig. 16.2) below a frequency

$$\boxed{f_B \simeq c/2\pi B} \tag{16.1}$$

characteristic of the baffle size. For noncircular baffles a conservative estimate would be that B should represent the distance from the speaker center to the nearest edge. For instance, with a square baffle 1 m across we would take B = 0.5 m and estimate f_B as (340 m/s)/(3.14 m) = 110 Hz.

We see that the larger the baffle is made, the lower will extend the frequency range over which it can prevent the inefficiencies of dipole-type radiation. For treble or midrange speakers this may be quite practical. But for low bass response it requires very large structures; to get f_B as low as 40 Hz, for example, requires a baffle diameter $2B$ of at least (340 m/s)/(3.14)(40/s) = 2.7 m, or about 9 ft. Enclosed cabinets will usually be a more attractive alternative in obtaining good bass response.

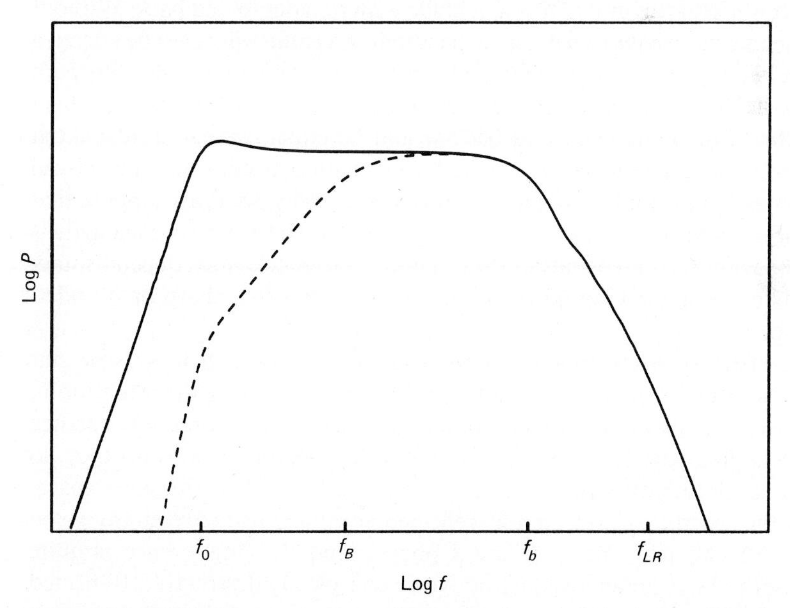

Figure 16.2 Idealized power output of a speaker in an infinite baffle as in Fig. 15.3 (solid line), and 6 dB/octave bass rolloff below f_B for baffle radius B (dashed line).

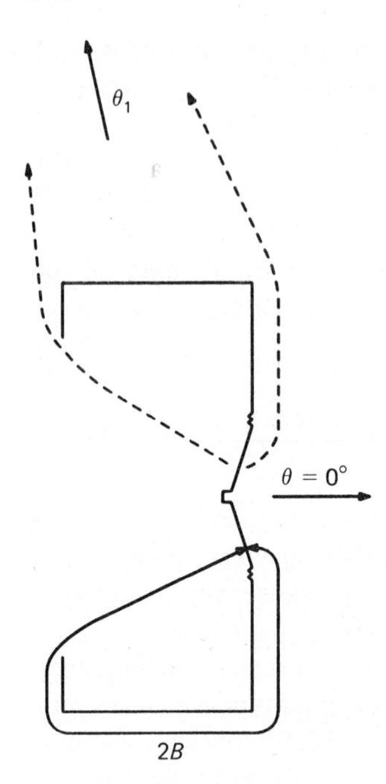

Figure 16.3 An open-back cabinet, with solid path suggested to play the role of $2B$ and dashed paths suggesting direction θ_1 of minimum intensity.

Open-backed mountings as shown in Fig. 16.3 are common in television consoles, where there is no intention of trying to reproduce very low bass frequencies. It is useful to think of this as a baffle with its sides folded back. Without attempting an exact analysis, we may expect that the baffle will again be effective above a frequency given by (16.1) if we take $2B$ to be roughly the shortest path around the baffle from one side of the cone to the other. Simply by asking where path lengths to the cone front and back would be equal, we can also make it plausible that the angle θ_1 where greatest cancellation occurs may be moved around somewhat beyond 90° from the cone axis. Finally, we must suppose that near the lower end of its useful range at f_B, where the sound wavelength is comparable with the dimensions of the structure, the effects must depend somewhat on its exact shape—round or square, shallow or deep, sharp or rounded corners, etc.

One further question might still bother you about open baffles: Now that the sound from the front and back sides of the cone is entering the same room, don't the two signals eventually combine and cause troublesome interference patterns? The first part of the answer is that after reflecting from walls they do indeed end up both contributing to the reverberant sound in the same space. But as for the second part, numerous reflected components arrive at any given point in the room with many different phases, and the interference is quite unlikely ever to be completely constructive or completely destructive. Reflected back-side waves are no more of a problem here than are reflected front-side waves.

16.2 ENCLOSED CABINETS

Consider now a complete enclosure of the space behind a speaker (Fig. 16.4). If this enclosure is built carefully enough that no significant amount of air can leak in or out when the cone moves, there will be no opportunity for any signal from the back side of the cone to interfere with the front-side output. That output will be about like that for an infinite baffle if $kB > 1$, where B represents the distance to the nearest edge of the cabinet face. That is, the radiation will go into the half space in front of the cabinet, and its load on the speaker will be represented reasonably well by the infinite-baffle formulas from Chap. 11. At low frequencies, where $kB \ll 1$, this will act as an unbaffled monopole source. That is, the radiation will cover a full 360° range of angles, and the radiation resistance will tend toward only half as much as for the infinite baffle. For kB close to 1 we would expect details of cabinet shape, such as asymmetry or rounded corners, to have their greatest effect.

One of the most important things about this sealed enclosure will be its loading effect on the cone. For such high frequencies that $kb \gg 1$ it would be proper to say that the back side of the cone radiates nearly as it would into an infinite space and that the resulting sound simply reverberates inside the enclosure until it is absorbed, in the same fashion we studied for room acoustics. But we are usually not concerned with that picture because the principal range for the speaker will have $kb < 1$. In the upper part of this range we might have $kb < 1$ and yet $kB > 1$, for which the picture is rather complicated. For a detailed theory we would have to specify the exact cabinet shape in order to find out (either by calculation or by measurement) what the first several normal-mode frequencies would be and would have to ask how the exact location of the speaker in the cabinet face affects its ability to excite each of those modes. A likely conclusion would be that the cabinet interior should be loosely filled with damping material to prevent these modes from making resonant peaks in the speaker behavior.

Let us focus our attention on the low-frequency limit, however, where $kB \ll 1$. If all dimensions of the cabinet are smaller than a quarter wavelength, the enclosed volume acts simply as a lumped compliance $C_a = V/\rho c^2$. Note that

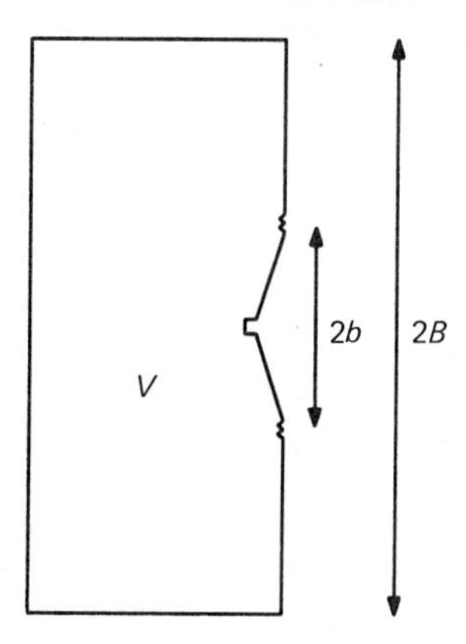

Figure 16.4 Enclosure of a speaker of diameter $2b$ in a cabinet with face width $2B$ and volume V.

this compliance is *not* an addition onto a resistive "radiation load" from the back side of the cone; it takes the place of that load. The cone motion must effectively drive airflow not only into the external radiation impedance load but into the stiffness of the enclosure as well. We should not rush too carelessly to simply draw the circuit of Fig. 16.5*a,* however, because whenever the cone moves outward this drives current U into the radiation load but $-U$ into the cabinet. Furthermore, the pressure on the back side of the cone will generally differ from that on the front not only in phase but in amplitude as well. It is safer to regard the cone as simply driving two independent acoustic circuits (Fig. 16.5*b*), where both have the same transformation factor A (the cone area) to relate their acoustic properties to the mechanical load they place on the cone. Let p_f and p_b be the pressure on the front and back sides of the cone. Then $p_f = Z_{a,\text{rad}}(+U)$ and $p_b = Z_C(-U)$. The net reaction force required from the cone upon the combined loads is $F = A(p_f - p_b) = AU(Z_{a,\text{rad}} + Z_C)$. Then the equivalent mechanical impedance of this load is $F/v = F/(U/A) = A^2(Z_{\text{rad}} + Z_C)$, so that the naive picture of Fig. 16.5*a* does turn out to be correct.

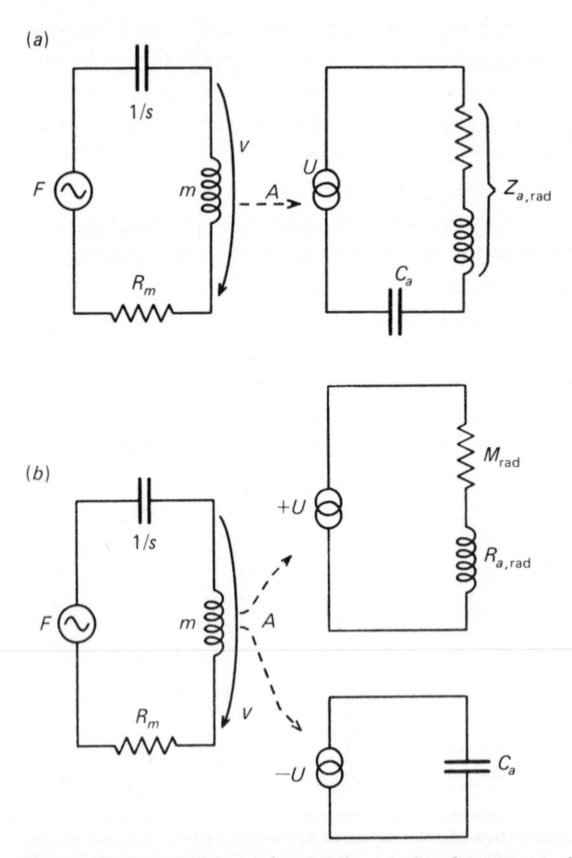

Figure 16.5 Addition of compliance C_a of enclosed air to acoustic load on a speaker; compare Fig. 15.2. Naive argument for simple series circuit (*a*) is justified by consideration as two separate loads (*b*).

The final effect of the cabinet, then, is to put an additional stiffness s_V in the mechanical circuit (Fig. 16.6). The total stiffness

$$s + s_V = s + A^2/C_a = s + A^2\rho c^2/V \qquad (16.2)$$

should be used in place of s in all the discussion of the previous chapter about the resonant frequency f_0. For a given cone mounting, the added stiffness will always raise f_0 compared with what it would be in an infinite baffle and thus will further limit the speaker's useful range. In order for this increase to be unimportant, the volume V would have to be large compared with $A^2\rho c^2/s$.

For the example used in Sec. 15.2 with s = 1200 N/m, b = 12 cm, and A = 0.045 m², the cabinet volume would have to be much greater than

$$V_0 = (0.002 \text{ m}^4)(1.2 \text{ kg/m}^3)(1.16 \times 10^5 \text{ m}^2/\text{s}^2)/(1200 \text{ N/m})$$
$$= 0.23 \text{ m}^3$$

in order for it to act like an infinite baffle and keep f_0 close to 34 Hz.

An alternative attitude is to plan from the beginning that s_V will take the primary role and deliberately use a very weak mounting with $s \ll s_V$. Then the resonant frequency will be

$$\boxed{f_0 \simeq (1/2\pi)\sqrt{s_V/m} = (Ac/2\pi)\sqrt{\rho/mV}.} \qquad (16.3)$$

This is the idea behind the **acoustic suspension** speaker. For a given cabinet size, this will keep f_0 as low as possible; or for a given f_0 it will minimize the necessary volume at the value

$$V_{\min} = (\rho/m)(Ac/2\pi f_0)^2. \qquad (16.4)$$

For instance, acoustic suspension of a speaker with b = 10 cm and m = 12 g (including m_{rad}) to attain f_0 = 80 Hz would require

$$V_{\min} = [(1.2 \text{ kg/m}^3)/(0.012 \text{ kg})][(0.031 \text{ m}^2)(340 \text{ m/s})/(500/\text{s})]^2$$
$$= 0.045 \text{ m}^3.$$

This could be done with dimensions 30 × 30 × 50 cm.

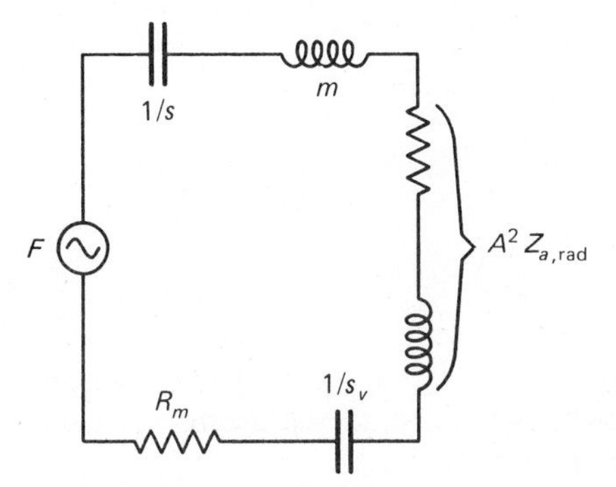

Figure 16.6 Incorporation of the equivalent mechanical impedance of the acoustic load, $Z_m = A^2 Z_a$, into the mechanical circuit; compare Fig. 15.4.

You might be concerned about one other detail of sealed cabinet mountings. Since we are considering wavelengths large enough to spread around the cabinet corners, the radiation inertance of the air on the front side is not necessarily equal to $0.85\rho b/A$ as we have been supposing for infinite baffling. We might expect the actual value to lie somewhere between that and the smaller load $0.61\rho b/A$ for an unbaffled pipe end. (See Eqs. 14.23 and 14.24.) In practice the effective value will depend not only on the cabinet itself but also on where it is placed; a nearby wall or floor is usually exploited to provide some baffling for a bass speaker.

16.3 VENTED CABINETS

Bass speakers are commonly mounted in bass-reflex or tuned-port enclosures that do have a rather large opening in front (Fig. 16.7). Let us try to understand the basic idea behind this approach by considering Figs. 16.8 and 16.9. The open port may well have a rectangular shape with area A_p, but let us model it as a circular opening with radius a chosen so that $\pi a^2 = A_p$. That approximation should have some validity as long as we confine our conclusions to wavelengths larger than the port dimensions.

There are two main points to consider: For a given volume flow U, how much power will be radiated; and for a given force F applied to the speaker coil, what motion U will in fact occur? An exact and general analysis would appear to be very complicated, but there is one thing that simplifies it greatly: The purpose of the port is to modify and extend low bass response, but it is supposed to make little or no change in how radiation occurs in the speaker's middle or high range. To see that this is possible, consider what happens to the current $-U$ entering the box with volume V. For any given midrange frequency and port area A_p, we can always ensure that it is easier to compress the air into V

Figure 16.7 A bass-reflex speaker cabinet. (Courtesy of JBL Inc.)

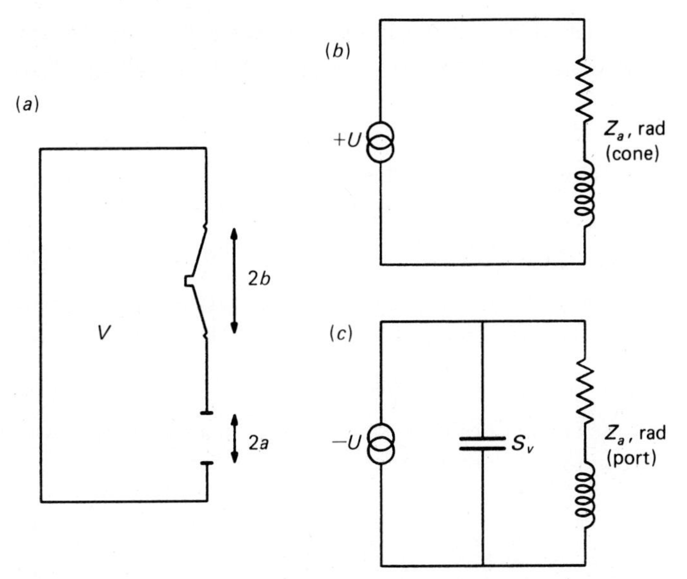

Figure 16.8 (*a*) Cabinet of volume V with speaker of cone diameter $2b$ and port of diameter $2a$. (*b, c*) The acoustic loads on the front and back of the cone.

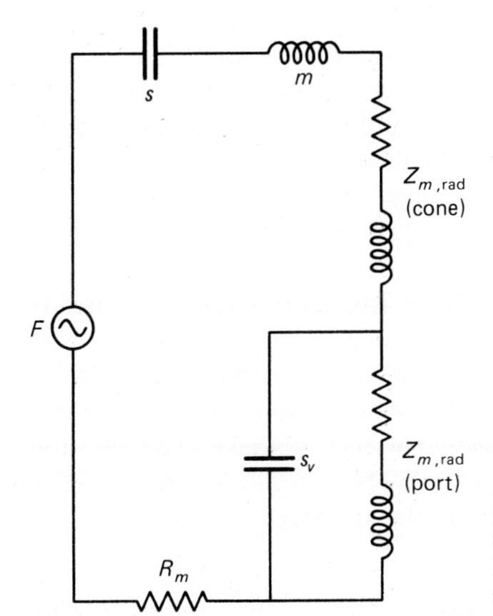

Figure 16.9 Equivalent mechanical circuit for a bass-reflex speaker, incorporating the mechanical equivalent of the acoustic loads. Compare Fig. 16.6.

than to drive it through A_p simply by making V large enough. That is, the admittance of the "capacitive" branch of the cabinet equivalent circuit (Fig. 16.8*c*) can be much larger than that of the "inductive" branch so that it will carry nearly all the current. Then very little air moves through the port, the radiation is primarily from the cone, and the cabinet acts nearly like the sealed cabinet discussed in the preceding section. This argument justifies that the port can be designed to contribute appreciably to the radiation only at such low frequencies that the distance between cone and port is small compared with a wavelength. Then the total radiated power can be approximated by simply using $U_{\text{cone}} + U_{\text{port}}$ in place of U_{cone} in baffled-monopole radiation formulas such as (11.41):

$$P = (\rho c k^2/2\pi)(U_{\text{cone}} + U_{\text{port}})^2. \tag{16.5}$$

Now consider more closely how these flows are related. In the port circuit the fraction $Y_p/(Y_p + Y_V)$ of $-U$ will go through the port. Thus

$$U_{\text{port}} = [Y_p/(Y_p + j\omega V/\rho c^2)](-U_{\text{cone}}). \tag{16.6}$$

At low frequencies the inertance of the air in the port opening is more important than its resistance; so $Y_p \simeq 1/j\omega M_p$. In this approximation, (16.6) simplifies to

$$U_{\text{port}} \simeq [1/(1 - \omega^2 M_p V/\rho c^2)](-U_{\text{cone}}). \tag{16.7}$$

This makes it clear that the behavior of the system changes in an important way in the vicinity of the Helmholtz resonance frequency of the cabinet,

$$\boxed{f_H = (c/2\pi)\sqrt{\rho/M_p V}.} \tag{16.8}$$

In terms of f_H, we can rewrite (16.7) as

$$U_{\text{port}} \simeq -U_{\text{cone}}/(1 - f^2/f_H^2). \tag{16.9}$$

For $f \gg f_H$, U_{port} is very small, as we already argued above. For $f \ll f_H$, $U_{\text{port}} \simeq -U_{\text{cone}}$ and the radiation is greatly reduced; the system has in effect been allowed to become a dipole radiator. This means we would design the cabinet with f_H less than the cone resonance f_0, so that this reduction in power would occur only below the point where the speaker is not expected to operate anyhow.

We see that the frequencies for which U_{port} could make an appreciable in-phase addition to U_{cone} are those just a little above f_H. If that were above f_0 it would only create an unwanted hump where the frequency response is already reasonably flat. It would be most useful to have that addition at the point where the cone output is dropping off; so again we are forced to the choice that f_H should be somewhat below f_0. The range of flat response can be extended roughly from f_0 on down to f_H, though it is not reasonable to try to make that extension exceed an octave. The apparently infinite U_{port} in (16.9) at $f = f_H$ is removed, of course, if we restore a finite value of R_{port} into the analysis.

As an example, consider again the speaker with $m = 16$ g, $s = 1200$ N/m, $b = 12$ cm, and $A = 0.045$ m^2, which had $f_0 = 34$ Hz. Suppose we wanted to put it in a cabinet with $V = 0.6$ m^3 with a port that would extend the response down half an octave to 24 Hz. Setting f_H equal to 24 Hz requires then that

$$\begin{aligned} M_p &= \rho c^2/(2\pi f_H)^2 V \\ &= (1.2\ \text{kg/m}^3)(1.16 \times 10^5\ \text{m}^2/\text{s}^2)/(151/\text{s})^2(0.6\ \text{m}^3) \\ &= 10.2\ \text{kg/m}^4. \end{aligned}$$

As a crude model for M_p take the radiation reactance of a baffled circular opening of radius a (Fig. 14.6):

$$M_p \simeq (\rho/\pi a^2)(2)(0.785a) = 0.50\ \rho/a.$$

This gives

$$a \simeq (0.50)(1.2\ \text{kg/m}^3)/(10\ \text{kg/m}^4) = 6\ \text{cm}$$

or a port area

$$A_p \simeq 0.011\ \text{m}^2.$$

An alternative picture (Fig. 16.10) would be a short length d of pipe with area πa^2, which would have inertance

$$M_p \simeq (\rho/\pi a^2)[d + 2(0.61a)].$$

The desired M_p of 10 kg/m^4 could be produced, for instance, with d = 14 cm and a = 10 cm.

Our other concern is whether the port action makes any important change in the mechanical load seen by the force F in Fig. 16.9. For $f \gg f_H$ the port admittance is low compared with that of the cabinet and the analysis for an enclosed cabinet is not substantially changed. For $f \ll f_H$ the port admittance is high, but the cone stiffness s limits U to very small values. The interesting range is just above f_H, where the total admittance $Y_p + Y_V$ may be much smaller than either one alone, since their reactive components have opposite sign. Here the presence of the port makes it harder to drive current around the equivalent circuit. Just as before when we considered the motional electrical impedance near f_0, the very fact that large motions are being driven imposes a greater load on the driver and tends to reduce the prominence of what otherwise would be a sharp peak in the system response.

These points are illustrated in Fig. 16.11, where the total output as given by Eq. 16.5 is calculated for (a) the unrealistic assumption of constant U_{cone}, (b) $U_{\text{cone}}(f)$ as it would be for a constant force F driving the speaker in an unvented

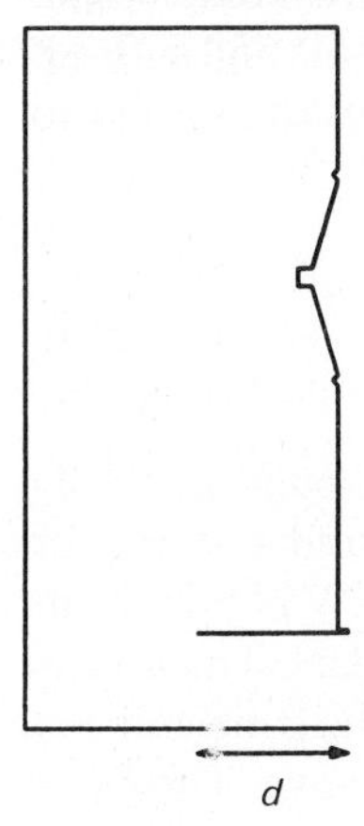

Figure 16.10 A short narrow passage offers an alternative way of achieving the desired port inertance.

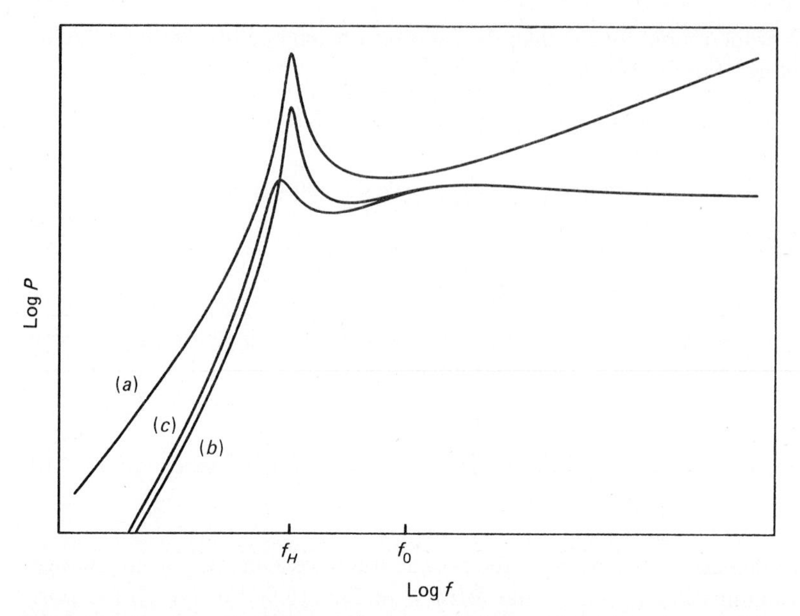

Figure 16.11 Acoustic power radiated according to Eq. 16.5, (*a*) assuming same U_{cone} at all frequencies, (*b*) using the U_{cone} that would result from constant force in Fig. 16.6, and (*c*) calculating U_{cone} for constant F with Fig. 16.9. This example has $f_0 = 2f_H$, $Q_m = 0.8$, and a judicious choice of port inertance.

cabinet as in Fig. 16.6, and (*c*) $U_{cone}(f)$ obtained as $AF/Z_m(f)$ from the circuit of Fig. 16.9 including the effect of the open port. In each case a finite port resistance $R_p = 2\pi\rho f^2/c$ has been included to keep the resonant effects finite. Neither this expression for R_p nor that for M_p above should be taken too seriously, because with less than a wavelength separation between port and cone each of them is partially loaded by the flow from the other. Any attempt at a more accurate theory would need to be checked empirically against laboratory measurements of effective radiation impedances for such complex configurations.

Another feature you will find on some cabinets is an additional diaphragm, or "drone cone," blocking the port entrance. Clever choices of mass and stiffness of this element could give the opportunity for further tuning of the system to improve the smoothness of its frequency response.

16.4 HORN LOADING

One of the drawbacks of the cone dynamic speaker is its low efficiency, generally only a few percent. This problem is more serious in large sound systems for auditoriums or stadiums, where the difference between 5 and 50 kW of minimum amplifier power required may represent many thousands of dollars. The underlying reason for the low efficiency is that the solid materials of the cone and coil are several thousand times more dense than the air they must move. There is a poor mechanical impedance match between the driver and the fluid, which would not be the case if we were trying to radiate sound into water instead of air. The speaker has a high mechanical internal impedance, just like a battery with high internal resistance that delivers only a fraction of its emf to an external load.

To illustrate this point, a speaker with total moving mass $m = 14$ g has a mechanical reactance at $f = 200$ Hz of

$$X_m = \omega m = (1260/\text{s})(0.014 \text{ kg}) = j17.6 \text{ kg/s},$$

supposing that this frequency is well above resonance. The air resistance load at this frequency, for a cone with radius $b = 10$ cm, is only

$$\begin{aligned} R_{m,\text{rad}} &= 2\pi^3 \rho f^2 b^4 / c \\ &= (62)(1.2 \text{ kg/m}^3)(4 \times 10^4/\text{s}^2)(10^{-4} \text{ m}^4)/(340 \text{ m/s}) \\ &= 0.9 \text{ kg/s}. \end{aligned}$$

Thus only a small fraction of the applied force (about 5% in this case) is available to do useful work.

Is there some way that the air load could be made to appear as a predominantly resistive mechanical impedance to the driver? Consider how the mechanical radiation load of a diaphragm in an infinite baffle was written in Chap. 11 as

$$Z_{m,\text{rad}} = \rho c \pi b^2 Z_1(2kb). \tag{16.10}$$

Our problem is not only that $|Z_1| \ll 1$ but also that the real part of Z_1 is much smaller than the imaginary part. If the diaphragm were feeding an infinite cylindrical tube instead of a half space, the impedance would simply be $\rho c \pi b^2$. This is both larger and entirely real; for instance, a pipe with $b = 10$ cm would present a pure resistance of 13 kg/s. Unfortunately, we need to get the sound out of the pipe to be of any use. If the pipe extends for a length L and then simply terminates in an open room, it will obviously introduce unwanted resonances at its own normal-mode frequencies.

We can get some of the advantages of the pipe without that drawback if we let it spread gradually instead of ending abruptly. That is the basic idea of horn-loaded speakers: The sound is launched into a narrow pipe with high input impedance for efficiency's sake, but a judiciously flared shape can then let the traveling wave out into the room. Suppose the wave carries an unchanging total power down a pipe whose cross-sectional area depends on distance x along its axis as $A(x)$, and that A changes so slowly with x that the wave is always practically plane. Since $P = pU$ remains constant while the impedance $p/U = z_c/A$ becomes smaller and smaller as A increases, p and U must change as

$$p(x) = p(0)\sqrt{A(0)/A(x)}, \qquad U(x) = U(0)\sqrt{A(x)/A(0)}. \tag{16.11}$$

Thus the horn acts as an impedance-matching device that can accept a high-p, low-U wave and gradually transform it into a low-p, high-U wave suitable for radiation into the low impedance presented by the room to the mouth of the horn. We expect that the pipe cross section must expand very smoothly, for any sudden change would constitute a discontinuity in impedance and create reflections. Those reflections would mean standing waves, and resonances at particular frequencies, which must be avoided.

Many possible horn shapes have been studied, including conical, exponential, and hyperbolic. Let us present here only the limited information we can handle without doing the exact details for any of those cases, since modern horn design has in any case progressed to still more complicated shapes.

First let us be more specific about the requirement that the horn not flare too abruptly. If the increase ΔA were an appreciable fraction of $A(x)$ within a distance Δx much smaller than $\lambda/4$, the wave would see a fairly sudden impedance change and be reflected from it. Thus a successful horn needs to have $\Delta A/A$ remain always small compared with $\Delta x/(\lambda/4)$, or

$$(1/A)\, dA/dx \ll 4/\lambda. \tag{16.12}$$

But we can turn this around the other way: For any given horn, this equation can only be satisfied for sufficiently small wavelengths and correspondingly high frequencies. The horn will act as a high-pass device, and the **cutoff frequency** below which it is ineffective can be estimated very roughly from (16.12) as

$$f_c \sim (c/4A)(dA/dx). \tag{16.13}$$

An exponential horn, for instance, would be described by $A(x) = A_0 \exp(x/d)$, where d is some fixed distance, and would have $f_c \sim c/4d$. A conical horn would have $A(x) = A_0(1 + x/d)^2$, where d can be interpreted as the distance back to

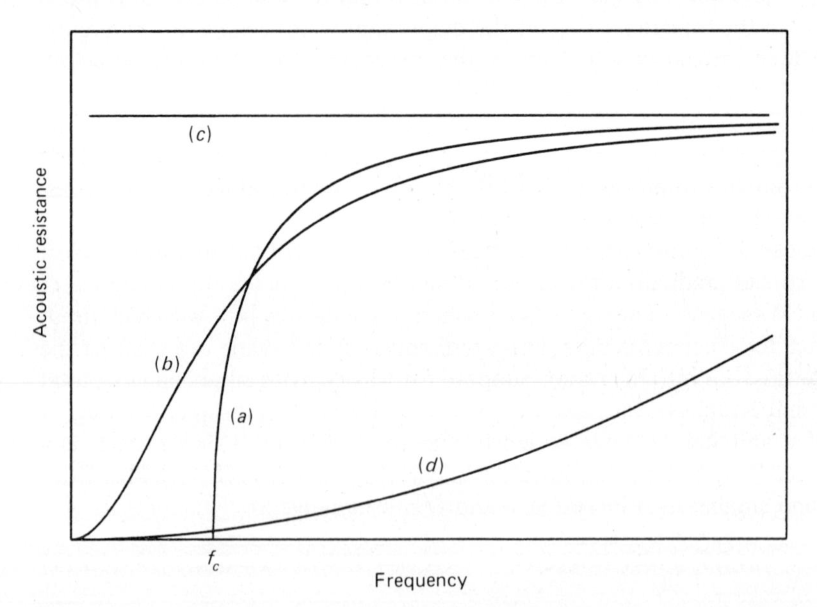

Figure 16.12 Series input resistance at the throat of long horns, (*a*) exponential and (*b*) conical. For comparison, input resistance of (*c*) a long cylindrical tube and (*d*) a bare diaphragm, both of the same diameter as the horn throat.

the apex of the missing tip of the cone, and we would estimate $f_c \sim c/2d$. Exact theory shows (Fig. 16.12) that the transition in behavior is quite gradual for the cone, whereas the exponential horn maintains nearly constant resistive loading for $f > c/4d$, but this plunges to a sharp and total cutoff (purely reactive loading) for $f < c/4\pi d$.

There are two other things we may expect about horn dimensions. One is that the total horn length L will exceed the wavelength of the sounds it is supposed to handle, so that the mouth area can be much larger than the throat while still not violating (16.10). The other is that the mouth width will be about a quarter wavelength or more, so that most of the wave will continue traveling forward without realizing that the guiding walls have suddenly ended. This very fact means that horns will tend to cover a frequency range in which they create a more or less directional beam of radiation. Traditionally this has meant that a simple shape such as the popular exponential horn effectively radiates with a pattern that becomes narrower at higher frequencies. [Recall how beam width for piston radiation in Chap. 11 was given by arcsin $(3.83/kb)$, and imagine that the horn mouth width can now approximately play the role of $2b$.] But in recent years more complex horn shapes have been developed (Fig. 16.13) that can give nearly constant directivity over a range of several octaves in frequency. These are commercially available in a variety of models, enabling the user to choose coverage patterns tailored to the desired application. The horizontal and vertical coverage angles need not be the same; patterns such as $90° \times 40°$ or $40° \times 20°$ may be selected.

16.5 SPEAKER SYSTEMS

We have seen that it may take considerable care to design a speaker and mounting that will have fairly flat response over three or four octaves. But the audible

Figure 16.13 Some constant-directivity horns. (Courtesy of Altec Corporation.)

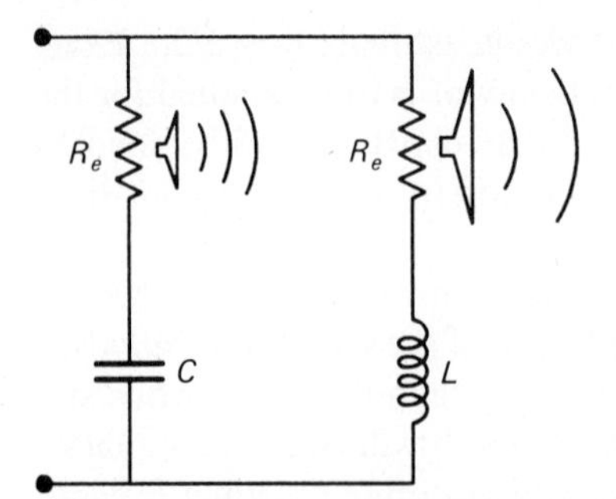

Figure 16.14 A first-order crossover network. Relations among R_e, L, C, and the crossover frequency are developed in Prob. 16.11.

range of sound encompasses close to 10 octaves; so it is too much to ask that any single speaker provide really high-fidelity sound reproduction over that entire range. Instead, we usually use speaker systems that divide the task among two or three speakers mounted in a common cabinet (in home-entertainment systems) or several sets of speakers (in large auditorium sound reinforcement).

There are several disadvantages to simply applying the total electrical signal indiscriminately to all the speakers, either in series or in parallel. A speaker may still radiate sound beyond its intended frequency range, but with an undesirable directional pattern or uneven frequency response. Two speakers located a short distance apart and both contributing output at a given frequency may cause uneven sound distribution through constructive or destructive interference of their outputs. Simultaneous strong signals in different frequency ranges in the same speaker may produce nonlinear distortion. Currents associated with the bass signal may produce heating of the coil of a small treble speaker without serving any useful purpose.

For all these reasons, it is important that we use filter circuits called **crossover networks** to send to each speaker only that part of the total signal which it is supposed to radiate. As a naive illustration of what a crossover network is supposed to do, consider Fig. 16.14. High-frequency signals will go preferentially through the small speaker and capacitor rather than the inductor coil. Low-frequency signals find a low-impedance path through the inductor but cannot easily get through the capacitor, so are presented mainly to the bass speaker.

If you do Prob. 16.11, you will see that this particular crossover circuit makes the transition between low- and high-frequency behavior only very slowly; there is a two- or three-octave range in the middle where both speakers share appreciably in the total power output. Any high-quality commercial speaker system will have a more sophisticated crossover network than this. Since the design of such networks is primarily a problem in electric-circuit theory, we will not pursue it further here.

In large systems the signal may be split according to frequency range before the final amplification stage. These "bi-amp" and "tri-amp" systems have separate power amplifiers to feed their woofers and tweeters.

PROBLEMS

16.1. Roughly how large must a baffle be to imitate the effect of an infinite baffle for a speaker whose intended range is entirely above **(a)** 2 kHz and **(b)** 300 Hz.

16.2. **(a)** For a bass speaker with $b = 12$ cm mounted in a baffle with an effective B of 30 cm, above what frequency is this baffle effectively infinite?
(b) How many decibels is the response at 90 Hz reduced compared with an infinite baffle?
(c) But how many decibels higher is it than for the unbaffled speaker alone?

16.3. Discuss the baffling of a small speaker mounted in an automobile dashboard without complete enclosure behind it. Estimate acoustic path lengths and cutoff frequencies.

16.4. Suppose an enclosed cabinet with volume V and walls of thickness t has a small air leak. Model the leak as a tube of length t and diameter $2a$ whose outer end radiates into the outside world. What equivalent circuit is seen by the airflow from the back side of the cone? What fraction of this flow goes through the leak, and is it more important for high or low frequencies? For illustration, use $V = 0.02$ m^3, $t = 1.5$ cm, and $f_0 = 60$ Hz. How big then could $2a$ be without allowing more than 10 percent leakage?

16.5. For a speaker of diameter 25 cm mounted in a cabinet $30 \times 30 \times 40$ cm, estimate the ranges of frequency in which **(a)** the cabinet is well modeled as a simple compliance, **(b)** the cabinet interior could be treated like a reverberant room with smooth frequency response, and **(c)** individual cabinet mode resonances might be a problem. What difference does it make whether the cabinet is lined with damping material, say with $\bar{\alpha}$ of 0.5 instead of 0.05 unlined? (*Hint:* Recall Eq. 12.11.)

16.6. If a speaker has $b = 8$ cm, $m = 6$ g (cone and coil alone), and $s = 300$ N/m, what enclosed cabinet volume V will give it a resonant frequency $f_0 = 120$ Hz?

16.7. Instead of requiring mounting in a large cabinet, tweeter units often are manufactured as sealed units already enclosed in back. Show that this is reasonable by estimating how much enclosed volume is required to allow a cone with $m = 2$ g and $b = 4$ cm to operate down to $f = 600$ Hz.

16.8. For a speaker with $b = 10$ cm, $m = 10$ g (cone and coil alone) and $s = 7200$ N/m, propose dimensions for a bass-reflex cabinet to extend its response down to about 80 Hz.

16.9. Roughly estimate minimum length and mouth width for an exponential horn if it is to handle frequencies above **(a)** 2 kHz, **(b)** 200 Hz, and **(c)** 20 Hz. If in each case the range is to extend to a highest frequency 10 times as great, what is the maximum throat diameter that will satisfy $k_{\max} b_{\text{th}} < 2$? Are the answers compatible with the requirement that $f_{\min}$ be well above the cutoff f_c?

16.10. What would you estimate as a minimum mouth width for a horn that is to radiate a beam 60° wide at **(a)** 2 kHz, **(b)** 200 Hz?

16.11. Suppose the crossover network of Fig. 16.14 drives two idealized speakers that both have the same purely resistive input impedance R_e at all frequencies, and that the desired crossover frequency is $\omega_0 = 2\pi f_0$. Choose L so that its impedance will be greater than R_e for $f > f_0$ but less than R_e for $f < f_0$, and vice versa for C_e.
(a) Show that this in effect is a choice that the circuit will have a Q of 1.
(b) Show that the impedance seen by the amplifier is simply R_e, regardless of frequency.
(c) Show that the power delivered to the bass speaker is $(\mathcal{E}^2/R_e)/[1 + (\omega C_e R_e)^2]$, and find the corresponding formula for the treble speaker.

(d) Show that the power is split half-and-half at $f = f_0$ and that the transition from 90-10 to 10-90 split extends from $f_0/3$ to $3f_0$. Discuss the drawbacks of this circuit.

LABORATORY EXERCISES

16.A. Measure acoustic output for the same speaker when it is baffled or enclosed in different ways. Interpret the results with concepts from this chapter.

16.B. Obtain a bass speaker and determine its resonant frequency in the lab. Define a reasonable target f_H for a bass-reflex cabinet suited to this speaker, and generate a specific design. For a more extended project, proceed to build the cabinet, measure the frequency response of the resulting system, and discuss how well the theory explains its behavior.

16.C. For one or more horn loudspeakers, measure the frequency response, directional pattern, and input impedance, and estimate the acoustic efficiency. Compare the results with typical cone-speaker characteristics.

Chapter 17

Microphones

The accuracy of sound measurements for scientific or technological purposes depends crucially on the instruments used to detect the sound. Similarly, the fidelity of music and speech reproduction can be no better than that of the microphones employed. Ideally we would insist on near perfection in our microphones; in practice, we should at least document the actual behavior of various microphones so that they can be intelligently chosen for different types of use.

The first section of this chapter reviews the main properties that describe a microphone's performance, and how they may be measured. We then discuss some idealized types of microphone behavior, which can serve as comparisons for realistic cases. As we did before with the cone dynamic loudspeaker, we use the electrostatic microphone as a case study to present more detail than for other types. A more general approach that serves to put all kinds of microphones (and loudspeakers as well) into a common framework will be reserved for the following chapter.

17.1 DESCRIBING MICROPHONE PERFORMANCE

The assigned task of any microphone is to accept an acoustic input and transform it into an electrical output. Specifically, the exact description of the sound occurring at some point in space is contained in the time history of the air-pressure fluctuations $p(t)$. We look for the electrical output voltage waveform $V(t)$ to be an exact replica of $p(t)$. Upon closer inspection, we find that we need to talk about at least four specific aspects of the relation between p and V.

First, is the output directly proportional to the input? Does a doubling of amplitude for a sine-wave input precisely double the output, and does the output remain precisely sinusoidal? If so, the microphone has the extremely important property of **linearity.** Any nonlinearity in the system would mean that input signals containing a mixture of components with frequencies f_1 and f_2 would generate new components in the output with frequencies $f_1 \pm f_2$, which have no counterparts in the input. (See Prob. 17.1.) Any such distortion products ought to be much weaker (50 dB or more) than the original components. A high-quality microphone will operate over a wide dynamic range of input levels, as much as 80 or 100 dB, responding accurately to both weak and strong inputs without introducing noticeable distortion. Measurement of distortion components can in principle be done with an electronic spectrum analyzer, but it requires an accompanying demonstration that the distortion did not arise in the test-signal generator or loudspeaker.

Second, how strong is the microphone output? Insofar as the response is linear, the ratio

$$M = V/p \tag{17.1}$$

will be independent of signal amplitude and thus characteristic of the microphone. This **sensitivity** M may be compared with a standard M_{ref} and expressed on a logarithmic scale as a **sensitivity level**

$$\boxed{\mathrm{ML} = 20 \log (M/M_{ref}).} \tag{17.2}$$

For instance, a microphone with $M = 2.5$ mV/Pa would be described as having ML = −52 dB with respect to the standard $M_{ref} = 1$ V/Pa. Before directly comparing specifications on different microphones, remember that you may sometimes encounter other reference standards, such as 1 V/μbar, in older literature. Sensitivity levels for hydrophones may be given with respect to 1 V/μPa.

But no real microphone responds equally well to signals of all frequencies and all directions of arrival; so we ought to represent M as a function of those variables. This is best done by separating it into three factors, just as we did for loudspeakers in (15.2):

$$\boxed{M = M_0(f_0)\, g_0(f)\, h(\theta, \phi; f).} \tag{17.3}$$

If a microphone is to be represented by the single number M_0, a reference frequency and reference direction must be specified. If the reference frequency were $f_0 = 1$ kHz, for example, g_0 would be defined as unity at that frequency. Where an axis of symmetry exists, it will be used as the reference direction; otherwise, whatever direction has maximum sensitivity can be chosen. Then h is taken to have the maximum value unity at $\theta = 0°$.

The **on-axis frequency response** $g_0(f)$ is our third very important property for microphones, just as it was for speakers. A flat response, independent of frequency, would normally be considered ideal for measurement purposes. But that does not preclude the possibility that a different response would better suit some other purpose (Fig. 17.1). For speech reproduction, for example, reduced response below 100 Hz would forestall problems picking up low-frequency back-

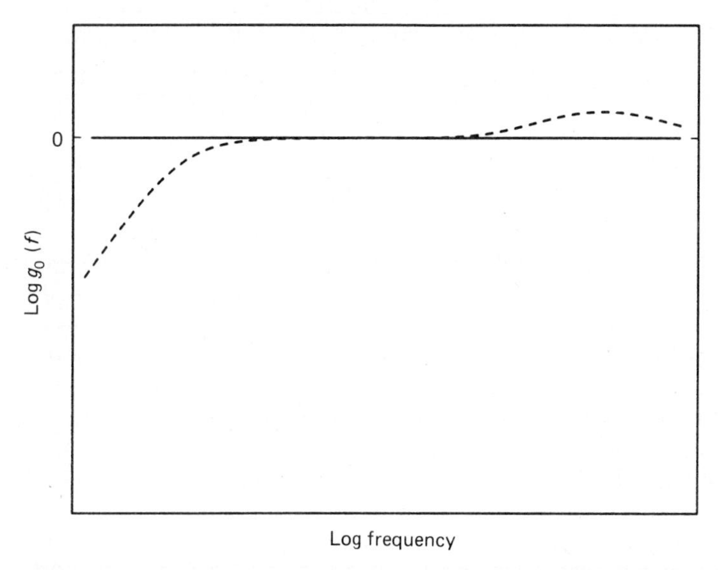

Figure 17.1 Hypothetical flat frequency response of a microphone, and departures from flatness that might be useful as described in the text.

ground noise while not losing any of the desired information. Or a lavalier microphone (worn hanging around the neck or clipped to a shirtfront) might be designed with enhanced response above 4 kHz to compensate for the blocking effect of the speaker's chin upon high-frequency components.

Measurements of frequency response are usually presented on a logarithmic scale. They are not trivial to perform, since we have no perfect loudspeakers available to provide the test signal. The most straightforward method is side-by-side comparison, with curves being recorded for each of two microphones located in the same sound field. Uneven loudspeaker response will at least have the same effect on both curves, and this can be removed if the frequency response of one microphone (flat or otherwise) is already known, either on theoretical grounds or from more fundamental measurements. A more elegant version of this uses the signal from a standard microphone (known to have flat response) in a feedback loop so that the driving voltage on the loudspeaker will be adjusted at every frequency in such a way as to always produce the same SPL at the microphones. Then a chart of the response of the second microphone will require no further correction.

Finally, consider the **directional response** patterns described by $h(\theta, \phi; f)$. From a series of frequency response curves, each taken with the microphone pointing at a different angle away from the line toward the source, information about directional response can be inferred. But it is more attractive to directly record a polar chart while a microphone is slowly rotated on a turntable at some distance from a fixed source in an anechoic environment, in quite the same way as with loudspeaker directional radiation patterns. The parameter f emphasizes that the directional pattern changes from one frequency to another, so that several charts may be needed to fully show its nature. Insofar as the directional pattern changes markedly with frequency, that means the frequency response depends strongly on direction. This can be a problem in music reproduction, because it means the microphone introduces different amounts of **off-axis coloration** for sources located in different directions.

In some applications, especially noise measurements in reverberant rooms, the relevant sound comes in simultaneously from all directions. Then instead of the **free-field** (single oncoming plane wave) frequency response g_0, it is the **random-incidence** (direction-averaged) response $g_0\langle h \rangle$ that we would like to be flat. Typically, $\langle h \rangle$ falls off once the frequency is higher than roughly $f_d = c/\pi d$, where d is the microphone diameter. In the detailed design of the microphone, this may be compensated by a clever choice of diaphragm resonance f_0 somewhat above f_d, together with an appropriate degree of damping of that resonance. It is then possible to keep either type of response fairly flat for an octave or so above f_d, but only at the expense of the other type (Fig. 17.2). Thus a well-equipped laboratory will offer a choice of microphone models suitable to both kinds of measurement.

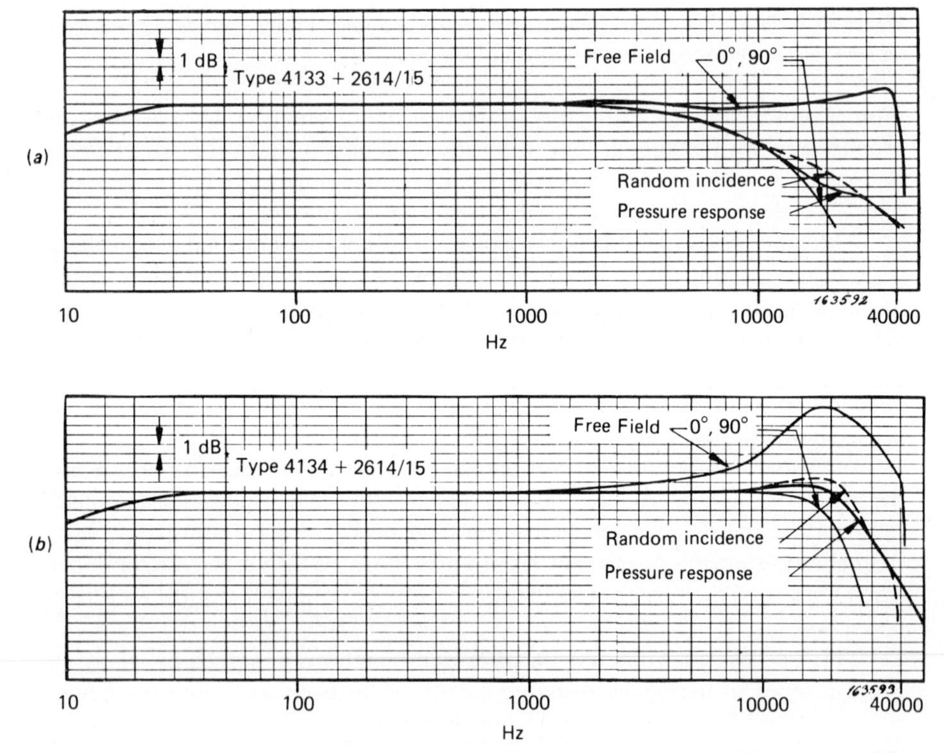

Figure 17.2 Frequency response of two microphones, (*a*) designed for free-field use and (*b*) designed for measuring reverberant sound fields. For each, the upper curve (labeled 0°) is the on-axis free-field response g_0 and the lower curve (90°) is $g_0 h(\theta = 90°)$, for a free field incident from the side. Other curves show random-incidence response $g_0\langle h \rangle$, and response to pressure in an enclosed calibration device. (Courtesy of Bruel and Kjaer Instruments, Inc., Marlborough, MA.)

17.2 SOME IDEALIZED MICROPHONE RESPONSES

To provide standards of comparison for real microphones, it is useful to describe some idealized cases that you will never actually measure in the laboratory. One helpful concept in classifying microphones is that of the **pressure-sensitive** microphone. Ideally this means that only the sound pressure amplitude matters, not the direction of wave travel. Such a microphone would have an **omnidirectional** pattern (Fig. 17.3). Conceptually, you may think of it as having a membrane exposed to sound waves on one side but completely enclosed on the other, like a speaker with a sealed cabinet. If such a device is sufficiently small, it can act as an ideal probe to measure a sound wave without disturbing that wave. In reality, the finite size of the device means that waves of sufficiently high frequency are altered in two ways: Arriving from the back side they cannot diffract around the obstacle well enough to exert as much pressure on the diaphragm, and arriving from the front they generate reflected waves that increase the total pressure on the diaphragm above that in the original incident wave. Furthermore, off-axis waves with half wavelength shorter than the diaphragm diameter can result in simultaneous positive pressure on some parts of the diaphragm and negative pressure on others, for little net effect. Thus at frequencies higher than $f_d = c/\pi d$ you should not expect any microphone to be precisely omnidirectional, and the larger the microphone (including its mounting capsule) the greater the departures will be.

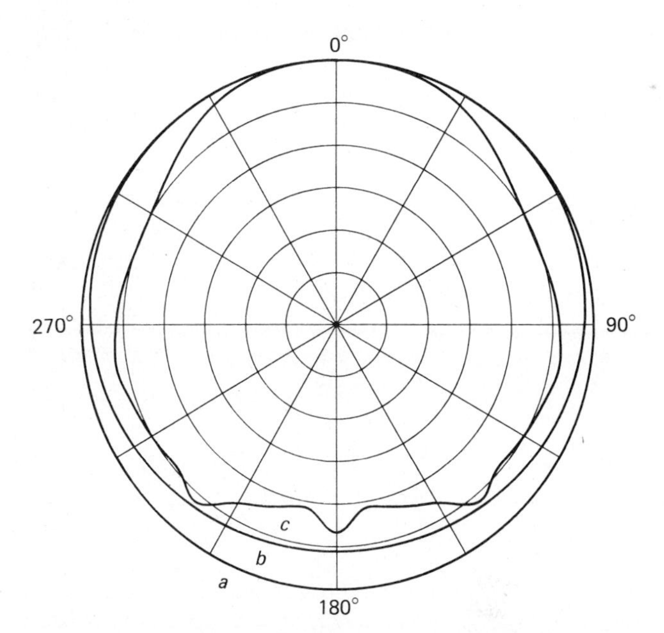

Figure 17.3 Directional response $h(\theta)$ for an omnidirectional microphone (5-dB increments radially). (*a*) For frequencies low enough that the device is much smaller than a wavelength, response is accurately equal in all directions. (*b, c*) The circular pattern deteriorates for high frequencies, for example, about 5 and 10 kHz for a 1-in microphone.

An alternative type is the **pressure-gradient-sensitive** microphone. In this case, envision a diaphragm exposed to the surrounding air on both sides, just as when we discussed a bare speaker cone in the preceding chapter. Insofar as a sound wave creates equal pressure on both sides, there is no net force, no diaphragm motion, and no output voltage. Only differences between front and back pressure give rise to an output signal. The mathematically ideal version of this device is so tiny that (1) it does not alter the wave that otherwise would have been present, and (2) the front and back are effectively an infinitesimal distance apart. The infinitesimal pressure difference, and resulting infinitesimal voltage, are truly proportional to that component of the gradient of pressure, ∇p, along the front-to-back direction. The pressure and pressure-gradient devices could be called monopole and dipole detectors, respectively, for they are very closely analogous to the monopole and dipole sources presented earlier.

Suppose a pressure-gradient microphone faces along the z axis, meaning that its diaphragm lies in the xy plane. Then a plane wave traveling in the direction of some wave vector $\mathbf{k}$ pointing at an angle θ away from the z axis has $p(\mathbf{r}, t) = p_0 \exp[j(\omega t - \mathbf{k} \cdot \mathbf{r})]$ and

$$\nabla p = -j\mathbf{k}p_0 \exp[j(\omega t - \mathbf{k} \cdot \mathbf{r})]. \tag{17.4}$$

Recalling that $k = \omega/c$, we can write the amplitude of the z component of this vector as

$$(\nabla p)_z = (\omega/c)p_0 \cos \theta. \tag{17.5}$$

This tells us that (1) something in the mechanical or electrical design of the

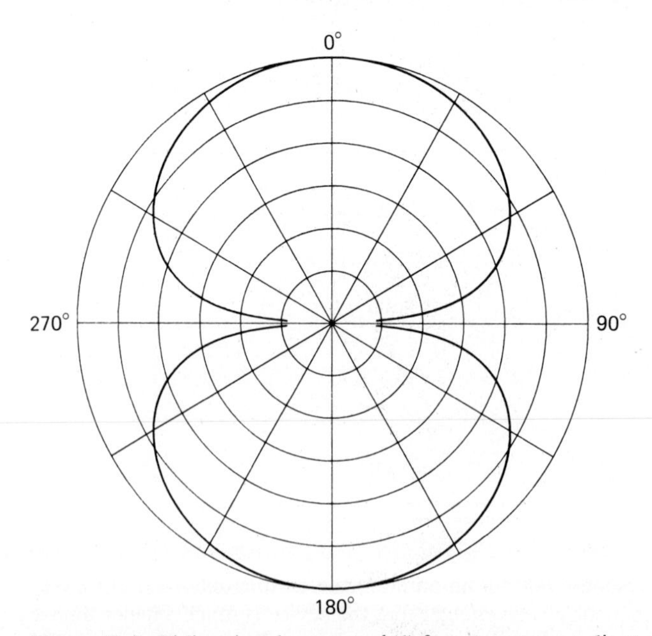

Figure 17.4 Bidirectional response $h(\theta)$ for a pressure-gradient microphone. As with Fig. 17.3, the pattern must be expected to become more irregular at high frequencies.

microphone is needed to compensate for the high-frequency boost represented by the factor ω, and (2) the directional pattern (Fig. 17.4) is given by

$$h(\theta) = \cos\theta. \tag{17.6}$$

This is called a **bidirectional** microphone. Again, shadowing by the diaphragm as well as its mounting means that real microphones must depart somewhat from this ideal pattern, especially at high frequencies.

It would clearly be of practical interest to have a more or less **unidirectional** pattern also. One way this has been achieved is to mount one omnidirectional and one bidirectional element together in a single capsule, with their electrical outputs added. We are free to adjust the relative strength of the two signals before adding; so the total output can be described by

$$h(\theta) = a + (1 - a)\cos\theta. \tag{17.7}$$

Notice how we have defined the parameter a such that $h(0)$ is forced to be unity. If $a = 0.5$, this takes the particularly simple form

$$h(\theta) = \cos^2(\theta/2); \tag{17.8}$$

this has zero response at 180° and is called a **cardioid** pattern. (The word means heart-shaped.)

The simple cardioid has a very broad lobe of acceptance, with sensitivity only 6 dB less at 90° than at 0° (Fig. 17.5). A narrower main lobe can be attained, at the expense of introducing a small secondary lobe in the backward direction, merely by adding larger proportions of bidirectional signal. If we use $a = 0.375$

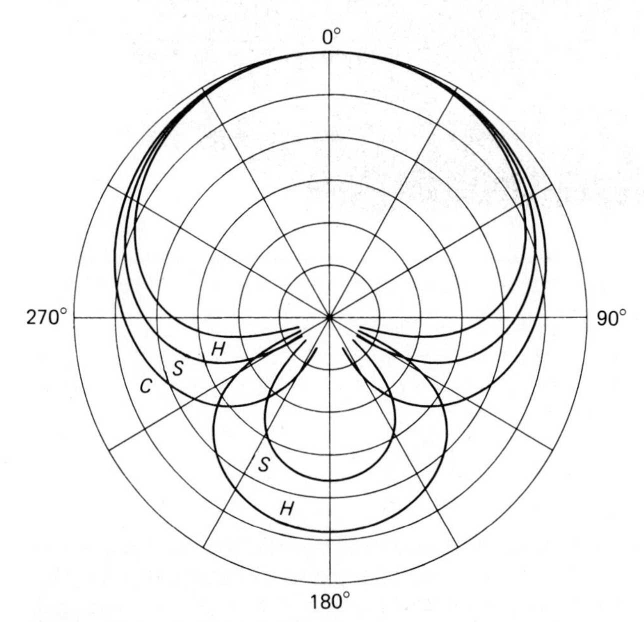

Figure 17.5 Unidirectional microphone responses from Eq. 17.7: C = cardioid, S = supercardioid, H = hypercardioid.

in (17.7), for example, we obtain the "supercardioid" response with 8.5 dB rejection at 90°. The "hypercardioid" corresponds to $a = 0.25$, with 12 dB rejection at 90°.

Response patterns in the cardioid family can also be obtained with a single microphone element, by mounting it in a nonsymmetrical way (Fig. 17.6). With the proper size and shape of baffle, signals arriving from the back can create nearly equal pressure on both sides of the diaphragm, while a phase delay on the order of 180° for a signal arriving from the front to reach the back side enhances the pressure difference. Very sophisticated designs are required, however, to make these phase relations right over a wide range of frequencies. You will find in practice that many microphones with a "cardioid" label have very little directionality at low frequencies.

Another interesting combination is a cardioid unit mounted in the same capsule with a bidirectional element, with their axes perpendicular. If the sum and difference of the outputs are both formed, they will provide sensitivity patterns along both diagonal directions (Fig. 17.7). Thus this acts as a stereo microphone.

To get really narrow unidirectional patterns, it is necessary to build devices with dimensions larger than the wavelength involved. One approach is the line or shotgun microphone discussed in Sec. 4.3. Another is a large parabolic reflector to focus sound onto an ordinary small microphone, just as a reflecting telescope does with light. Many readers will have seen these used by television crews on

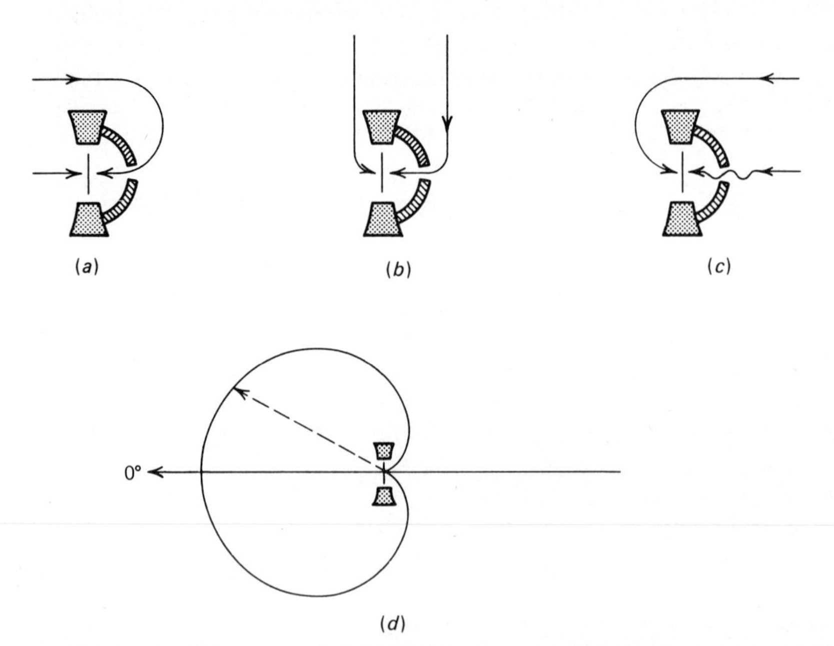

Figure 17.6 Addition of an unsymmetrical phase-shifting structure to a pressure-gradient microphone can delay waves approaching the back side (on the right) in such a way that net force on the diaphragm can be large for $\theta = 0°$ (*a*), finite for $\theta = 90°$ (*b*), and small for $\theta = 180°$ (*c*). (From D. E. Hall, *Musical Acoustics: An Introduction,* Wadsworth, 1980.)

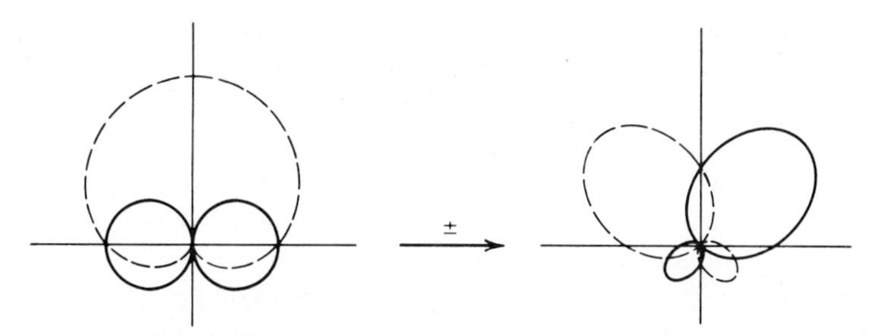

Figure 17.7 Addition and subtraction of cardioid and bidirectional signals to obtain a stereo coverage pattern. Remember that the two lobes of the bidirectional pattern $h(\theta) = \cos\theta$ have opposite phase; so one must be called negative in performing the addition. (From D. E. Hall, *Musical Acoustics: An Introduction,* Wadsworth, 1980.)

the sidelines at a football game, aimed onto the playing field to pick up the quarterback's signals and the linemen's grunts. The effective beamwidth of such a reflector, with diameter D, is roughly D/λ radians. (For justification of that statement, you may consult the section on diffraction theory in any optics text.) Thus a reflector with $D = 1$ m will have significant directionality for wavelengths less than about a meter or frequencies above several hundred hertz. That is good enough to pick up intelligible speech.

How much advantage does a directional microphone have in excluding noise arriving randomly from all directions while still effectively hearing a speaker located on-axis? A measure of this is provided by integrating the response pattern over all angles to obtain the average response to background noise:

$$\langle h^2 \rangle = (1/4\pi) \iint h^2(\theta, \phi) \sin\theta \, d\theta \, d\phi. \tag{17.9}$$

For axially symmetric patterns, this simplifies to

$$\langle h^2 \rangle = (1/2) \int_0^{\pi} h^2(\theta) \sin\theta \, d\theta. \tag{17.10}$$

For both the bidirectional and ordinary cardioid patterns this results in $\langle h^2 \rangle = 1/3$ for the noise, whereas $h(0) = 1$ for the on-axis signal. Then we may say either that the microphone has a **directivity** factor of 3 or that it has a **signal-to-noise ratio** which is 10 log (3) = 4.8 dB better than an omnidirectional microphone. The greatest directivity available in the cardioid family of patterns is a factor of 4, corresponding to 6 dB, for the hypercardioid.

17.3 THE ELECTROSTATIC MICROPHONE

As an example for detailed analysis, let us consider the electrostatic microphone. Since the heart of its operation lies in the changing separation between two charged conducting sheets, it has also been popularly called the condenser (i.e., capacitor) microphone. Since the diaphragm must be electrically conducting, it may be made of either metal foil or metal-coated insulator (such as aluminized Mylar). Traditionally this type of microphone was more expensive, more fragile, and more inconvenient to use than others, so was limited mainly to situations where its superiority in flat frequency response was crucial. The inconvenience had to do with the necessity of providing extra leads from a dc power supply to maintain a large voltage V_0 (as much as 100 V or more) across the capacitor (Fig. 17.8). But an alternative way of maintaining a steady electric field between the capacitor plates is to have one of them coated with a permanently polarized material called an electret (the electrical analog of a permanent magnet). In recent years the quality of thin electret films (commonly made from the fluoropolymer Teflon FEP) has developed to the point where electret capacitor microphones are now competitive with other types in both scientific and musical applications.

When pressure $p = p_1 \cos \omega t$ acts on a circular diaphragm, what output voltage is generated? The answer depends on the frequency, because the diaphragm is a system with its own natural resonance. If the tension $\mathcal{T}$ applied at the edge is the only restoring force and the mass per unit area is σ, the resonant frequency for a diaphragm of radius a has been given in Eq. 8.14 as $f_1 = (0.383/a)\sqrt{\mathcal{T}/\sigma}$. The normal range of operation is for $f < f_1$, where the motion is stiffness-controlled and the membrane displacement nearly the same as if the pressure were static. Setting the acceleration equal to zero in (8.6) and adding another term for the externally applied pressure requires the displacement $\xi(r)$ to satisfy

$$0 = \mathcal{T}(\xi_{rr} + \xi_r/r) - p. \qquad (17.11)$$

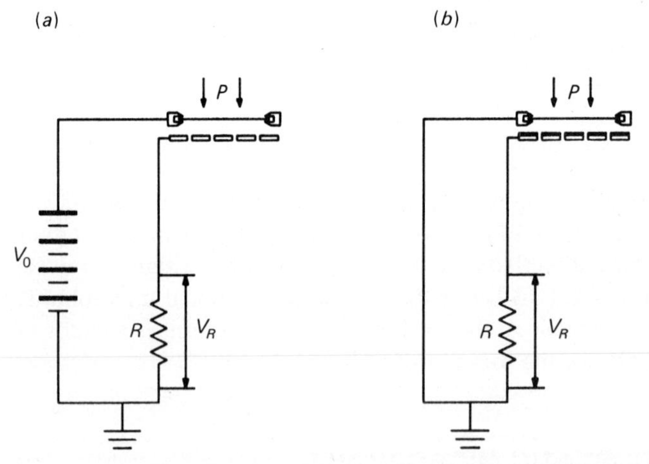

Figure 17.8 Essential elements of an electrostatic microphone. (*a*) Acoustic pressure p acts on a diaphragm that is one plate of a capacitor with dc voltage V_0 maintained across it. The back plate is perforated so that air is not trapped in the thin space between the two plates. AC voltage output appears across the resistor R. (*b*) If either the back plate (as shown here) or the diaphragm has a layer of electret material with permanent polarization pointing across the gap, the external power supply is unnecessary.

This equation has a very simple solution obeying the requirement that ξ be zero at $r = a$, namely,

$$\xi(r) = -(p/4\mathcal{T})(a^2 - r^2). \tag{17.12}$$

Once the membrane is bulged in this way, we no longer have a parallel-plate capacitor. Each small element dA of plate area, however, may be viewed as a small capacitor with $C = \epsilon_0\, dA/(s + \xi)$, where s is the equilibrium plate separation and $\epsilon_0 = 8.85 \times 10^{-12}$ F/m. All these elemental capacitances are connected in parallel; so the effective capacitance of the combination is simply their sum:

$$\begin{aligned} C &= \epsilon_0 \int dA/(s + \xi) \\ &= (\epsilon_0/s) \int_0^a 2\pi r\, dr[1 - \xi/s + (\xi/s)^2 - \cdots]. \end{aligned} \tag{17.13}$$

Here the binomial expansion of $(1 + \xi/s)^{-1}$ has been used so that the integral will be quite easy to carry through when (17.12) is used for ξ. The result is

$$C = C_0(1 + pa^2/8\mathcal{T}s + p^2a^4/48\mathcal{T}^2s^2 \cdots), \tag{17.14}$$

where $C_0 = \epsilon_0\pi a^2/s$ is the original capacitance. Now the resistance R is chosen to be so large that the response time RC is longer than the period of the sound waves, so that the amount of charge CV_C on the plates must remain nearly constant. But since that charge is equal to C_0V_0, the voltage across the capacitor must then vary as

$$\begin{aligned} V_C &= V_0/(1 + pa^2/8\mathcal{T}s + p^2a^4/48\mathcal{T}^2s^2 \cdots) \\ &= V_0(1 - pa^2/8\mathcal{T}s - p^2a^4/192\mathcal{T}^2s^2 \cdots). \end{aligned} \tag{17.15}$$

Application of Kirchhoff's loop rule tells us that there must be just enough current flowing through the resistor to keep $V_C + V_R = V_0$. Thus the voltage drop V_R presented to the output amplifier is

$$V_R \simeq V_0(pa^2/8\mathcal{T}s + p^2a^4/192\mathcal{T}^2s^2). \tag{17.16}$$

Now we may use $p = p_1 \cos \omega t$ to write this as

$$V_R(t) = V_0[(p_1a^2/8\mathcal{T}s) \cos \omega t + (p_1^2a^4/384\mathcal{T}^2s^2)(1 + \cos 2\omega t)]. \tag{17.17}$$

The leading term shows that the microphone sensitivity should be

$$\boxed{M = V/p = V_0a^2/8\mathcal{T}s = V_0C_0/8\pi\epsilon_0\mathcal{T}.} \tag{17.18}$$

Notice that ω does not appear in this expression, indicating that the frequency response should be flat for all $f \ll f_1$. We see that a desire for more sensitivity will encourage the use of higher bias voltages, larger diaphragms, and smaller electrode spacings. Yet the electric field strength V_0/s cannot be made so large as to cause arcing across the air gap, so must remain rather less than 8×10^5 V/m. Neither can the diameter be made too large without either lowering f_1 or pushing beyond the tearing strength of the diaphragm material.

For $f \gg f_1$ the diaphragm motion is limited by mass rather than stiffness; so its response amplitude must fall off in proportion to f^{-2}, that is, 12 dB/octave. Near f_1 the mechanical resistance becomes important; just as with the resonance of a dynamic speaker, a choice of Q near unity can keep the response nearly flat in this range instead of having a sharp peak. Adjusting the size of the perforations in the back plate can allow the right amount of airflow through them to provide this desired amount of damping.

The last term in (17.17) represents an output at a different frequency than the input, that is, harmonic distortion. We could characterize the relative strength of this undesired signal by the ratio

$$V(2\omega)/V(\omega) = p_1 a^2/48\mathcal{T}s = p_1 M/6V_0. \qquad (17.19)$$

Since this needs to be kept much smaller than unity, we can see that the same changes in a, $\mathcal{T}$, or s that would increase the sensitivity tend at the same time to increase the distortion.

As an example consider the diaphragm described in Sec. 8.2, with $a = 6$ mm, $\mathcal{T} = 2860$ N/m, and $f_1 = 12$ kHz. If we require that the harmonic distortion not exceed 1 percent in amplitude (a 40-dB margin) for a maximum SPL of 140 dB (or $p_1 = 200$ Pa), then (17.19) gives

$$p_1 a^2/48\mathcal{T}s < 0.01.$$

This means the plate separation must not be less than

$$\begin{aligned} s_{\min} &= p_1 a^2/0.48\mathcal{T} \\ &= (200 \text{ Pa})(36 \times 10^{-6} \text{ m}^2)/(0.48)(2860 \text{ N/m}) \\ &= 5.2\ \mu\text{m}, \end{aligned}$$

for which the capacitance is

$$\begin{aligned} C_0 &= \epsilon_0 \pi a^2/s \\ &= (8.85 \times 10^{-12} \text{ F/m})(3.14)(36 \times 10^{-6} \text{ m}^2)/(5.2 \times 10^{-6} \text{ m}) \\ &= 200 \text{ pF}. \end{aligned}$$

If we limit the applied electric field to $E = 4 \times 10^5$ V/m, the maximum V_0 is then $Es = 2$V and the corresponding sensitivity

$$\begin{aligned} M &= (2 \text{ V})(36 \times 10^{-6} \text{ m}^2)/(8)(2860 \text{ N/m})(5.2 \times 10^{-6} \text{ m}) \\ &= 0.63 \text{ mV/Pa}, \end{aligned}$$

so that

$$\text{ML} = 20 \log (0.63 \times 10^{-3}) = -64 \text{ dB } (re \text{ 1 V/Pa}).$$

(A more generous spacing such as $s = 50\ \mu$m could still have the same ML if a proportionately larger V_0 of 20 V were applied, and in that case C_0 would be only 20 pF.) At the low-frequency end (say 40 Hz) the electrical impedance of a 200-pF capacitor is

$$1/\omega C = 20 \text{ M}\Omega;$$

choosing R somewhat larger than this will keep the RC leakage time scale relatively long. A smaller R would mean a drop-off in response at low frequencies.

We see that the electrostatic microphone is a high-impedance device and ought to be connected to a preamplifier whose input impedance is at least as high. Any appreciable length of connecting cable between the microphone and the preamplifier would provide a shunt capacitance that would prevent the full voltage predicted above from appearing across R. For cable with 100 pF/m, for instance, even one meter is enough to seriously degrade the low-frequency response for most microphones of this type. Therefore, a small preamp is generally attached directly to the microphone unit inside its housing, and only the amplified signal is sent through any cable. It is to power this preamp that small batteries are still needed even when the microphone itself has its bias voltage provided by an electret.

17.4 SOME FURTHER COMPARISONS

How many basically different ways might there be to design a microphone with flat frequency response? Let us try to take the same sort of general view we did in Sec. 15.2 when asking how flat speaker response could be achieved. If we consider only the case where the device is small compared with all wavelengths of interest, there still remain three important choices. First, will the total force on the diaphragm be proportional to the sound pressure (and thus independent of frequency) or to the pressure gradient (with the gradient contributing a factor proportional to f)? Second, what will be the dominant impedance in the mechanical system consisting of the diaphragm and whatever else is attached to it? The velocity amplitude created per unit force may be proportional to f if it is stiffness-controlled, to f^{-1} if mass-controlled, or independent of f if resistance-controlled. Finally, how much output voltage will be generated per unit velocity? If the device depends on magnetic fields (either coil motion through a field as for a simple dynamic transducer or changing magnetic fluxes in moving-armature or magnetostrictive devices), the voltage is directly proportional to velocity and no further frequency dependence is introduced. But if it depends on electric fields (as for capacitive or piezoelectric devices, as well as for carbon microphones), the voltage is proportional to displacement and thus to velocity multiplied by f^{-1}.

There are $2 \times 3 \times 2 = 12$ possible combinations of these three choices, and you may easily verify that only four of them have an overall sensitivity independent of frequency. First, a pressure-sensitive microphone based on any electrical effect should have stiffness-dominated motion. That will naturally be the case for crystal or carbon microphones since the diaphragm must bend or squeeze a solid material in order to move. And that is why an underlying assumption of the electrostatic design was that enough tension must be applied to the membrane so that it, rather than the mass, controls the motion. Second, a pressure-sensitive magnetic device (and specifically the unidirectional moving-coil dynamic microphone) should be resistance-dominated, with its own (highly damped) natural resonance right in the middle of the operating range. That is why this kind often does not have as flat a response as a good electrostatic microphone, with strong rolloff at both low and high ends of the spectrum. The presence of strong damping means this is a very inefficient device in the sense of not converting much of the incident acoustic energy into electrical output; but that is not nearly so important here as it was for the dynamic speaker, since the microphone output is routinely amplified anyhow. The third possibility would be an electric-effect pressure-gradient device with resistance-dominated motion, again subject to some disadvantages as a highly damped system.

The final case is that of a magnetic-effect pressure-gradient device, which needs to have mass-controlled motion in order to achieve flat frequency response. The classic example of this bidirectional type is the ribbon dynamic microphone (Fig. 17.9), in which current simply flows once down a small strip of foil instead of round and round a coil. This makes it possible to put the magnetic pole pieces one at each edge of the ribbon so that its two faces are left equally free to receive sound. Our argument above shows why this ribbon should not be under any appreciable tension. Let us give without formal proof an expression for the sensitivity of a ribbon microphone:

$$M_0 = (BL)(LW)D/mc. \tag{17.20}$$

Here the magnetic field strength B times ribbon length L gives the emf per unit velocity, the length L times width W gives the force per unit pressure applied, m is the ribbon mass, and c the speed of sound. D is an effective front-to-back distance, finite for the real device instead of infinitesimal as for an ideal pressure gradient. Theoretically, D may be ill-defined, but empirically it can be determined

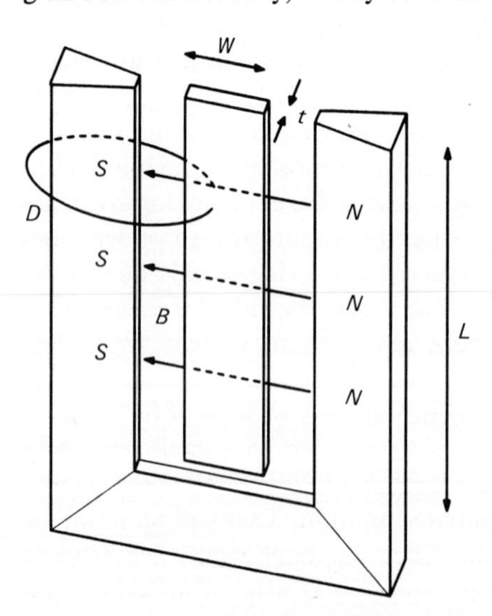

Figure 17.9 Essential elements of a ribbon microphone. The ribbon R moving between the magnetic poles N and S has both sides exposed and develops an induced emf along its length.

by measuring M_0 and turns out to be similar to the typical dimension of the housing that baffles the ribbon. If the ribbon mass is written in terms of its thickness t and material density ρ_m, the sensitivity can be expressed as

$$\boxed{M_0 = BLD/\rho_m tc.} \tag{17.21}$$

This suggests that maximum sensitivity is to be obtained by making the ribbon as long and light as possible, but that its width is not very important.

An interesting thing that can happen with pressure-gradient microphones is called the **proximity effect.** This is not really the fault of the device so much as of the way it may be used. Suppose one pressure and one pressure-gradient microphone have been designed in accordance with the argument above, so that they have identical flat frequency responses when operated side by side under free-field conditions. But hidden under that "free-field" label is the assumption that the source is far enough away that the approaching waves are practically plane, so that $\nabla p = -jkp$. What if the microphones are at a finite distance r from the source and the waves are admittedly spherical? From

$$p(r, t) = (A/r) \exp[j(\omega t - kr)] \tag{10.6}$$

we calculate

$$(\nabla p)_r = \partial p/\partial r = -(jk + 1/r)p. \tag{17.22}$$

The last term reminds us that the spherical wave's amplitude gradients can drive a pressure-gradient microphone even when the phase differences are negligible. At sufficiently high frequency that will not matter, but at low frequency the microphone is being driven by a signal stronger (in relation to the amplitude p) than was assumed in its design. Therefore, the pressure-gradient microphone output will be enhanced by the factor

$$|1 + 1/jkr| = \sqrt{1 + 1/(kr)^2} \tag{17.23}$$

over its free-field response. This represents a boost of 6 dB/octave for frequencies below $kr = 1$, that is, $f < c/2\pi r$. The closer the microphone is to the source, the greater the portion of the bass range that is affected, as illustrated in Fig. 17.10. This effect underlies the old radio broadcast tradition of lending more warmth to a male singer's voice by placing him very close to a ribbon microphone.

In closing this chapter, let us remember two important limitations on any simple theory of microphone design. First, the frequency and directional response are not entirely determined by the elementary transducer properties but depend also on the baffle or housing in which it is mounted, especially for high frequencies. In particular, the response may change radically according to what kind of openings may be present to allow sound signals to reach the back as well as the front of the diaphragm. Second, even a standard model complete with calibration curves to show its performance as packaged can still yield results of widely differing quality according to how wisely it is placed with respect to the sound source. To cite just one example that we are prepared to understand here, the theory in Chap. 12 shows how different proportions of direct and reverberant sound components will be detected depending on whether the signal is picked

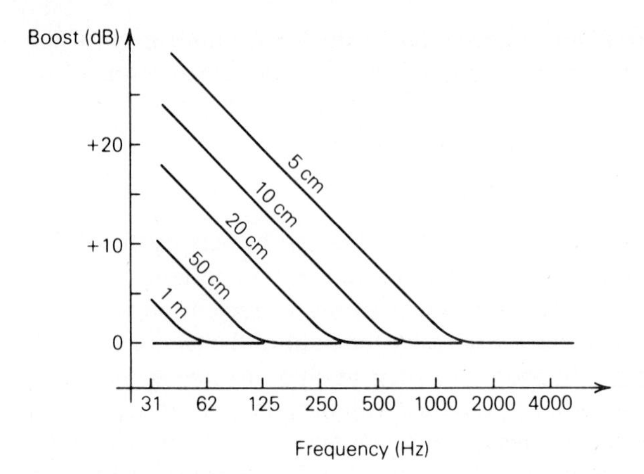

Figure 17.10 Proximity effect: the excess response of a pressure-gradient microphone over that of a pressure microphone when both are placed close to a monopole source. (From D. E. Hall, *Musical Acoustics: An Introduction,* Wadsworth, 1980.)

up closer to or farther away from the source. The relative importance of clarity and of room ambience, which would be enhanced by close or distant placement, respectively, depends on the type of signal involved. The art of good microphone placement for music and speech reproduction is worth a book in itself, and we will not pretend to cover its details here.

PROBLEMS

17.1. Suppose the output of a microphone is related to its input by $V(t) = M[p(t) + \beta p^2(t)/p_0]$, where $p_0 = 1$ Pa and β is a parameter describing nonlinearity of response. For $p(t) = p_1 \cos(\omega_1 t) + p_2 \cos(\omega_2 t)$, what is $V(t)$? (Use trigonometric identities to express it as a sum of sinusoidal components.) Identify the terms that represent distortion products. If you require the amplitudes of those extra components to be at least 60 dB below the strongest true signal component for all values of p_1 and p_2 in a working range between 50 and 130 dB SPL, what is the maximum allowable value of β?

17.2. In going through a collection of microphones and their manufacturers' specification sheets, you find X with ML = −93 dB ($M_{\text{ref}} = 1$ V/μbar) and Y with ML = −47 dB ($M_{\text{ref}} = 1$ V/Pa). Which is more sensitive? What voltage output will each provide when exposed to a wave with SPL = 70 dB?

17.3. A set of precision laboratory microphones all employ the same basic pressure-sensitive design but have diameters of 2.52, 1.26, and 0.63 cm. In what frequency range would you have doubts in each case about departure from flat and omnidirectional response? What other advantage might lead you nevertheless to sometimes choose the larger diameter?

17.4. At what angle does null response occur for an ideal supercardioid, and how many decibels down is the response at 180° compared with 0°? Answer the same questions for the hypercardioid. At what angle is the response 6 dB down for each of these, as well as for cardioid and bidirectional?

17.5. Use ideas from Chap. 14 to propose qualitatively how you might design a pressure-gradient microphone housing to create an appreciable phase delay between front- and back-side signals to the diaphragm, even when the entire device is quite small compared with a half wavelength of the incident sound.

17.6. Consider an electrostatic microphone whose nominal 1-in size describes the outer casing and whose diaphragm radius is 1.2 cm. For an electrode spacing $s = 30$ μm, what is its capacitance? If the diaphragm has surface density 0.05 kg/m^2, what tension must be applied to put the resonance at $f_1 = 8$ kHz? For applied voltage $V_0 = 12$ V, what is the sensitivity level ML? If you chose deliberately to make the response drop off for frequencies below 100 Hz, what resistance R might you use? What amplitude of diaphragm motion would occur for incident sound with SPL = 120 dB? Estimate how much harmonic distortion is in the output.

17.7. Suppose a ribbon microphone employs an aluminum foil with length 1 cm, width 1 mm, and thickness 20 μm in a field $B = 0.6$ T. What is the ribbon mass? If the housing has an effective D of 2.6 cm, what is M_0 and the corresponding ML? For incident sound with SPL = 80 dB what is the output voltage? Using a resistivity 2.8×10^{-8} Ω-m, what is the resistance of this ribbon from end to end? Would you say this device has a high or a low output impedance? Over what frequency range can you reasonably expect the directional pattern to be well described by $h(\theta) = \cos\theta$?

17.8. Suppose the housing of a pressure-gradient microphone were designed so as to compensate for the proximity effect and give it a flat response at a distance $r =$ 10 cm from a point source. What would happen to its response when moved out to $r = 50$ cm?

17.9. Suppose other considerations dictated that a particular hydrophone be pressure-sensitive, magnetostrictive, and stiffness-dominated. How would its sensitivity vary with frequency? Can you propose a way of compensating for this after the signal is in electric form?

17.10. A pressure-sensitive microphone is placed at distance r from an isotropic point source providing signal power $P = 10^{-4}$ W. There is a general background noise level of 65 dB SPL. If the microphone is to detect signal at least 10 dB stronger than noise, how close to the source must it be? How much greater could r be if the microphone were a hypercardioid instead of omnidirectional?

Chapter 18

Transducers: General Theory

For the sake of concreteness, it has been advantageous in the preceding chapters to study the specific examples of dynamic loudspeakers and electrostatic microphones in some detail. But this is not the place to present a separate theory for each of a dozen more types; there are other sources where you can find information of an encyclopedic kind. The main purpose of this book is supposed to be a more general and fundamental insight; so let us ask about the possibility of a basic framework in which you can later fit all the individual cases as you need them. For this purpose we first introduce the important concept of reciprocity, then give a brief glimpse of how such a general theory of transducers can be formulated in mathematical terms, and finally note some consequences that are useful in providing absolute calibration of electroacoustic devices.

18.1 SOME BACKGROUND

A valuable organizing concept for dealing with sound waves and transducers is that of **reciprocity.** To see what this means, consider Fig. 18.1*a*, where some alternating current i is supplied to speaker S, and a resulting sound pressure p is received by microphone M. Under a broad range of conditions, the same current i in S after an exchange of positions, as in Fig. 18.1*b*, will still produce the same pressure p at M. When that is true for all positions of S and M and for an entire range of frequencies, we say a reciprocity condition has been obeyed. Many readers will be reminded of the reciprocity theorem for electric circuits, which also involves exchanging a source (of voltage) and a receiver (of current) and finding that the detector still registers the same current as before.

On closer examination, we may see that there are two different aspects of

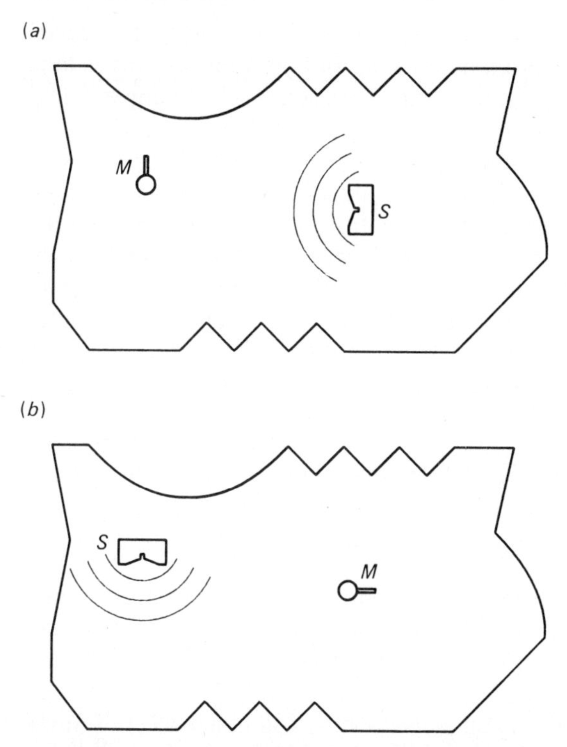

Figure 18.1 (*a*) A specified input signal to speaker *S* produces an output from microphone *M* in the presence of arbitrary surroundings. (*b*) If conditions for reciprocity hold, exchanging locations of *S* and *M* while keeping a constant input will leave the output unchanged.

reciprocity to be considered. One is the reciprocal nature of the transducers themselves, which we take up in Sec. 18.2. The other is the reversibility of the sound waves propagating through the space between *S* and *M*. It may be helpful for you to think of this reversibility as a kind of symmetry that can be seen in several areas of physics. It closely resembles the time-reversal symmetry of Newton's laws of motion. For a certain class of forces, namely, nondissipative ones, classical mechanics makes no distinction between $+t$ and $-t$; that is, a movie film of particles moving through electric, magnetic, or gravitational fields and colliding elastically with each other will appear equally realistic whether run forward or backward. Symmetries such as this provide very powerful tools for describing and classifying physically possible motions.

Not only particles but fields as well have important time-reversal symmetries. Most familiar to people unversed in science is the reversibility of light rays: roughly speaking, "if I can see you, then you can see me too." Though it may appear that this reversibility breaks down in cases such as partial reflection at an interface between air and glass, there is still an underlying symmetry that can be expressed in more sophisticated ways. That symmetry ultimately stems from the time-reversal properties of the fundamental natural law represented by Maxwell's equations.

But the mention of Maxwell's equations is an admission that so-called geometrical optics is only an approximate theory of how light behaves as long as its wavelength is very short compared with all other relevant distances. Long-wavelength behavior requires the study of further phenomena characteristic of wave propagation, and especially diffraction. It becomes important to admit that the wave is inherently smeared out in space; unlike ideal point particles or rays, what ultimately happens to one part of a wave front is not independent of the other parts around it. Proper statement of reciprocity relations for fields then involves what those fields do throughout an entire three-dimensional region of space, not just along the one-dimensional continuum of a single ray. The mathematical proofs become rather abstract and will not be given here.

For sound fields, let us be content to express reciprocity this way, again without formal proof:

> If the positions X and Y of a simple source and a simple receiver are interchanged, but all the surroundings remain as before, an unchanged source strength will give an unchanged signal strength at the receiver.

The word simple is being used here in the specific technical sense (from Chap. 11) of a source (1) very small compared with a wavelength and (2) involving net volume changes, that is, having a monopole moment. (Dipole sources or pressure-gradient detectors require more complicated statements of their reciprocity properties.) Notice the implication that somehow the combined effect of waves propagated along all possible paths from one point to the other, not just along a single ray, is identical in either direction. To emphasize that we are concerned here only with the sound waves and not with the transducers, let us make a more quantitative statement entirely in terms of the acoustic quantities p (pressure at the receiver) and Q (source strength in the sense of volume flow rate):

> The ratio Q/p is determined entirely by the shape and nature of the boundaries of the region and the location of the points X and Y; it does not depend on other details of the devices or on which one is sending and which receiving.

A rather surprising illustration of how this principle can be applied is suggested by Pierce (in his book cited in the Preface, p. 199): Suppose a noise source and a listener are symmetrically located on opposite sides of a barrier and it is desired to reduce the received noise level by putting absorbing material on one face of the barrier. Which side should it be on to be most effective? Whichever side you put it on, reciprocity tells us that the source and receiver could trade positions and the sound level would remain the same; thus it must make no difference which side has the absorber!

18.2 REVERSIBLE TRANSDUCERS

Reciprocity concepts can be useful not only for the sound fields themselves but perhaps even more so for the transducers with which they interact. We indicated in Chap. 2 that most of these devices are **reversible transducers,** that is, that a microphone is just a miniature loudspeaker operating in reverse. Before presenting the theory for such devices, we must mention an exception that is not covered by that theory, namely, the **parametric transducer.** The name indicates that the input signal controls a parameter of some system that generates the output signal. That system has energy resources of its own, so that the energy in the output does not have to come entirely from that in the input. The device may not work that way at all if we attempt to exchange the roles of input and output. The carbon microphone described in Sec. 2.1 provides one illustration, where the dc source that drives the current through the carbon granules provides the energy for the output signal. This is a nonreciprocal device; so the following remarks should not be applied to it.

Here then is a very brief version of the sort of theory that can be developed to describe the reciprocal actions of all kinds of passive linear transducers. By describing them as passive, we are choosing to deal only with those cases where the transducer has no internal source of energy. The requirement that the output energy must all be provided by the input places stringent limits on the characteristics of such transducers, and the main goal of this section will be to show one such limit.

Figure 18.2 shows two different ways in which we might try to view the transducer as a black box with electric terminals on one side and acoustic terminals on the other. From a mathematical viewpoint, any calculation based on one of these pictures has an exact counterpart for the other; so they simply offer two entirely equivalent ways of expressing the same theory. In Fig. 18.2*a* the acoustic volume velocity U is presented as analogous to electric voltage V, and pressure p as corresponding to current i. This is appropriate in an extended set of circuit analogies (alternative to the set given in Chap. 14), in which U has the role of driving the system (like V) and causing p (like i) to occur. The ratio U/p, which is the acoustic admittance, is then defined to be analogous to the

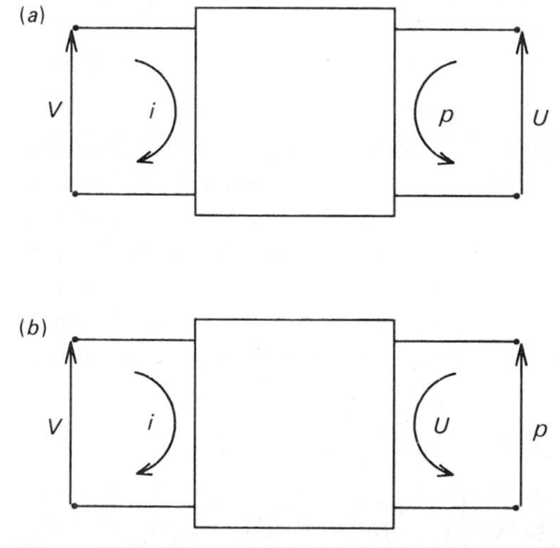

Figure 18.2 Two possible black-box representations of an electroacoustic transducer, using (*a*) mobility analogy and (*b*) impedance analogy.

electric impedance V/i. This may be described as a **mobility analogy,** or as an analogy using a standard technique called **duality** in electric-circuit theory to interchange the roles of p and U. In my experience, this analogy can be quite difficult to accept intuitively when first encountered, and we will not use it here. Nevertheless, it sometimes offers advantages in particular cases, and you should be aware that you may encounter it in the literature.

We do the following theory only with the **impedance analogy** of Fig. 18.2*b*. This is the one we have used in previous chapters, in which pressure p plays the same "pushing" role as V, and U (like i) is viewed as a resulting flow. Our two-port device will have the property that either V *or* p may cause a current i to flow, and likewise both V and p may drive U. Now let us build on the assumed linearity of the transducer: The two causes must produce additive effects, each of which is linearly proportional to its cause. That is,

$$i = Y_{ee}V + Y_{ea}p, \tag{18.1}$$

$$U = Y_{ae}V + Y_{aa}p. \tag{18.2}$$

The subscripts e and a denote whether the particular cause and effect are electric or acoustic, and the matrix Y is a generalized admittance. The four matrix components must be constant in time and independent of signal amplitude, although in general they will depend on signal frequency. The values of these four functions of frequency must completely characterize how such a transducer will act. It is more common to solve these two equations for V and p and write them in terms of an impedance matrix Z:

$$V = Z_{ee}i + Z_{ea}U, \tag{18.3}$$

$$p = Z_{ae}i + Z_{aa}U. \tag{18.4}$$

The impedance Z is the inverse of the admittance Y. But remember that you cannot just invert the individual components without using matrix algebra. Component Z_{ee}, for instance, is

$$Z_{ee} = Y_{aa}/(Y_{ee}Y_{aa} - Y_{ea}Y_{ae}), \tag{18.5}$$

not just $1/Y_{ee}$!

Can we identify these coefficients with easily measurable quantities? The electrical input impedance for loudspeaker operation, for instance, is V/i. But this quantity depends on the relation between p and U, and thus on the acoustic load presented to the speaker by its surroundings. The load impedance can be changed, for instance, by the way a speaker is baffled. The parameters involved here are most directly related to two extreme cases. First, (18.3) shows that if the speaker cone is clamped (or "blocked") so that no diaphragm motion can occur, the electric input impedance with $U = 0$ will simply be

$$(V/i)_{U=0} = Z_{ee}; \tag{18.6}$$

such a situation may be rather difficult to attain in practice, however. The other extreme would be to drive the speaker with its output going into an "acoustic short circuit" where p remains identically zero, for which (18.1) gives

$$(V/i)_{p=0} = 1/Y_{ee}. \tag{18.7}$$

This could be realized by testing the speaker inside a vacuum chamber. Conceptually similar are the two extremes of acoustic input impedance for microphone operation, according to whether the electric output operates into an open circuit,

$$(p/U)_{i=0} = Z_{aa}, \tag{18.8}$$

or a short circuit,

$$(p/U)_{V=0} = 1/Y_{aa}. \tag{18.9}$$

The off-diagonal matrix elements, or **transduction coefficients,** are of particular interest, for they tell how effective the device is at converting electrical signals into acoustic or vice versa. From (18.3), Z_{ea} is the ratio V/U when $i = 0$, that is, for an open-circuit microphone output. This can be related to the usual measure of microphone sensitivity by dividing (18.3) by (18.4):

$$M_0 = (V/p)_{i=0} = Z_{ea}/Z_{aa}. \tag{18.10}$$

Similarly, the "acoustic open circuit" of a blocked speaker would have $p/i = Z_{ae}$ or

$$S_b = (p/V)_{U=0} = Z_{ae}/Z_{ee}. \tag{18.11}$$

Unfortunately this differs from the practical S used in Chap. 15 both because $U \neq 0$ in ordinary speaker operation and because we were concerned there with the value of p some distance away rather than right at the transducer face; but at least it gives some idea of the physical meaning of Z_{ae}.

We might wish in some application to have a very sensitive microphone and thus a large ratio Z_{ea}/Z_{aa}, and at the same time to independently specify some particular values of Z_{ae} and Z_{ee}. Is it possible to design a transducer in which the four components of the impedance matrix have any numerical values we please? Not if the transducer is passive, for there are thermodynamic limits (among others) that must be obeyed. Let us examine the simplest such limit, by noticing that the arrows have been drawn in Fig. 18.1 in such a direction that Vi and pU both represent positive rates of work done by the outside world upon the transducer. Now during an individual cycle of oscillation, energy might be stored briefly in a capacitance or a compliance and then given back a little later in the cycle. But the long-term average of the energy absorbed by the device can only be positive (for a **dissipative** device) or zero (in the ideal nondissipative case). If we use the phasor representation for a sinusoidal signal, the total power absorbed by the transducer is

$$P = (1/2)\ \mathrm{Re}(Vi^* + pU^*) \geq 0. \tag{18.12}$$

Using (18.3) and (18.4), this becomes

$$\mathrm{Re}(Z_{ee}ii^* + Z_{ea}Ui^* + Z_{ae}iU^* + Z_{aa}UU^*) \geq 0. \tag{18.13}$$

But ii^* is itself real, and $\mathrm{Re}(Z_{ee})$ is just the resistance R_{ee}, and similarly for the last term. So, using the abbreviation

$$W = (1/2)(Z_{ea} + Z_{ae}^*), \tag{18.14}$$

we can write (18.13) as

$$R_{ee}ii^* + WUi^* + W^*iU^* + R_{aa}UU^* \geq 0. \qquad (18.15)$$

Since (18.15) must hold for every possible combination of i and U, it must hold in particular if either one is zero, that is, when either side of the device is left open-circuited. This gives

$$\boxed{R_{ee} \geq 0 \quad \text{and} \quad R_{aa} \geq 0,} \qquad (18.16)$$

which state the eminently reasonable requirement that no passive device can have negative input resistance. Other operating conditions with i and U both nonzero are best considered by writing (18.15) in terms of the (possibly complex) ratio $H = i/U$ in order to make UU^* a common factor in all four terms:

$$R_{ee}HH^* + WH^* + W^*H + R_{aa} \geq 0. \qquad (18.17)$$

The minimum of this quadratic function of H represents the conditions most nearly able to violate the energy requirements and can be found by setting equal to zero the partial derivatives with respect to both the real and imaginary parts of H. It turns out that the minimum occurs for $H = -W/R_{ee}$, and its value is $R_{aa} - WW^*/R_{ee}$. So we can guarantee that (18.15) is satisfied for a given transducer regardless of what is connected to it only if

$$\boxed{WW^* \leq R_{ee}R_{aa}.} \qquad (18.18)$$

Thus if either side of a transducer is nondissipative, that is sufficient to require that W be zero, or

$$\boxed{Z_{ea} = -Z^*_{ae}.} \qquad (18.19)$$

For such a transducer, knowledge (either from theory or from measurement) of either transduction coefficient is enough to immediately and uniquely determine the other one. That is, the two coefficients must be equal in magnitude, and with their complex phase angles related as shown in Fig. 18.3. The simplest transducers fall into two special categories, those for which phase angle α is either 0° or ±90°. By connecting together one of each kind you can, if desired, construct a compound transducer with any other phase angle.

In the case where $\alpha = -90°$, a passive dissipationless transducer must have

$$Z_{ea} = Z_{ae} = -jK, \qquad (18.20)$$

where K is a positive real function of frequency. Such a transducer is called **reciprocal** in a more specific sense than we have used up to now. The imaginary multiplier j denotes a quarter-cycle time lag between velocity and applied voltage, or between electric current and applied pressure. This is the behavior we expect in all ideal transducers based on electrical effects and may be understood in terms of voltage being in phase with displacement, which in turn lags 90° behind velocity.

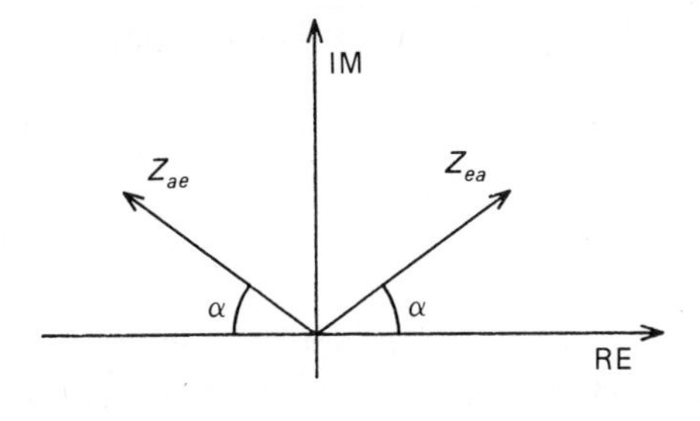

Figure 18.3 For a transducer with $R_{ee}R_{aa} = 0$, the transduction coefficients must be located in the complex plane so that the two angles α shown are equal.

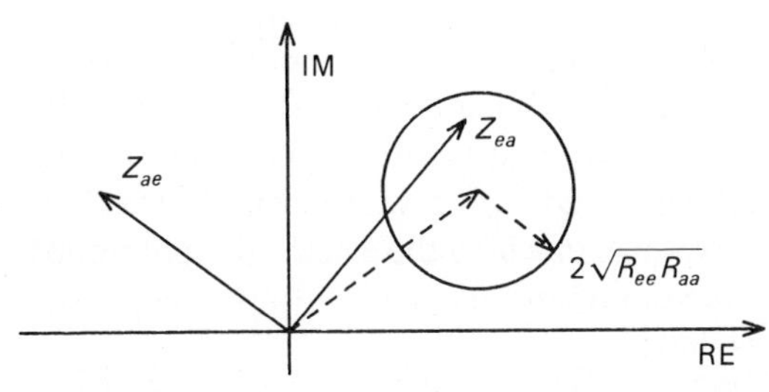

Figure 18.4 For a transducer with $R_{ee}R_{aa} \neq 0$ and any specified Z_{ae}, the condition $P > 0$ requires only that Z_{ea} lie somewhere within the circle shown, with radius $2|W|_{\max}$ and centered on the tip of $-Z^*_{ae}$. The same diagram can be used with the roles of Z_{ae} and Z_{ea} reversed.

In the case with $\alpha = 0°$,

$$Z_{ea} = -Z_{ae}, \tag{18.21}$$

but both are real. This occurs for ideal transducers based on magnetic effects, for which velocity and voltage (or current and pressure) peak at the same time. The minus sign in (18.21) may be thought of as related to our discussion of back emf at (15.16), where two applications of right-hand rules representing vector cross products gave successive 90° rotations in space around B. Because of the minus sign, such transducers are sometimes called **antireciprocal.**

What limitations would remain if we allowed both R_{ee} and R_{aa} to be nonzero in order to escape the condition (18.19)? Then according to (18.18) W may also be nonzero, so long as

$$|W| < \sqrt{R_{ee}R_{aa}}. \tag{18.22}$$

Together with (18.14) rearranged in the form

$$Z_{ea} = -Z^*_{ae} + 2W, \tag{18.23}$$

this may be interpreted as in Fig. 18.4 to mean that, for a given Z_{ae}, Z_{ea} could lie anywhere within the circle shown without violating energy conservation. If the radius of the circle is large enough, it could encompass the origin. Thus it is apparently not ruled out that we could devise ways of making a **nonreciprocal** transducer, that is, one that would transduce normally in one direction but be totally "deaf" in the other direction. But such a transducer would have to be dissipative and in fact would need enough pure resistance to make

$$R_{ee}R_{aa} > |Z_{ae}|^2/4. \tag{18.24}$$

(The analogous statement would still hold if it were Z_{ea} rather than Z_{ae} that was to be nonzero.)

18.3 APPLICATIONS OF RECIPROCITY

There are two interesting ways in which this theory may be applied. One is the use of information about one type of operation (e.g., as a loudspeaker) to predict how a transducer would behave when used in reverse (as a microphone). The other and somewhat less obvious application is in absolute calibration procedures.

As an illustration of the first type, consider a dynamic loudspeaker. We are prepared to consider reciprocity only for simple sources; so we must assume that the entire device is quite small compared with a wavelength. Let us limit this illustration further to a frequency range in which the electrical input impedance is just the coil resistance and the dominant mechanical impedance is that of the coil and cone mass. Then we have

$$Z_{ee} = R_e \tag{18.25}$$

and, using the mechanical impedance $Z_m = j\omega m$ together with $Z_a = p/U = (F/A)/(vA) = Z_m/A^2$ for diaphragm area A,

$$Z_{aa} = -j\omega m/A^2. \tag{18.26}$$

An extra minus sign has been introduced to account for U being counted positive for inward diaphragm motion, in accord with Fig. 18.2. Note that the real part R_{aa} is zero, so that Eq. 18.19 will apply.

The analysis of Chap. 15 in this case boils down to $U = vA = (F/j\omega m)A = (A/j\omega m)(Bli)$. The air load on the cone is relatively negligible; so we are looking at a transducer effectively shorted on the output (acoustic) side. Thus we treat the external pressure p in (18.4) as being practically zero, so that we can identify

$$(U/i)_{p=0} = -Z_{ae}/Z_{aa} \simeq ABl/j\omega m. \tag{18.27}$$

(18.26) can then be used to deduce

$$Z_{ae} = Bl/A. \tag{18.28}$$

Note that this is real, as claimed in the preceding section for any magnetic transducer.

This is enough information to tell without any separate analysis how this device would work as a microphone, for (18.19) immediately requires that

$$Z_{ea} = -Bl/A. \tag{18.29}$$

If the microphone works into a relatively high output impedance so that i is very small, it will have sensitivity

$$M_0 = (V/p)_{i=0} = |Z_{ea}/Z_{aa}| = BlA/\omega m. \tag{18.30}$$

The appearance of the factor ω suggests that we might connect the output to a finite electrical impedance that would compensate for this frequency dependence and provide a flat response (as in Prob. 18.3). Such need for input or output processing is hard to avoid if a transducer is to be used as both loudspeaker and microphone in broadband applications, as, for instance, in intercom systems.

Another important application of reciprocity is in microphone calibration. As long as a primary standard microphone is available, with its sensitivity certified by the National Bureau of Standards, it is often enough to calibrate another microphone simply by comparing the two side by side, both exposed to the same test signal. But if you were made responsible for providing the primary standard, how could you achieve an absolute calibration of it? The straightforward way would be to make a direct measurement of the pressure in a sound wave, together with the voltage output this wave causes in the transducer. Yet, aside from a microphone, other ways of measuring small pressure fluctuations (or, equivalently, producing known amounts of sound pressure) are few and difficult. We would usually find it much more convenient if we could confine ourselves to electrical measurements and avoid any need ever to make any other kind.

It is easiest to see how reciprocity could help here in the case where we have two identical reversible transducers, so that one can be used to create a signal and the other to detect it. The quantities to be measured are the current input i_1 to the first transducer and the open-circuit voltage output V_2 from the other. Now in general V_2 will be proportional to i_1, yet their ratio will depend on the acoustical environment, so that environment must be precisely known. Two ways present themselves to accomplish that end. The first is to place the transducers a distance d apart in an anechoic chamber, so that only the direct wave is detected. It is a simple spherical outgoing wave whose amplitude p_2 is given in terms of the source strength U_1 by (11.1) and (11.10):

$$p_2 = \rho f U_1/2d. \qquad (18.31)$$

If the transmitter is small enough for the air load on it to be negligible compared with its internal mechanical impedance, we can again use

$$(U_1/i_1)_{p_1=0} = -Z_{ae}/Z_{aa}. \qquad (18.27)$$

And since no significant current is to be drawn from the receiver, we have also

$$M_0 = (V_2/p_2)_{i_2=0} = Z_{ea}/Z_{aa}. \qquad (18.10)$$

But the reciprocity of the transducer assures that these two quantities both have the same magnitude, and the difference in phase can be ignored. Thus if they are multiplied together we can eliminate both p_2 and U_1 with (18.31):

$$\boxed{M_0^2 = V_2 U_1/p_2 i_1 = 2dV_2/\rho f i_1.} \qquad (18.32)$$

For example, measurement of $i_1 = 5$ mA and $V_2 = 0.3$ mV for $d = 3$ m and $f = 5$ kHz gives

$$M_0^2 = (6\ \text{m})(3 \times 10^{-4}\ \text{V})/(1.2\ \text{kg/m}^3)(5000/\text{s})(0.005\ \text{A})$$
$$= 6 \times 10^{-5}\ \Omega/(\text{kg/m-s}^4).$$

Then $M_0 = \sqrt{6 \times 10^{-5}} = 7.7\ \text{mV/Pa}$

and $\text{ML} = 20 \log (0.0077) = -42\ \text{dB}\ (re\ 1\ \text{V/Pa}).$

Problem 18.7 shows how the calibration can still be accomplished even if the two transducers are not identical.

An alternative method that avoids the need for an anechoic chamber would be some other acoustical environment whose nature is precisely controlled and calculable. Such a case occurs when the microphone manufacturer provides a coupler, a small chamber of known volume with hard walls in which holes are provided to mount the two transducers. When the coupler is much smaller than a wavelength, its volume V is a simple acoustic compliance, so that (18.31) can be replaced by

$$p_2/U_1 = 1/2\pi f C_a = \rho c^2/2\pi f V \tag{18.33}$$

and (18.32) by

$$\boxed{M_0^2 = 2\pi f V V_2/\rho c^2 i_1.} \tag{18.34}$$

18.4 CONCLUDING REMARKS

This chapter, even more than those before, has quickly sketched the outlines of a broad area while omitting a great deal of detail. That is in keeping with the fundamental purpose of this book, which has been to acquaint you with basic concepts and give some practice in their use so that you would be able to profitably read the broad spectrum of sources that contain all those other details. I sincerely hope that your interest has been whetted and that you will go on from here to study those other sources and learn far more than could have been packed into any single introductory text.

PROBLEMS

18.1. Suppose the coil resistance of a dynamic microphone is 2 Ω and its open-circuit sensitivity is 0.3 mV/Pa; it also has $Bl = 2$ T-m and $m = 0.2$ g. Show that this sensitivity is reduced by about 3 dB if it operates into a load resistance of only $R_L = 6\ \Omega$ at a frequency of 1 kHz. Do this by using (18.3) and (18.4) together with $V = -iR_L$. Can you draw an analogy with the effect of motional impedance in a loudspeaker, discussed in Chap. 15?

18.2. Could you use a typical home hi-fi loudspeaker (say a 20-cm cone dynamic speaker in a 50-liter cabinet) as a microphone? Why would the large cone area by itself tend to make this much more sensitive than the typical small microphone? But what other physical property prevents the sensitivity from being so high after all? What sensitivity M would you predict on the basis of reciprocity at 200 Hz if in loudspeaker operation a current of 0.2 A produced a signal with SPL = 80 dB at a distance of 4 m? Why would such a prediction be dubious at 4 kHz?

18.3. Consider a mass-dominated dynamic transducer with $Bl = 0.1$ T-m, $m = 0.5$ g, and coil resistance $R = 3\ \Omega$. Let it be connected to a load resistor $R_L = 50\ \Omega$ through a series capacitor C_e. Show that the voltage across R_L alone could provide an output V/p of $BlAR_LC_e/m$ to replace (18.30). How small should the capacitor be to get this flat frequency response over a range from 100 Hz to 5 kHz? What then is the sensitivity M?

18.4. For an electrostatic microphone, Z_{aa} is due to the membrane stiffness; integrate (17.12) over the circular area of radius a to show that Z_{aa} is $j(8\mathcal{T}/\pi\omega a^4)$. From the discussion in the preceding chapter, identify the coefficients Z_{ee} and Z_{ea}. Can you now write Z_{ae} and use it to show what U/i will be when this device is operated as a loudspeaker under acoustic-short-circuit conditions? For the example with ML = −64 dB described in Sec. 17.3, what current i, source strength U, and SPL at a distance of 2 m would be produced by applying a voltage input of 0.2 V (rms) at 2 kHz? In what ways would you change this device in order to make it more effective as a loudspeaker?

18.5. Two identical small transducers are placed 1.2 m apart in nonreflective surroundings. An alternating current of 15 mA at 2 kHz in one of them generates sound that is detected by the other; its open-circuit output is 50 μV. **(a)** What is the sensitivity M and sensitivity level ML of each transducer? **(b)** If you change the frequency to 1 kHz while still supplying 15 mA, what change do you expect in the output voltage? **(c)** Why will this calibration method fail if you use too high a frequency?

18.6. Two identical transducers face into a coupler whose volume is 20 cm^3. Current of 20 μA and frequency 1 kHz into one produces open-circuit output 10 μV from the other. What is the sensitivity M and sensitivity level ML of each transducer?

18.7. You wish to perform an absolute calibration using two reversible transducers, but you suspect they are not identical. Show that M^2 in (18.32) must be replaced by M_1M_2 and that merely reversing the roles of transmitter and receiver does not produce any further information. Show how a further experiment in which both are used as receivers, placed side by side in a sound field produced by some third device (whose properties need not be known), can provide M_1/M_2 so that both individual sensitivities can be calculated. As an example, suppose in the first experiment a current 2 mA at 500 Hz in transducer A produces output 5 μV in transducer B located 2 m away. In the second experiment, output voltages 25 and 40 μV come from A and B, respectively. What are M_1, M_2, and the corresponding ML's?

Appendix A

Problem Solving

Physics and engineering courses generally include large numbers of problems for you, the student, to solve. It is appropriate to stress this activity, for it is the ultimate test of whether you really understand the material being presented. To a large extent, the best way to learn how to handle these problems is simply to do as many as possible. It can also be helpful, however, to spell out the techniques you should be trying to use.

A.1 PRELIMINARIES

Probably the most common weakness in problem-solving technique is a failure to appreciate the importance of preliminary thinking. Being in great haste to get an answer—any answer—and go on to the next problem is a sure way to generate all kinds of errors. Whenever physics teachers get together and gripe about engineering students, the standard caricature that always comes up is the student who simply grabs the nearest formula, relevant or not, and starts plugging numbers into it. Let us list several suggestions that could help users of this book avoid living up to that sad expectation:

1. John Wheeler is supposed to have advocated that you "never do a calculation until you know the answer." That is, never get involved in actual computing until you have thought about why you are going to do it that particular way and already have some *physical* idea what the answer should be like. Once the calculations get rolling, they tend to take on a life of their own, and it becomes all too easy to forget the original physical question.
2. Draw a picture. Label the parts. It is amazing how often this can save you from silly errors, like substituting a given quantity of 30° as the value of an angle θ in an equation when in fact the given angle was $90° - \theta$.
3. Don't be afraid to spell out the obvious. A clearly stated list of "what is known" can be followed by "what is to be found" and then "physical connections relating

these to each other." Such lists can help you see whether any particular equation offers a prospect of progress toward finding the unknowns. They can also spare you from wasting hours on impossible tasks such as trying to milk the values of four unknown quantities from only three equations.

4. Always carry the units along with the numbers when you calculate. It is astounding how often the units can forcefully deliver the message, "This answer cannot possibly be correct, so you had better go back and search out an error."
5. Chip away one piece at a time. In elementary texts, many lengthy problems are split into small parts; if you take (*a*), then (*b*), (*c*), etc., each will be very simple. But eventually you must learn through experience to accept more generally stated problems and provide the piece-by-piece approach yourself.
6. Ask whether your answers are reasonable. What sets good students apart from poor ones is not so much whether they make mistakes in their work but that the good students will recognize the odor of an error and go back to correct it. An answer of 42 seconds for an exam problem about a falling body may be just another number to a poor student, while it can be a clue to a good student that he has dropped a couple of powers of 10.
7. Check limiting cases. What happens when a parameter of a problem takes on an extreme value, such as zero or infinity or an angle of 90°? If the formula you are using does not produce sensible results in such limits, you should examine very carefully whether you are making proper use of it.

A.2 MATHEMATICAL METHODS

At high school level, it is reasonable that a standard problem might boil down to this: having just learned that $F = ma$, you are asked how much force is required to give acceleration 5 m/s^2 to a mass of 3 kg. Many students have the false expectation that college physics will just be more of the same, whereas in reality the simple "plug-in" problem becomes less and less important the further you go. There are several distinct directions in which more sophistication must be added.

Bolder Use of Simple Equations First, you must feel equally comfortable using an equation forward or backward. If given $F = 15$ N and $m = 3$ kg, you must take responsibility for turning $F = ma$ around into $a = F/m$. Beginning engineering students can usually handle this, but it is the point at which, sadly, the typical liberal-arts person remains stuck for life.

More significantly, there is simply not going to be a numbered equation with a box around it for every situation that comes up. You must take the responsibility for joining together two or more equations to get a desired result. This is probably responsible for the greater share of the grief experienced by the typical engineering student studying general physics.

Specific vs. General Results Students are usually much too eager to start punching numbers into a calculator. As soon as you take a specific number and start doing arithmetic, you are bound to get an answer that applies only to that specific problem. If someone asks, "Well, what if m had been 4 kg instead of 3 kg?" you are stuck with repeating the whole calculation. If you could sometimes get an answer that applies to a whole family of problems at once, that would be much more powerful. And that, of course, is what algebra was invented for.

Though there are occasional exceptions, I urge as a general rule that you carry calculations as far as possible in algebraic form. Wait until the last possible moment before substituting specific numbers for the situation at hand. This will give the greatest chance for insight into why the answer is what it is and how it would change in another situation. It may also save unnecessary labor in that a particular variable may completely cancel out by the time the algebra is done. If numbers were substituted too soon, you might have multiplied by a number only to turn around and divide later by that same number.

Computer Aid in Presenting Results The more interesting and realistic the situations we try to analyze, the more often our final results require fairly lengthy or complicated calculations. Especially when we wish to see a result that is not just a specific number but a function showing how one thing depends on another (for instance, how a loudspeaker's response changes with frequency), the labor of calculating and plotting points on a graph by hand may be prohibitive. Fortunately, personal microcomputers and programmable calculators are now so readily available that this is no longer an excuse for skipping the most interesting and significant problems.

Analytic vs. Numerical Techniques It has been traditional to consider a physics problem "soluble" if its answer could be written in the form $x = f(y, z, \ldots)$, where the unknown x was isolated on one side of an equation and the other side was written entirely in terms of known quantities $y, z, \ldots$. Furthermore, the function f on the right-hand side was to involve only a finite number of analytic operations (addition, subtraction, multiplication, division, and evaluating known functions such as square root, exponential, or cosine). This analytic approach is centered on the symbol-manipulating techniques of algebra and calculus. It is extremely elegant and powerful, but that can blind us to the fact that it is not the only possibility.

In cases where analytic techniques encounter roadblocks, it can save the day to turn to number-crunching; that is, there are times we should turn our backs on the approach advocated under Specific vs. General Results. By giving up elegance and generality and moving ahead with specific numbers instead of algebraic symbols, we may find solutions to additional types of problems. At a more advanced level, you will find this to be very important in such areas as solution of nonlinear differential equations. For our purposes in this book, however, let us discuss only the simpler issue of transcendental equations.

Consider the equation

$$1 + e^{-x} = x, \tag{A.1}$$

which arises in studying how a piano string is set in motion by its hammer. No amount of effort in moving symbols around can ever put this equation in the form $x = f(e)$. It is rather generally true that when an unknown x occurs both inside and outside a function such as exp, log, or sin, purely algebraic solutions do not exist. But let us point out several ways in which you *can* obtain a solution.

Guessing In spite of its inelegance, this is a perfectly legitimate way to find a solution. Pick any number you like for x, and try substituting it. For instance, $x = 1$ leads to $1.368 = 1$, and $x = 2$ to $1.135 = 2$. Both are clearly incorrect. But after one or two guesses, the next one can be an educated guess rather than random. Since $x = 1$ made the left side too large compared with the right, and $x = 2$ the other way around, you ought to strongly suspect that somewhere in between lies a value of x that will obey the equation. Try $x = 1.4$ and you get $1.246 = 1.4$, which should make you confident that the correct value lies below 1.4. Try 1.2 and get $1.301 = 1.2$; closer, but now too low. What is your next guess, and how nearly does it satisfy the equation? Are you convinced that by repeating the procedure enough times you can get an answer as accurate as you wish?

Once you know roughly where an answer lies, you can speed up the process by using interpolation to guide your next educated guess. For instance, since $x = 1.2$ made the left side 0.1 too large and $x = 1.4$ made it 0.15 too small, try going a fraction $0.1/(0.1 + 0.15)$ of the way from 1.2 to 1.4. This is $x = 1.28$, which gives $1.278 = 1.280$ and is indeed much closer to the final answer.

Graphing Suppose you plot a graph of the left side of (A.1) as a function of x. Then, using the same axes, plot a graph of the right side also. The solution of the equation $LHS = RHS$ must be represented by the point where the two curves intersect (Fig. A.1). If you have taken care to plot the graph very accurately, you will be able to read off the solution immediately by simply noting the coordinates of the intersection.

Iterating This technique is not widely taught in elementary mathematics courses, but it can be very useful. It is well adapted to fast and accurate solutions with a pocket calculator. Suppose you put the equation in the form $x = f(x, \ldots)$, with no limitation on how the right side depends on x along with other, known quantities. Substitute any value you like for x on the right side, carry out all the operations indicated by the function f, and take the result as a new guess at x. Put that new x into the right side and repeat the procedure. That is, starting from the first arbitrary guess x_0 generate a sequence of values $x_1, x_2, \ldots$ from

$$x_{i+1} = f(x_i, \ldots). \tag{A.2}$$

In the example above, $x_{i+1} = 1 + \exp(-x_i)$ and an initial guess $x_0 = 1$ produces the sequence

1.36788
1.25465
1.28518

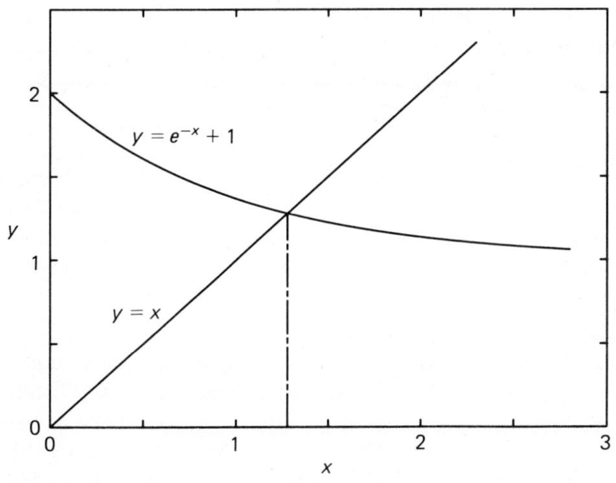

Figure A.1 Graphical solution of example $1 + e^{-x} = x$.

1.27660
1.27898
1.27832
1.27850
1.27845
1.27847
1.27846

which rapidly converges to the correct answer.

If you had the misfortune to rearrange (A.1) in the alternate form $x_{i+1} = -\ln(x_i - 1)$, the sequence would diverge instead. For instance, an initial guess as good as 1.280 would produce 1.273, 1.298, 1.209, 1.564, 0.573, and then an error flag for the logarithm of a negative number. Whenever that happens, just rearrange the equation to isolate the other x on the left side. Usually, one or the other version will converge to the right answer.

All of the Above What is likely to be most helpful in practice is not to spend half an hour on a painstakingly accurate graph but half a minute on a quick sketch. That can be enough to tell you whether it is reasonable to expect any solution at all, whether there will be one or several solutions, and roughly what their values will be. The rough graph, then, can serve as a guide to increase the efficiency of accurate calculations using guesses or iterations.

Let us illustrate all that with one more example. Suppose we wish to solve the equation

$$1 + \beta x = \ln(x),$$

where β is a parameter. It is easy to make a quick sketch of a graph showing the left- and right-hand sides separately (Fig. A.2). And doing this makes it immediately

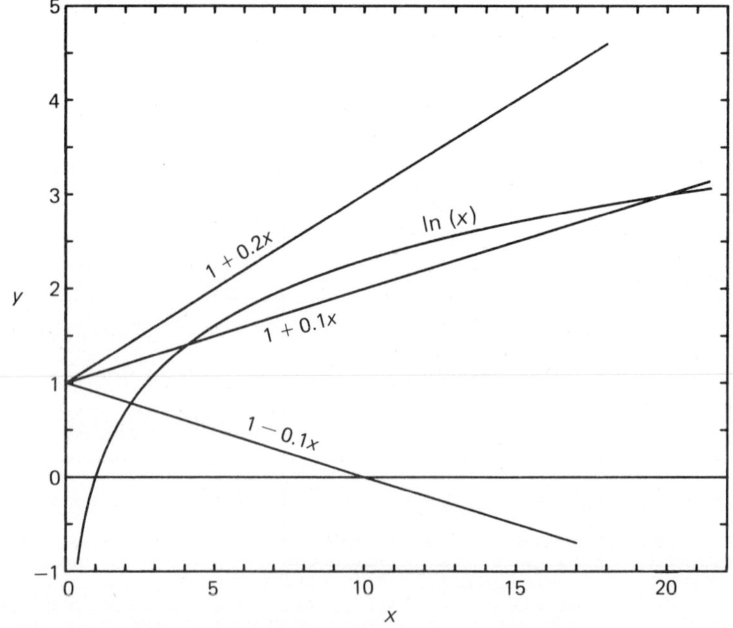

Figure A.2 Graphical solution of example $1 + \beta x = \ln(x)$.

apparent that the nature of the problem depends in a very important way upon β: (1) for $\beta < 0$ there is sure to be one and only one solution, and that solution will lie somewhere in the range $0 < x < e$; (2) for small positive β there will be two solutions, both with $x > e$, and we must ask whether the physical problem provides grounds for choosing one or the other; (3) for sufficiently large positive β there is clearly no solution and we know not to waste time looking for one. In this example it happens to be quite easy to determine the critical value β_c that divides cases 2 and 3: note that in the borderline case the straight line would be tangent to the curve, meaning their slopes would be the same so that $\beta = 1/x$. But then $1 + 1 = \ln(1/\beta)$, so

$$\beta_c = e^{-2} = 0.135 \quad \text{and} \quad x(\beta_c) = e^2 = 7.4.$$

As an illustration of case 2 let $\beta = 0.1$. The two obvious ways to rearrange the original equation to try an iterative solution are

$$x_{i+1} = \exp(1 + \beta x_i)$$

and

$$x_{i+1} = [\ln(x_i) - 1]/\beta.$$

The first of these converges to $x_1 = 4.09$ (although only if the initial guess for x is less than x_2) and the second converges to $x_2 = 19.9$ (but only for initial x greater than x_1); so both solutions are available.

To illustrate case 1, consider $\beta = -0.1$. Here the first form converges to the correct solution $x = 2.185$ while the second form always diverges. For values of β less than -1 neither attempt at iteration converges (although something else takes place that is fascinating—try it). But educated guessing guided by the graph can still produce correct solutions fairly quickly, for instance, $x = 0.6874$ when $\beta = -2$.

Appendix B

Phasors

The theory of waves and vibrations entails many calculations involving physical quantities that oscillate as a function of time (and perhaps also position). Sinusoidal functions in particular play a very important role. The algebra and calculus involved can often be shortened by using the **phasor representation** of simple harmonic motion. We outline here the basic properties of this technique so that we can use it at any point where it would be advantageous.

B.1 PHASORS IN LINEAR EQUATIONS

The basic idea of a phasor is the temporary replacement of a real physical quantity, for purposes of calculation, by a complex number (or function). Complex numbers may be conveniently thought of as having two components and representing points in a plane (Fig. B.1). The components are called the real and imaginary parts of the number; for instance, if $z = 3 + j4$, then $\text{Re}(z) = 3$ and $\text{Im}(z) = 4$. Here j stands for $\sqrt{-1}$. Complex numbers add just like vectors: if $z = x + jy$ and $w = u + jv$, then $z + w = (x + u) + j(y + v)$. But it is in multiplication that they are entirely different from vectors: multiplying out the binomials term by term and using $j^2 = -1$ gives $zw = (xu - yv) + j(xv + yu)$. Multiplication is much more convenient when each number is represented by polar rather than cartesian coordinates in the complex plane. If we write $|z|$ for the magnitude of z (which is $\sqrt{x^2 + y^2}$) and α for its angle from the real axis, and similarly for $|w|$ and β, then zw simply has magnitude $|zw| = |z||w|$ and angle $\gamma = \alpha + \beta$.

The usefulness of phasors depends on two fundamental things: (1) the oscillating physical quantity in question obeys a linear differential equation (such as $d^2x/dt^2 + \Omega^2 x = 0$ for a simple harmonic oscillator), so that solutions may be combined in accordance with the superposition principle; and (2) imaginary exponentials can be evaluated with the identity

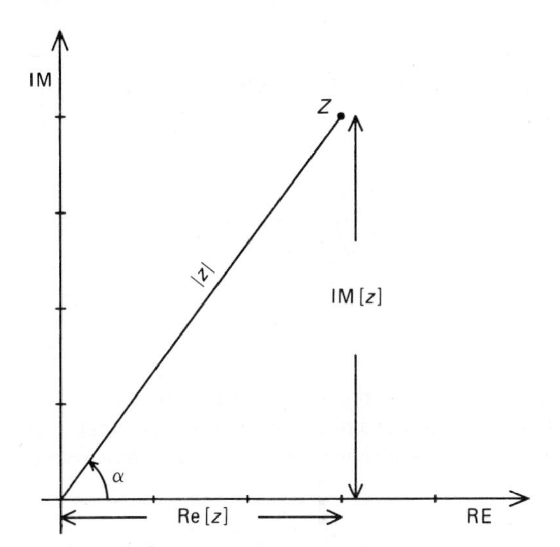

Figure B.1 Representation of the complex number $3 + j4$ as a point in the complex plane.

$$\boxed{e^{j\xi} = \cos \xi + j \sin \xi.} \tag{B.1}$$

This enables us to do many calculations using $e^{j\xi}$ as a mathematical representation of the oscillating quantities, yet declare at the end that the real part alone is itself a solution and can be used to represent the actual physical variable. The exponential notation has the advantage of making it unnecessary to remember or derive various trigonometric identities. The identities $\cos(\alpha + \beta) = \cos \alpha \cos \beta - \sin \alpha \sin \beta$ and $\sin(\alpha + \beta) = \sin \alpha \cos \beta + \cos \alpha \sin \beta$, for instance, can be seen as simple direct consequences of the trivial equation $\exp[j(\alpha + \beta)] = [\exp(j\alpha)][\exp(j\beta)]$.

Let us spell out what (B.1) means in terms of the plane of complex numbers. Since the real and imaginary components of $e^{j\xi}$ are $\cos \xi$ and $\sin \xi$, this complex number lies on a unit circle around the origin. Its magnitude is 1, and its phase angle is $\arctan(\sin \xi/\cos \xi) = \xi$. In the case where $\xi = \omega t$, this says the complex number $e^{j\omega t}$ sweeps round and round the unit circle at the uniform rate ω radians per second. More generally, $e^{j(\omega t - \phi)}$ represents a phasor that would not happen to be passing through the real axis at $t = 0$ unless ϕ were zero. Figure B.2 illustrates how ϕ can be interpreted as a **phase lag** telling how far the occurrence of the peak value lags behind time zero.

We are particularly interested in writing phasors such as

$$F = F_0 e^{j(\omega t - \phi_F)}, \tag{B.2}$$

which is used in Chap. 5 to represent a sinusoidal driving force. We find that the resulting velocity can be written as

$$v = v_0 e^{j(\omega t - \phi_v)}. \tag{B.3}$$

Though both of these phasors rotate rapidly in the complex plane, they do so at the same frequency and thus always maintain the same relative orientation. This is expressed by the fact that their quotient, the mechanical impedance, is independent of time:

$$Z = F/v = (F_0/v_0)e^{-j(\phi_F - \phi_v)}. \tag{B.4}$$

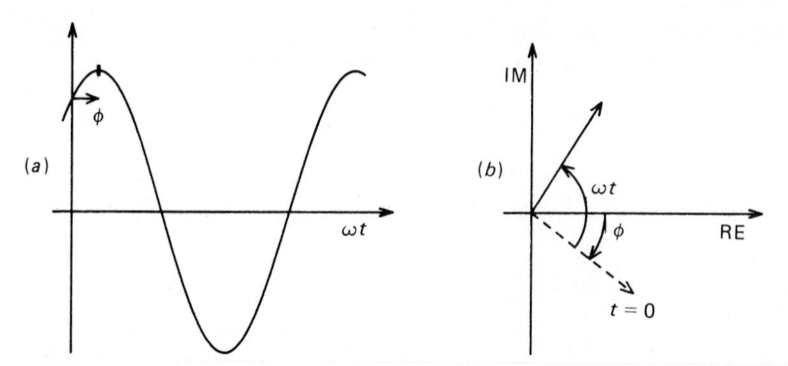

Figure B.2 (*a*) Graph of $\cos(\omega t - \phi)$, illustrating interpretation of ϕ as a phase lag of the waveform peak relative to $t = 0$. (*b*) Representation as a phasor beginning from angle $-\phi$ at $t = 0$ and spinning counterclockwise, whose projection on the real axis reproduces part (*a*).

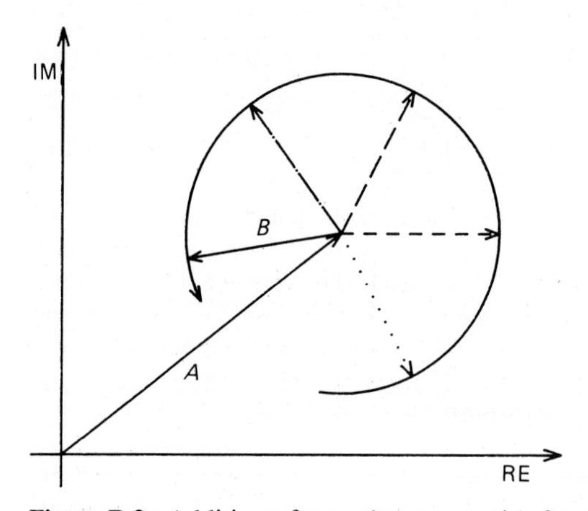

Figure B.3 Addition of two phasors, A with frequency ω_1 and B with frequency ω_2. It is helpful to imagine them illuminated by a strobe flashing at the same frequency as A rotates in order to make it seem stationary. Then when B is added onto A, its tip will appear to slowly trace out a circle that is at least $A - B$ and at most $A + B$ from the origin.

When we go on to write such expressions as

$$Z = R + j(\omega m - s/\omega), \tag{B.5}$$

we do not mean that a physical property of the system has values that are imaginary in the sense of unreal. The j signifies nothing more or less than a tendency for the oscillating quantities F and v to be out of phase with each other in time.

An example involving phasors with different frequencies is provided by the phenomenon of beats. If two sinusoidal sources are each creating sound at a certain location, the total acoustic pressure will have the form

$$p(t) = Ae^{j(\omega_1 t - \phi_1)} + Be^{j(\omega_2 t - \phi_2)}. \tag{B.6}$$

Even before doing any formal calculation, we can see in Fig. B.3 the suggestion that the combined disturbance can have an amplitude at most $A + B$ and at least $A - B$. Furthermore, if the individual phasors rotate at rates ω_1 and ω_2, one must keep overtaking the other at the beat rate

$$f_b = (\omega_1 - \omega_2)/2\pi. \tag{B.7}$$

That is, this is how often the total disturbance will go from maximum to minimum amplitude and back to maximum again. (To verify this in detail, do Prob. B.3.)

B.2 PHASORS AND ENERGY

Phasors are most useful for representing oscillating quantities obeying linear equations. Displacement, velocity, acceleration, and force in mechanical systems; current and voltage in electric circuits; and pressure, density, and volume velocity in acoustic systems are all examples of physical variables whose behavior may appropriately be represented by oscillations with a certain frequency, amplitude, and phase.

But we must also deal with the energy associated with these motions. Sometimes it may specifically be an energy density, at other times a power (rate of delivery of energy) or intensity. The various forms of energy depend quadratically on the vibration amplitudes. We must be careful not to be misled by a simple expression such as $K = mv^2/2$ for kinetic energy, since substituting $v = v_0e^{j\omega t}$ would give $K = (mv_0^2/2)e^{2j\omega t}$. That is clearly wrong, since it suggests that K would sometimes have negative values. The reason this does not work is that $\text{Re}(v^2) \neq [\text{Re}(v)]^2$. Since the actual physical velocity is supposed to correspond only to the real part of the phasor v, the safe thing to do would be always to extract the real part first and then square to find the energy. In this case the result would be $K = (mv_0^2/2)\cos^2 \omega t$, which can also be written as $(mv_0^2/4)[1 + \cos(2\omega t)]$. Although the energy does in fact vary with frequency $2\omega t$, its minimum value is now correctly given as zero.

We often want only to know the time-averaged energy or power; for instance, in the case above this would be $\langle K \rangle = mv_0^2/4$. Let us derive a convenient way to extract such an average in the general case. Suppose any two physical quantities have the values

$$f = \text{Re}[f_0 e^{j(\omega_1 t - \phi_1)}] \tag{B.8}$$

and

$$g = \text{Re}[g_0 e^{j(\omega_2 t - \phi_2)}]. \tag{B.9}$$

Then

$$fg = f_0 g_0 \cos(\omega_1 t - \phi_1)\cos(\omega_2 t - \phi_2) = (f_0 g_0/2)\{\cos[(\omega_1 + \omega_2)t - (\phi_1 + \phi_2)] + \cos[(\omega_1 - \omega_2)t - (\phi_1 - \phi_2)]\} \tag{B.10}$$

so that $\langle fg \rangle$ will be nonzero only if $\omega_1 = \omega_2$. And in that case its value is

$$\langle fg \rangle = (f_0 g_0/2)\cos(\phi_1 - \phi_2). \tag{B.11}$$

But this same expression will appear if we consider the phasors F and G whose real parts are used in (B.8) and (B.9) and calculate

$$F^*G = f_0 g_0 e^{-j(\omega_1 - \omega_2)t} e^{j(\phi_1 - \phi_2)}. \tag{B.12}$$

Here the complex conjugation denoted by the asterisk is defined by $(x + jy)^* = (x - jy)$, from which it follows that $[\exp(j\alpha)]^* = \exp(-j\alpha)$. Again there is a nonzero average only if $\omega_1 = \omega_2$. Comparing (B.11) with (B.12), we find that the average of the product of any two sinusoids can be obtained very simply from their phasor representations as

$$\boxed{\langle fg \rangle = (1/2)\,\text{Re}(F^*G).} \tag{B.13}$$

The most familiar example should be the power delivered when a voltage drop V drives a current i through any electric-circuit element. If V and i are phasor representations of these quantities, then instead of the familiar $P = iV$ learned for dc circuits we should use $\langle P \rangle = (1/2)\,\mathrm{Re}(i^*V)$. If i and V in turn are related by the (complex) impedance of the circuit element, $V/i = Z = R + jX$, this becomes

$$\langle P \rangle = (|V|^2/2)\,\mathrm{Re}(1/Z) = (|V|^2/2)(R/|Z|^2) \tag{B.14}$$

or

$$\langle P \rangle = (|i|^2/2)\,\mathrm{Re}(Z) = |i|^2 R/2 \tag{B.15}$$

instead of just V^2/R or i^2R.

PROBLEMS

B.1. If $w = 3 - j4$ and $z = 8 + j6$, calculate **(a)** $w + z$, **(b)** $w - z$, **(c)** wz, **(d)** w/z, **(e)** z/w, **(f)** w^*z, and **(g)** $\mathrm{Re}(w^*z)$. Show the representation of each of these in the complex plane. [*Hint:* Consider using polar representation for **(c)** to **(g)**.]

B.2. Show that the real part of any complex number $z = x + jy$ can be found by calculating $(z + z^*)/2$. Use this to show that (B.14) is correct in replacing $\mathrm{Re}(1/Z)$ by $R/(R^2 + X^2)$.

B.3. Suppose that $A > B$ and $\omega_1 - \omega_2 \ll \omega_1$ in Eq. B.6. Factor out $e^{j\omega_1 t}$, and interpret everything else that multiplies this factor as a slowly varying amplitude. Find the magnitude of this amplitude, and thus justify Eq. 4.8.

B.4. If the intensity (power per unit area) in a sound wave is equal to the product of the actual physical pressure p and velocity v, and if these are represented by the phasors

$$p_c = (0.02\ \mathrm{Pa})e^{j(500t-40°)} + (0.01\ \mathrm{Pa})e^{j(800t+30°)}$$

and

$$v_c = (50\ \mu\mathrm{m/s})e^{j(500t-60°)} + (20\ \mu\mathrm{m/s})e^{j(800t-20°)},$$

what is the average intensity $\langle I \rangle$?

Appendix C

Bessel Functions

Many kinds of waves, when they occur in cylindrical geometry, are governed by a differential equation that can be reduced to the form

$$x^2\, d^2y/dx^2 + x\, dy/dx + (x^2 - m^2)y = 0. \tag{C.1}$$

Here the independent variable x is proportional to radial distance from the axis of a cylindrical coordinate system, and so is limited to $0 < x < \infty$. Like all second-order differential equations, this has two independent solutions. Only one of them is well behaved at $x = 0$, however, and it is only this "Bessel function of the first kind" with which we are concerned. Like the trigonometric functions, the Bessel functions have been tabulated, graphed, and otherwise studied at great length. It is very useful, then, to treat them as known functions that will be regarded as a satisfactory way to express answers for any problems we wish to solve.

Let us summarize here just a few of the most important properties of Bessel functions. Much more detailed information can be found in various mathematical handbooks, such as the National Bureau of Standards *Tables of Functions and of Zeros of Functions* (U.S. Government Printing Office, 1954) or E. Jahnke and F. Emde, *Tables of Functions* (4th ed., Dover Publications, New York, 1945).

C.1 FUNCTION VALUES AND ROOTS

Each different value of the parameter m in Eq. C.1 will produce a different solution; that is, there is actually a whole family of functions $y(x) = J_m(x)$ depending on the label m as well as on x. J_m is called the Bessel function of order m. We will be concerned only with integer values of m.

The Bessel functions may be constructed with standard techniques for solving linear differential equations, and we will not go through the details here. The overall nature of these functions is best appreciated first with a graph (Fig. C.1). We see there that J_0 and

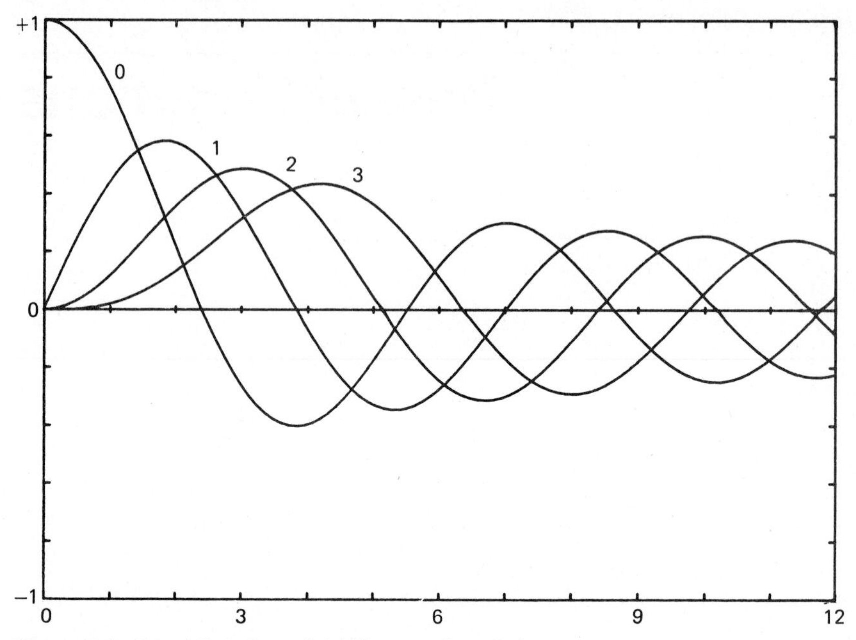

Figure C.1 Bessel functions $J_m(x)$ for $m = 0, 1, 2, 3$.

J_1 resemble cosine and sine, except that the amplitude of the fluctuations decreases with increasing x instead of remaining constant.

We can write some simple asymptotic expressions for very large or small values of x, which make it unnecessary to consult a table or graph to obtain the function value. For small x, the leading term in a power-series expansion is

$$J_m(x) \simeq (1/m!)(x/2)^m \quad \text{if} \quad x \ll 2\sqrt{m+1}. \tag{C.2}$$

Somewhat larger values of x can be handled by retaining another term:

$$J_m(x) \simeq (1/m!)(x/2)^m[1 - (x/2)^2/(m+1)] \quad \text{if} \quad x \ll 2\sqrt{2(m+2)}. \tag{C.3}$$

For sufficiently large x, we have instead

$$J_m(x) \simeq \sqrt{2/\pi x}\,\cos(x - \pi/4 - m\pi/2) \quad \text{if} \quad x \gg (m+1)\pi/2. \tag{C.4}$$

Although in principle this duplicates information already contained in the graph (Fig. C.1), we give in Table C.1 a brief listing of values that will cover the range where neither of the limiting forms shown in Eqs. C.3 and C.4 is accurate. Along with J_0 and J_1 we also tabulate the related piston radiation functions R_1 and X_1, which are discussed in Chap. 11.

Sometimes it is useful to know the values of x for which J_m is zero. An accurate reading of Fig. C.1 or interpolation in Table C.1, for example, shows that $J_0(x)$ first crosses the axis for $x = 2.405$. Since these are oscillating functions, they have many zero crossings, and we designate the nth root of J_m as j_{mn}:

$$J_m(x = j_{mn}) = 0. \tag{C.5}$$

A few of the smallest roots are listed in Table C.2. The first row of the table can be extended to the right by the approximate formula

TABLE C.1 SELECTED VALUES OF BESSEL FUNCTIONS J_0 AND J_1, AND OF THE PISTON RADIATION IMPEDANCE FUNCTION*

x	$J_0(x)$	$J_1(x)$	$R_1(x)$	$X_1(x)$
1.0	+0.765	+0.440	0.120	0.397
1.2	0.671	0.498	0.170	0.462
1.4	0.567	0.542	0.226	0.521
1.6	0.455	0.570	0.288	0.571
1.8	0.340	0.582	0.354	0.613
2.0	+0.224	+0.577	0.423	0.647
2.2	0.110	0.556	0.495	0.671
2.4	+0.003	0.520	0.567	0.686
2.6	−0.097	0.471	0.638	0.693
2.8	0.185	0.410	0.707	0.690
3.0	−0.260	+0.339	0.774	0.680
3.2	0.320	0.261	0.837	0.662
3.4	0.364	0.179	0.895	0.638
3.6	0.392	0.096	0.947	0.608
3.8	0.403	+0.013	0.993	0.573
4.0	−0.397	−0.066	1.033	0.535
4.2	0.377	0.139	1.066	0.492
4.4	0.342	0.203	1.091	0.450
4.6	0.296	0.257	1.109	0.408
4.8	0.240	0.299	1.122	0.364
5.0	−0.178	−0.328	1.131	0.323

* For values of x not shown here, asymptotic forms such as (C.3) or (C.4) may be used.

TABLE C.2 VALUES OF j_{mn}, THE nTH ROOT OF $J_m(x)$

n \ m	0	1	2	3	4
1	2.405	3.832	5.136	6.380	7.588
2	5.520	7.016	8.417	9.761	11.065
3	8.654	10.173	11.620	13.015	14.372
4	11.792	13.324	14.796	16.223	17.616

$$j_{m1} \simeq m + 1.86\sqrt[3]{m} \qquad (m \gg 1). \tag{C.6}$$

Entries to the lower left of the diagonal can be estimated from

$$j_{mn} \simeq (n + m/2 - 1/4)\pi \qquad (n \gg m). \tag{C.7}$$

C.2 DERIVATIVES AND INTEGRALS

Many identities are known that make it possible to write derivatives or integrals of Bessel functions in terms of other functions from the same family. We give here just a few of these, which will be sufficient for our purposes.

First, just as the derivative of the cosine function is the negative sine, so also

$$dJ_0(x)/dx = -J_1(x). \tag{C.8}$$

But the analogy is not complete, for unlike the derivative of sine we have an extra term in

$$dJ_1(x)/dx = J_0(x) - J_1(x)/x. \tag{C.9}$$

As examples of integrals involving Bessel functions, we may give

$$\int xJ_0(x)\,dx = xJ_1(x) \tag{C.10}$$

and

$$\int xJ_0^2(x)\,dx = (x^2/2)[J_0^2(x) + J_1^2(x)]. \tag{C.11}$$

Another example, where a Bessel function results from term-by-term integration of a power-series expansion of an exponential function, is

$$\int_0^{2\pi} e^{jx\cos\phi}\cos m\phi\,d\phi = 2\pi j^m J_m(x). \tag{C.12}$$

Finally, another integral needed in Chap. 11 is

$$2\int_0^{\pi/2} \frac{J_1^2(x\sin\theta)\,d\theta}{\sin\theta} = 1 - \frac{J_1(2x)}{x}. \tag{C.13}$$

Appendix D

Partial Answers to Selected Problems

1.1. Frequency ratio of order 10^{11} to 10^{13}.

1.5. 8.5 km, 60 mg/m^3.

1.6. 194 dB.

1.7. 0.32 Pa.

1.10. **(a)** 28 m, 40 m.

1.11. **(b)** 6 dB, 20 dB.

1.12. **(a)** 18 km, **(c)** 2.6 km.

2.1. **(b)** 19 μV.

2.2. **(a)** 61 dB, **(d)** 136 W.

2.3. **(c)** 1 s, **(f)** No.

2.5. **(b)** 16 sones.

2.6. **(b)** 0.095 A, **(d)** 0.94 A.

2.10. **(a)** 1.41, **(c)** 3.

3.1. **(b)** 81 dB.

3.2. **(b)** 50 dB.

3.3. **(a)** A function of f, not just a number, **(b)** 109 dB, **(d)** 120 dB.

3.4. 7×10^{-8} Pa^2/Hz.

3.5. **(b)** 50.5 dB.

3.7. $C_n = 2V/\pi n$.

3.8. $A_n = -4/\pi(n^2 - 1)$ for some n—which ones?

3.14. $(-j\tau/4\pi)[(\sin y)/y - (\sin z)/z]$, $y = (\omega - \omega_0)\tau/2$.

3.15. $(\tau/2\pi j)[1 + (\omega - \omega_0)^2\tau^2]^{-1}$ plus another relatively unimportant term.

3.19. **(b)** $\sqrt{TK/4\pi\omega}$.

4.2. **(b)** 79 and 68 dB.

4.3. **(a)** $I_N = NI_1$, **(b)** $0 \leqslant I_N \leqslant N^2 I_1$.

4.4. 5 s.

4.7. **(a)** arcsin $(0.61\ \lambda/Nd)$, **(b)** ±34°, ±8°.

4.8. $y = 4.49$, 13.3 dB; $y = 4.56$, 12.0 dB.

4.9. 94°, 27°.

4.11. 2.5 m, 0.25 m.

4.13. $p \sim 1 + 2 \cos y + 2\beta \cos 2y$.

4.15. 25 dB, 16 dB, 3.0 m.

5.2. 1.4 g/s.

5.3. $N = 730$.

5.6. 5.1 cm, 11°.

5.9. $m \geqslant 0.5$ g, $Q \geqslant 270$.

5.11. 30 s.

6.1. **(a)** 3.16 m/s, **(b)** 4.6 s.

6.2. **(a)** 1.87 kg/s at −84°.

6.3. **(a)** 0.53 s/kg at +84°.

6.4. **(a)** $0.5 - j1.28$ kg/s, 1.65 cm with 21° lag, 53 W.

6.10. You must show that $B^2 + (G - 1/2R)^2 = (1/2R)^2$.

6.12. 157 kg/s at −61°, 4 W.

6.13. Roughly 100 W for both—check against Fig. 6.11 or 6.12.

7.1. 0.9 mm.

7.3. $C_3 = 0.42$ mm

7.5. Constant times p^{-3}, odd p only.

7.8. 0.13 J, 1.3 W.

7.9. $m_s V^2/6$, $(16/n^4\pi^4) m_s V^2$, 12 dB/octave.

7.11. $s = T/L$.

7.12. 5.1 m/s, 0.95 mm, 1.46 mm.

7.13. 0.046 kg/s, 35 mm.

7.14. $0.028\ L/n$.

8.2. 153 Hz.

8.3. Tenth mode is $f_{13} = 274$ Hz.

8.5. This should lead you to prove Eq. 8.16.

8.6. 58 m/s; 9, 14, and 22 cm.

8.7. Comparable restoring forces would produce comparable f_1's; $\gg 70$ N/m; ×8.

9.5. 156 dB.

9.7. **(a)** 2.2 nm, **(b)** 70 μm/s, **(c)** 45 mg/m³, **(d)** 4.5 μJ/m³, **(e)** 6.7 mW/m².

9.8. **(a)** 240 m/s, **(b)** 1.1 mK, 3 μm, **(c)** 74 μW/m².

9.11. 173 dB, 133 dB.

10.1. **(b)** 50 Pa, 20 cm/s, 0.16 mm, 17 mJ/m³, 3 W/m².

10.2. **(b)** $147 + j198$ or 246 kg/m²s at 53.5°.

10.4. 430, 335 kg/m²s.

10.7. 72°.

10.8. **(b)** 83 Pa, 45 cm/s, 3 W/m²; $p_{rms}v_{rms} = 19$ W/m².

10.9. **(a)** 3, **(b)** 65 percent, 29 percent.

11.1. **(a)** 134 dB, 0.75 W, **(b)** 154 dB, 75 W.

11.2. 0.022 m³/s, 1.2 W, $(2.2 + j3.9) \times 10^3$kg/m⁴s.

11.6. **(a)** $f \lesssim 500$ Hz, **(c)** $f \lesssim 200$ Hz.

11.7. **(a)** 1.1×10^{-6} m⁴/s, 0.35 mW.

11.9. **(a)** 10^{-3}m³/s, **(b)** 33 mm/s, 10 μm, **(c)** $5 + j8$ kg/s.

11.10. **(a)** 1 mm, $0.05 + j1.0$ kg/s, **(b)** 0.1 μm, $13 + j1$ kg/s.

11.11. $13.6 + j6.9$ kg/s, 7.3 N.

11.13. **(a)** 6.3 kW, **(b)** 3.1 kW, **(c)** 31 W; **(d)** 6.3 kW, 3.9 kW, 31 kW.

11.14. 18 dB.

11.15. $kb > 3.2$ and 79.

12.2. 18, 23, 27, . . . , ms.

12.4. 1.6 s, 1.0 s.

12.5. **(a)** 17,000, **(c)** 250, **(f)** 21 Hz.

12.6. **(b)** 0.3 s, 66 dB.

12.9. **(a)** 2.2 s $\gtrsim T_r \gtrsim$ 1.9 s, **(b)** 1.3 mW, **(c)** 0.01.

12.10. $T_r \simeq 1.9$ s.

12.11. 1.14 s.

12.12. **(b)** 0.15 ± 0.09, 0.15 ± 0.05.

12.13. **(c)** 6 m.

12.14. In larger room: 67 and 97 dB.

12.16. **(b)** 1.60 vs. 1.43 s.

12.18. **(d)** 19.3, **(e)** 11.2 cm.

12.20. 100+ ms, 2000+ words, 4 cm.

13.1. **(b)** 77 dB.

13.2. **(a)** 59 dB, **(b)** 65 dB, **(e)** 8 to 10 dB reduction.

13.3. $L_{eq} = 63$ dB.

13.5. **(a)** 2.5 kg/m², **(b)** Recall Chap. 4.

13.6. **(a)** NR = 26 dB at 250 Hz, **(c)** 90 and 55 dBA.

13.7. **(a)** TL = 51 dB, **(c)** $\sigma > 360$ kg/m² or double wall.

13.8. **(a)** 66 dB, **(b)** 50 dB average, 54-dB peaks, 18 s separation.

13.9. **(b)** Cars 62 dB; trucks 66 dB average, 75 dB peaks.

13.10. **(b)** Cars 64 dB steady, trucks 71 dB peaks.

14.1. **(a)** 2 m.

14.2. **(a)** 0.1 Hz $\ll f \ll$ 12 kHz.

14.4. 124 dB.

14.5. 137 Hz.

14.6. 3-dB bandwidth 32 Hz, $V = 64$ cm^3 if neck is unflanged, $A/l_e \sim 2.2$ cm.

14.7. 210 Hz, $Q_1 \sim 2000$.

14.8. 0.18, −46°, −15 dB.

14.9. $1170 - j1690$ kg/m^{2}s.

14.12. 270 Hz or 240 Hz.

14.13. $T = 0.5$ at $f = 74$ and 82 Hz.

14.14. 4.9 mm.

15.1. 1.35×10^4 N/m.

15.2. A few watts.

15.4. Range 500 or 600 Hz to 2 or 3 kHz, efficiency up to 4 percent, 95 dB at 1 m for 1 V in.

15.5. 1.3 g, 3 kHz.

15.6. 200 Hz (counting air mass on both sides): $1.4 + j16.4$ kg/s, $0.2 - j2.2$ Ω, 18 mH, 370 μF, 26 Ω.

16.2. **(c)** 8 dB.

16.4. $2a < 8$ mm.

16.5. **(b)** $f \gtrsim 1.4$ or 4 kHz, depending on damping.

16.6. 0.0113 m^3.

16.7. Depth $\gg$ 2 cm is enough.

16.8. Answer not uniquely determined, but one possible example is $V \simeq 0.05$ m^3, $A_p \simeq$ 90 cm^2.

16.9. **(a)** 20 cm, 5 cm, 1 cm.

17.1. 3×10^{-5}.

17.2. Y: 0.28 mV.

17.4. S: 127°, 12 dB, 78.5°.

17.5. Consider inertance of a small hole.

17.6. 3100 N/m, −52 dB, 12 MΩ, 0.23 μm.

17.7. 0.5 mg, −102 dB, 1.7 μV, 14 mΩ.

17.9. Can you use R and C to make a 6 dB/octave low-pass filter?

17.10. 0.5 m, 1.0 m.

18.1. Show that correction factor is $1 + 1/Y_{ee}R_L$.

18.2. ML = −30 dB; what is λ at 4 kHz?

18.3. $C_e < 0.6$ μF, $M < 2$ μV/Pa.

18.4. 6.3×10^{-4} m^3/A-s, −40 dB.

18.5. **(a)** −55 dB, **(b)** 25 μV.

18.6. −63 dB.

18.7. −50 and −46 dB.

B.1. **(d)** $0.5j$, **(f)** $-50j$.

B.4. 1.07 μW/m^2.

Index